PLACE-BASED EVALUATION FOR INTEGRATED LAND-USE MANAGEMENT

Place-Based Evaluation for Integrated Land-Use Management

Edited by

JOHAN WOLTJER
University of Westminster, UK

ERNEST ALEXANDER
University of Wisconsin-Milwaukee, USA

ANGELA HULL
Heriot-Watt University, UK

MATTHIAS RUTH
Northeastern University, USA

Routledge
Taylor & Francis Group
LONDON AND NEW YORK

First published 2015 by Ashgate Publishing

2 Park Square, Milton Park, Abingdon, Oxfordshire OX14 4RN
52 Vanderbilt Avenue, New York, NY 10017

Routledge is an imprint of the Taylor & Francis Group, an informa business

First issued in paperback 2020

British Library Cataloguing in Publication Data
A catalogue record for this book is available from the British Library

The Library of Congress has cataloged the printed edition as follows:
Place-based evaluation for integrated land-use management / [edited] by Johan Woltjer, Ernest Alexander, Angela Hull and Matthias Ruth.
pages cm
Includes bibliographical references and index.
ISBN 978-1-4724-4548-3 (hardback)
1. Regional planning. 2. Land use--Planning 3. City planning. 4. Public works--Planning. I. Woltjer, Johan.
HT391.P5325 2015
307.1'2--dc23

2015000869

ISBN 13: 978-1-4724-4548-3 (hbk)
ISBN 13: 978-0-367-66874-7 (pbk)

Contents

PART II: UNDERSTANDING THE EVALUATION OF IMPACTS AND SPACE

PART III: SPATIAL ANALYSIS FOR INTEGRATED PROJECTS

List of Figures

List of Tables

Notes on Contributors

Ann Åkerskog is Senior Lecturer at the Department of Urban and Rural Development, Swedish University of Agricultural Sciences. Her main research interests are environmental impact assessment, landscape planning, landscape assessment, policy implementation, and planning practice.

Ernest Alexander is Professor Emeritus of Urban Planning at the University of Wisconsin-Milwaukee, USA, and now practices in Israel. His research focuses on decision making, organizations, institutions and institutional design, and design and evaluation in the planning process.

Jos Arts is Professor of Infrastructure Planning, University of Groningen, the Netherlands. He is also a strategic advisor at the Dutch Ministry of Infrastructure and Environment. His research interests include impact assessment and environmental, spatial and infrastructure planning.

Mark Baker is a reader in planning policy and practice within the School of Environment, Education and Development at the University of Manchester, UK. His research interests include the UK planning system, regional and strategic planning, and plan-making processes.

Matteo Basso is a PhD candidate in regional planning and public policy at the IUAV University of Venice, Italy. His research interests are spatial transformations analysis, public policy analysis, and planning theory.

Els Beukers was a PhD candidate at the Urban Planning Department, University of Amsterdam, the Netherlands, and is currently a policy consultant. Her interests are assessments of integrated infrastructure plans, 'soft' and local effects, and theory-practice interaction.

Like Bijlsma is a researcher at the Netherlands Environmental Assessment Agency. She is interested in contemporary urban dynamics, new towns, transformation sites, and urban areas.

Giorgio Borghelot is an urban planning practitioner working in the Venice region, Italy.

Dino Borri is Full Professor in Urban Planning Techniques, Polytechnic of Bari. His research interests include planning theory, futures research, and the application of information technology in spatial planning.

Emile Dopheide is Development Economist and Senior Lecturer at the Faculty of Geo-Information Science and Earth Observation (ITC), University of Twente, the Netherlands. His research interests are the use of assessment methods and indicators in spatial planning processes.

Sylvia Dovlén is Assistant Professor at the Department of Urban and Rural Development, Swedish University of Agricultural Sciences, Uppsala, Sweden. Her research interests involve the role of sustainable development in municipal and regional planning.

Ana Maria Esteves is Director with Community Insights Group, a boutique consulting company operating in the mining, oil, and gas sector. She is interested in local content, social performance, and social impact management. Ana Maria is also a visiting professor at the University of Strathclyde in Glasgow, Scotland.

John Gaber is a professor in the Department of Political Science at the University of Arkansas, USA. His research interests are in planning and program evaluation, community health, and planning research methods.

Maria Giannopoulou is Associate Professor in Urban and Regional Planning at the Department of Civil Engineering, Democritus University of Thrace, Greece. Her research focuses on spatial analysis and planning, multi-criteria decision analysis, GIS, and urban planning.

Laura Grassini is Adjunct Professor in Territorial and Environmental Planning at the Department of Civil, Environmental, Territorial Engineering and Chemistry of the Technical University of Bari. Her research interests are strategic planning, environmental planning, and technological change.

David Hamers is Senior Researcher at the Netherlands Environmental Assessment Agency. His research interests are (sub)urban development and planning, and the relationship between research and design.

Niels Heeres is a PhD candidate at the Faculty of Spatial Sciences, University of Groningen, the Netherlands. His research interests include road infrastructure planning, infrastructure-land use interaction, and instruments for planning and decision support.

Stephen Hincks is a senior lecturer in the School of Environment, Education and Development at the University of Manchester, UK. His research interests are conceptualising, measuring and analysing the dynamics of spatial development, and the analysis and evaluation of spatial and territorial policies.

Anton van Hoorn is Senior Researcher at the Netherlands Environmental Assessment Agency. He focuses on the development of cities, including themes such as airports, retail, energy, and Europe.

Angela Hull holds the Chair in Spatial Planning at Heriot Watt University in Edinburgh, UK. She carries out research on spatial planning and governance, and institutional triggers to sustainable urban management (transport, energy, housing, etc.).

Tom Kauko is the leader of the Urban, Rural and Regional research group at the Department of Geography, Norwegian University of Science and Technology. His main fields of research interest cover property value (mainly concerning the residential sector) and sustainable built environment (mainly urban regeneration).

Abdul Khakee is Professor Emeritus in the Department of Planning and Environment, Royal Institute of Technology, Stockholm. He studies the theory and practice of urban planning, investigating planning evaluation, the relationship between future studies and planning, planning ethics, public participation, and cultural policy.

Atif Kubursi is Professor Emeritus of Economics at McMaster University in Canada. His research interests are macroeconomics, applied econometrics, development economics, climate change, and risk management.

Nancy S. Lee is a research associate with the Dukakis Center for Urban and Regional Policy, Boston, USA. Her research interests are in local and regional economic development, sustainability, energy, and transportation.

Caterina De Lucia is Assistant Professor in Environmental Economics at the Department of Economics, University of Foggia, Italy. Her research interests are: vulnerability issues due to climate change, natural disasters, renewable energy policies and consumer behaviour, and economic growth-environment relationships.

Francesca Medda is Full Professor in Applied Economics and Finance at University College London. She is Director of the UCL QASER Laboratory. Her research focuses on innovative financial instruments, smart cities innovation, infrastructure investments, and risk analysis.

Sahar Mirzaee is a graduate student at Northeastern University, Boston. Her research interests are environmental economics and policy, economic modelling, optimization and decision making, climate change, and adaptation policies.

Vítor Oliveira is Senior Researcher at the Research Centre for Territory Transports and Environment, University of Porto, and Professor of Planning at the Universidade Lusófona do Porto, Portugal. His research areas are urban morphology, urban planning, and evaluation.

Tammy Overacker is an adjunct lecturer in the Department of Political Science at the University of Arkansas, USA. Her research interests are in community development, Health Impact Assessments (HIAs), and program evaluation.

Domenico Patassini is former Full Professor, IUAV University of Venice, Italy. His research interests are with evaluation theories and practices, landscape planning, development projects, and risk assessment.

Anastasia Roukouni is a PhD Candidate at the Department of Rural and Surveying Engineering, Aristotle University of Thessaloniki, Greece. Her research interests are transportation planning, land use, transportation interaction, evaluation methods, innovative financial instruments, and GIS.

Karsten Rusche is Researcher as ILS, the Research Institute for Regional and Urban Development in Dortmund, Germany. His research interests are valuation of public goods provision (e.g. green infrastructure), ecosystem service benefits, spatial econometrics, and statistics.

Matthias Ruth is Professor of Public Policy and of Environmental Engineering, and Director of the Resilient Cities Laboratory at Northeastern University in Boston, USA. His research focuses on dynamic modelling of natural resource use, industrial and infrastructure systems analysis, and environmental economics and policy.

Andreas Schulze-Bäing is Research Associate at the Department of Geography and Planning, University of Liverpool, UK. His research interests are land-use planning, strategic spatial policy, and rural–urban relationships.

Taede Tillema is Researcher at the Netherlands Institute for Transport Policy Analysis. His research interests include road infrastructure planning, accessibility studies, and pricing policies.

Frank Vanclay is Professor of Cultural Geography in the Department of Cultural Geography, Faculty of Spatial Sciences, University of Groningen, the Netherlands. His research interests are in the areas of social impact assessment and management, social performance, social licence to operate, community engagement, corporate social responsibility, human rights impact assessment, and place; as well as social aspects of natural resource management, environmental sociology, rural sociology, sociology of food and agriculture, and applied social research in general.

Athanasios Vavatsikos is Lecturer in Spatial Decision Support Systems at the Department of Production Engineering and Management, Democritus University of Thrace, Greece. His research focuses on multi-criteria decision analysis, GIS, and project management.

Brian Webb is a research associate in the Centre for Urban Policy Studies at the University of Manchester, UK. His research lies at the intersection between intergovernmental spatial policy coordination, spatial analysis, and the delivery of national and regional infrastructure.

Jost Wilker is Researcher at ILS, the Research Institute for Regional and Urban Development in Dortmund, Germany. His research interests are green infrastructure, ecosystem services, economic valuation, and participation.

Johan Woltjer is Professor of Urban Infrastructures, University of Westminster, UK. He was affiliated with the University of Groningen, the Netherlands, until 2014. He concentrates his research work on institutional innovations for urban regions and infrastructure development and, among other things, the role of capacity, governance and resilience.

Cecilia Wong is Director of the Centre for Urban Policy Studies and Co-Director of Cities@Manchester at the University of Manchester, UK. Her research areas include developing quantitative Indicators, spatial policy monitoring and evaluation, and spatial analysis and planning methodology.

Junming Zhu is a postdoctoral associate at the School of Forestry and Environmental Studies, Yale University, USA. His research interests are energy and resource policies, industrial ecosystems, and industrial environmental management based on case studies, econometrics, and network analysis.

Preface

Many of our activities to provide adequate infrastructures and develop liveable cities and regions involve evaluation. Planners, policy makers, researchers, politicians, and citizens worldwide need to estimate the success and understand the future impacts of policies and projects. These actors typically analyse their utilization, assess the social or economic opportunities that infrastructures, cities and regions provide, and project their projects' environmental effects. Evaluation involves tools and institutional arrangements for the interpretation of possible or actual impacts of proposals and existing results. Planning evaluation is about making judgments about the value of proposals and policies in urban planning and land use management. We need evaluation to make good decisions on future infrastructure, and to understand failures and successes of earlier decisions as well.

These kinds of evaluation-related activities have been central to research by the 'Planning and Evaluation Network', an extensive international network of academics from a variety of experiences and disciplines (urban planning, regional studies, policy sciences, environmental management). The members of this network aim to share research, exchange views and discuss the current state and future of evaluation in planning. This volume contains a set of original contributions that address an emerging theme focusing on place and local integration for evaluating infrastructure and planning projects. The chapters present some new lines of thought, policy innovations and cases from practice in Europe and the USA.

All the contributions were presented and discussed at the 8th International Workshop on Evaluation in Planning at the University of Groningen in 2013. Later the contributions were reviewed by expert policy practitioners as well: officials from the Dutch Ministry of Infrastructure and Environment, who provided constructive reflection and support. This book has been compiled following a tradition that owes a great deal to the driving force of Professor Abdul Khakee. This approach expresses international academic insights as well as policy practice, identifies policy innovations and employs case study illustration.

Johan Woltjer, Ernest Alexander, Angela Hull, Matthias Ruth

List of Abbreviations

AHP	Analytic Hierarchy Process
AICP	American Institute of Certified Planners
AIR	*Architectuurinstituut Rotterdam* (Rotterdam Architectural Institute)
AVI	Italian Vulnerability Area
BAD	Benefit Assessment District
BCA	Benefit-Cost Analysis
BID	Business Improvement District
BREEAM	Building Research Establishment Environmental Assessment Method
BRS	Business Rate Supplement
CAB	Citizen Advisory Boards
CBA	Cost-Benefit Analysis
CCF	Climate Challenge Fund
CEE	Central East-European
CGE	Computable General Equilibrium
CHF	Swiss Francs
CIA	Community Impact Assessment
CIL	Community Infrastructure Levy
CO_2	Carbon Dioxide
CSR	Corporate Social Responsibility
DBFM	Design-Build-Finance-Maintain
DCLG	Department for Communities and Local Government
DFB	Design-Finance-Build
DLR	Docklands Light Railway
DM	Decision Maker
DOD	US Department of Defence
DOE	US Department of Energy
EIA	Environmental Impact Assessment
EIS	Environmental Impact Statement
EPA	US Environmental Protection Agency
ESDP	European Spatial Development Perspective
EU	European Union
EUR	Euros
FPIC	Free, Prior and Informed Consent
GBP	British Pounds
GDP	Gross Domestic Product

GHG	Greenhouse Gas
GHGs	Greenhouse gas emissions
GIS	Geographic Information Systems
GLA	Greater London Authority
GTH	Green Road Infrastructure
GVA	Gross Value Added
HCfSE	Head of Centre for Sport and Exercise
HIA	Health Impact Assessment
I&M	*Infrastructuur en Milieu* (Dutch Ministry of Infrastructure and Environment)
ICLEI	International Council for Local Environmental Initiatives
IEA	International Energy Agency
IIED	International Institute for Environment and Development
IPCC	Intergovernmental Panel on Climate Change
IPM	Ideal Point Methods
ISO	International Organization for Standardization
KSH	Hungarian Statistical office
LULU	Locally Unwanted Land Uses
MTH	Mestre Through Highway (*Passante di Mestre*)
NCE	Normal Curve Equivalent
NGO	Non-Governmental Organization
NIMBY	'Not In My Backyard'
NIS	Negative Ideal Solution
NMVOCs	Non-methane Volatile Organic Compounds
NNDR	National Non-Domestic Rates
NO_x	Nitrogen Oxide
NPPF	National Planning Policy Framework
NPV	Net Present Value
OECD	Organisation for Economic Co-operation and Development
PBL	Netherlands Environmental Assessment Agency
PDM	*Plano Director Municipal*
PIE	Plan Implementation Evaluation
PIS	Positive Ideal Solution
PM_{10}	Particulate Matter <10 microns
POS	*Plan d'Occupation des Sols*
PPG	Planning Policy Guidance
PPP	People-Planet-Profit
PPP	public-private partnership
PPR	Plan-Process-Results
PPS	Planning Policy Statement
PUD	Planned Unit Development
R&D	Research and Development
RTPI	Royal Town Planning Institute
RWS	*Rijkswaterstaat* (Dutch Public Works Agency)

SEA	Strategic Environmental Assessment
SFMR	Regional Railway System
SHLAA	Strategic Housing Land Availability Assessment
SHMA	Strategic Housing Market Assessment
SIA	Social Impact Assessment
SIMP	Social Impact Management Plan
SOM	Self-Organising Map
SO_x	Sulfur Oxides
SRI	Socially Responsible Investment
TDR	Transferable Development Right
TEEB	The Economics of Ecosystems and Biodiversity
TEN	Trans European Network
TfL	Transport for London
THW	Transition Heriot-Watt
TIF	Tax Increment Financing
TOD	Transit Oriented Development
UN	United Nations
USD	US Dollars
VC	Value Capture
VCF	Value Capture Finance
VOC	Volatile Organic Compounds
WHO	World Health Organization
WTP	Willingness To Pay
WTPs	Water Treatment Plants

Chapter 1

Place-Based Evaluation for Infrastructure and Spatial Projects: An Introduction

Johan Woltjer, Ernest Alexander, Angela Hull and Matthias Ruth

Introduction

In recent years, there have been some major changes in the management of planning projects and infrastructure development, such as roads, rail and waterways. The emphasis is increasingly on local and regional integration of these projects. Besides the linkages between projects, their value and interactions with other related planning matters including environment, housing, industry, green and water have become more pertinent. In other words, land-use planning and infrastructure management have become spatially and thematically more integrated (e.g. Black, 2010; Bertolini, 2012; Geerlings et al., 2012; Hull, 2008; Hijdra et al., 2013; Busscher et al., 2014).

These changes have a profound influence on questions of evaluation: the qualities legitimate project proposals should have, the benefits and costs related to development initiatives, the complexity and effectiveness of integrated land-use management practice. These kinds of questions are central to planning evaluation. The assumption behind planning evaluation practices is that well-considered assessment and analysis help planners to clarify impacts of projects, make proposals more legitimate and make planning intervention useful given societal needs. Evaluation research has been involved in suggesting tools and designing rules and measures, not only for expressing levels of socio-economic progress and development, but also in terms of environmental and institutional realities (Khakee et al., 2008; Oliveira and Pinho, 2010).

The challenge now is to reinforce local and regional consideration of planning projects, and establish a stronger place-based understanding for their evaluation. Planning evaluation then implies looking at local and regional circumstances, and establishing an 'open eye' among evaluators for the specifics of cases in terms of local values, benefits, impacts, synergies, use, complexities and spatial change. This book, therefore, presents pointers for improving evaluation and the institutional design of evaluation processes for place-based infrastructure development and spatial planning.

This book brings together contributions from experts in the fields of spatial planning, regional science and infrastructure management to tackle an emerging agenda of spatially-oriented integrated evaluation. The book sets out to clarify the nature and roles of evaluation in the wider context of current planning and policy practices, presenting current academic thinking and concepts, case studies, methods, and policy and practice review to examine and assess integrated land-use management.

Place-Based Evaluation

The idea of incorporating and internalising various place-based factors into regional policy-making and planning evaluation has been a prominent theme in recent years (e.g. McCann et al., 2012). Emphasizing place in planning evaluation implies a broad definition and scope of projects, plans and programs. Local capacities such as levels of innovation, ecological resources, financial opportunity or political support are important, as are externalities between infrastructure and environmental factors. A place-based approach implies the integration of distinctive spatial circumstances into broader policy-making and evaluation practice. Evaluation tools, then, are area-oriented, and seek to express qualities at specific locations. Assessment of infrastructure and spatial projects requires less focus on generic indicators such as regional income, and relies more on contingent, specific markers for evaluation like local capacity.

A place-based approach raises several implications for planning evaluation research and practice. The emphasis on place and spatial context implies the need for distinct assessment items for evaluation such as co-benefits and co-costs, social impacts, individual value, long-term effects, and community engagement. Methodological improvements are also needed. Evaluation tools such as cost-benefit analysis (CBA), geographical information systems (GIS), scenario studies, institutional analysis and environmental assessment should express local geographies more clearly. Overall, the book has four kinds of implications: the need to express value and benefit, a focus on impacts in place, locally based spatial analysis, and the importance of institutional design for spatial change. These implications are briefly discussed below.

A first implication of a place-based approach is that evaluation practice emphasizes the importance of *expressing value and benefit* in land-use and infrastructure development. An important aspect is that evaluation can help clarify the values spatial plans and projects derive from infrastructure. Tools such as impact studies, economic assessments, and broader cost-benefit analysis can be helpful. Decisions on value-capture, for example, may then be better informed. Another aspect is that land-use projects typically generate mutual benefits and costs such as longer-term accessibility changes to green and urban space, which determine the quality of cities and regions. At the same time, little is known about these co-benefits and co-costs (Ruth, 2013). Evaluation practice, therefore, should also focus on the measurement and consideration of hidden and cumulative benefits from infrastructure use and the broader potential of infrastructure projects, and include these kinds of benefits and costs. Evaluation that includes co-benefits and co-costs requires decisions on the demarcation of the relevant area and time horizon.

The second implication involves a *focus on impacts in place*. Understanding local conditions and local capacities is increasingly important in planning evaluation. Place-oriented evaluation explores both the spatial and institutional integration of physical infrastructures with other uses. An important consequence of this approach is that recipients and users of infrastructure facilities and impacts are a key point of reference in evaluation. In other words: linkages between infrastructure supply and demand, and those affected (individuals, places, groups, users, communities), are central. The evaluation process involves looking at specific effects on certain groups in society, and estimating how impacts are accumulated over time, in space, and origin.

A third implication of a place-based approach to evaluation involves *locally based spatial analysis*. A local emphasis in evaluation implies that evaluators work with open source, contextualized, and community-oriented evaluation data. Professional and administrative data from specialist monitoring systems, for example for noise measurement or using transport modelling, would be supplemented by local insight and knowledge. Such an emphasis also implies the need for participatory processes in applying tools such as cost-benefit analysis (CBA). Local involvement in CBA can make evaluation situation-specific, and thus provide a better understanding of relevant environmental conditions and capacities. It also allows for learning processes aimed at generating knowledge, structuring options, identifying compliance, and perhaps improvement of the underlying plan or project.

The fourth kind of implications involves *institutional design for spatial change* and stakeholder opportunity. This means that evaluation should concentrate on how local institutions determine who is involved in spatial change and who might benefit from projects and potentially contribute to them. Benefits and contributions may range from issues of employment, the improvement of facilities, combining investments, or contributions in terms of knowledge and commitment. Such an involvement involves institutional design for value-capture, partnerships, social responsibility and 'buy-in'. Such an approach needs evaluative insight into the position of the parties involved in an infrastructure project, and the extent to which the infrastructure project or plan offers improved options and opportunities for local stakeholders.

Overview of the Book

The chapters of the book are organized into four parts:

- Part I: Evaluating Value and Benefit in Land-Use and Infrastructure Development;
- Part II: Understanding the Evaluation of Impacts and Space;

- Part III: Spatial Analysis for Integrated Projects;
- Part IV: Evaluating Planning Intervention, Institutions and Spatial Change.

Part I: Evaluating Value and Benefit in Land-Use and Infrastructure Development

The chapters in this section show how places of infrastructure projects are often associated with planned projects' direct economic value and costs. But the longer-term and more indirect user benefits from public values embedded in projects such as transit, highways and waterways also deserve attention. Evaluation practices, therefore, should better understand the relation between assets and users, and apply user-oriented criteria. Measurement of such values should be dynamic, ongoing, and include implicit and more indirect benefits and place-specific characteristics of a plan or project. New institutional arrangements are required to make these benefits explicit.

Ernest Alexander in his chapter discusses the important role of institutional design for planning and delivering infrastructure projects. Institutional design characteristics such as organizational structures, rules and procedures are essential to facilitate effective planning processes. This chapter asks the question what kind of institutions, organization and processes are best suited for effective planning, delivery and operation of a particular infrastructure project in its specific context. Value capture is seen as critical, as it ensures the funding needed to make projects feasible. The position of evaluation includes assessing alternative institutional designs' value-capture potential. The chapter, therefore, emphasises the need for considering alternative institutional designs for value capture, particularly special assessments, functional authorities, and specified forms of public-private partnership.

Matthias Ruth, Junming Zhu, Nancy S. Lee and Sahar Mirzaee call attention to a couple of innovative aspects for policy and planning – the co-benefits and co-costs of environmental planning, policy and investments, and the indeterminacy of causal relationships between system interventions and outcomes. Their chapter argues that plans, policies and investments generate co-benefits and co-costs (such as health benefits from policies proposing traffic congestion reductions to improve transport), and that their magnitude can easily be decisive for decision making. The chapter also explores how co-cost and co-benefit analysis may be used to help shape planning, particularly through institutional innovation needed for capture of co-benefits, and minimization of co-costs.

The chapter by *Karsten Rusche and Jost Wilker* starts from the principle that high quality green environments have a significant positive impact on the attractiveness of cities and regions, and deliver economic, social and environmental benefits. The role of evaluation in this chapter largely is to clarify and justify investments in green infrastructure. The chapter focuses on the economic value and individual benefits of a series of landscape parks in the city of Stuttgart. Results from the analysis in this chapter show that benefits from green infrastructure generally well exceed their costs. The most significant benefit gains are generated through recreation and leisure, improved river access, and health and well-being. A detailed analysis like this shows specific values from green, and the usefulness to specify benefits for use in strategic planning.

Anastasia Roukouni, Francesca Medda, Maria Giannopoulou and Athanasios Vavatsikos use the Crossrail project in London to show how evaluation can express the contribution of transport investment to sustainable economic growth. The focus is on land value capture as a tool for funding high cost public transport systems. In the case, a levy called the Business Rate Supplement is used to raise funds from infrastructure generated value. Special attention in the analysis is spent on issues of timing, as value capture strategies are based on dynamic development, and distance, given space infrastructure and their zones of impact. The chapter essentially highlights the idea of assessing added value and using value capture finance for large transport infrastructure investment at a wide level of scale.

Part II: Understanding the Evaluation of Impacts and Space

The understanding from this theme is that evaluation activities should express more clearly the place-based spatial characteristics within which planning and plans unfold, and the impacts plan implementation has on local economies, the communities in which these economies function, and the ecosystems within which all of them are embedded. These characteristics include, in particular, institutional capacities, local economic potential, social impacts, and benefits broadly defined. Therefore, evaluation work must be place-based, and should contribute to raising spatial awareness among public and private stakeholders. One way to generate such awareness is to develop participatory evaluation processes that include the dissemination of planning and decision support tools and results to the broader public, and the associated generation of a community-based 'evaluation vocabulary'.

One such evaluation process is the Social Impact Assessment (SIA) discussed by *Frank Vanclay and Ana Maria Esteves*, who emphasize recent trends in moving SIA from traditional ex ante prediction of negative impacts to a new paradigm of seeking to maximize positive outcomes to communities while minimizing harm. Since plans are established and investment and policy decisions are typically made under incomplete information, the SIA process is carried out as an adaptive management process in which all stages from pre-establishment of plans to outcomes post closure are monitored and evaluated to inform subsequent adjustments, learning, and re-intervention in the complex systems that plans try to shape.

Since communities are integral to the success of plans and the adaptive management that should guide them, engaging communities is essential to both the planning and evaluation process. Despite considerable experience with community engagement across a range of applied research fields, little systematic information exists in the planning literature that provides clear direction to inform practicing planners on their community engagement technique options, such as surveys, focus-group meetings or workshops, for example. Drawing on experiences in the related field of health impact assessments (HIAs), *John Gaber and Tammy Overacker* in their chapter distil information from 95 international projects on community engagement activities with the goal of better understanding the practices and experiences of community health planners with community engagement processes. These experiences, they argue, can provide valuable insight to the plan evaluation process.

In the following chapter, *Vitor Oliveira* presents the Plan-Process-Results (PPR) methodology developed to evaluate planning and plan implementation, and demonstrates its application to the Plano Director Municipal (PDM), the master plan for Porto, Portugal. A rich data set for the application of the PPR methodology includes the plan itself, other regional and strategic plans that affect or are affected by it, interviews, official statistics and cartographic material, as well as public accounts such as newspaper articles. This data set allows for a rich analysis applying a wide range of place-specific evaluation criteria, ranging from internal consistency of the plans and their relevance in the broader context of planning goals, to public participation, commitment of adequate resources for plan implementation, and plan effectiveness. With this chapter, Oliveira showcases how the PPR process can directly shape the design of plans and of planning practices that are being prepared, and identifies areas for future research in planning and evaluation.

Plans affect future realizations of local conditions, and as such are also based on the anticipation of such conditions. However, a wider range of futures will likely prevail than what is typically assumed in the planning process. Careful integration of future scenarios can therefore broaden the perspectives of planners and researchers concerned with both the planning and evaluation processes. *Abdul Khakee and Laura Grassini* attend to the methodological and practical challenges of using future scenarios in that manner and illustrate the approach with an application to a case study in Izmir, Turkey. That case study shows how future scenarios can provide deeper and richer appreciation of present space and thereby improve planning practice.

Another set of constraints on and synergies for current planning actually involves inconsistencies less between future scenarios but rather with the broader landscape of already existing plans and frameworks. To the extent that other plans and policy frameworks are not considered, conflicts may emerge, or opportunities to generate co-benefits may be missed in the planning process. This is the case discussed by *Cecilia Wong, Brian Webb, Andreas Schulze-Bäing, Mark Baker and Stephen Hincks*. These authors use GIS mapping overlays to identify the patterns of spatial synergies and conflicts that arise from sectoral government policies and programmes. They illustrate their approach for the case of housing delivery in England and highlight that even relatively simple mapping overlays can greatly inform policy debates and encourage enhanced partnerships among government policy makers and stakeholders. Such partnerships may result in enhanced coordination, management and delivery of complex spatial planning policies across different spatial levels.

Domenico Patassini, Matteo Basso and Giorgio Borghelot evaluate spatial changes of location patterns of economic activities generated by the development of large infrastructure systems, such as regional transport networks. Such infrastructures may serve as an important pull for economic activities, provide a source for agglomeration economies, economic multipliers and accelerators, and thus serve as a key factor of regional competiveness and have far-reaching social and environmental impacts. Their analysis showcases the large-scale and long-term impacts of the 'Mestre Through Highway' within the Venetian Metropolitan Area of Italy on spatial patterns of economic activity. The challenges associated with shaping the planning and implementation of the Mestre Through Highway demonstrates the limitations of good spatial governance when administrative procedures are characterized by inertia, when business interests motivate select communities and interest groups, and when adverse effects are diffuse and long-term.

Part III: Spatial Analysis for Integrated Projects

This section demonstrates how evaluation tools, such as cost-benefit analysis (CBA), can be made more 'spatial', and how efforts are under way to express synergies and benefits from projects, in a more distinct way. With this objective in mind, each of the four presentations explores how to adapt the strengths of CBA with its focus on single projects to more effectively assess integrated transport plans. All the chapters are written by scientists working in the Netherlands. Two chapters reflect on how CBA is used there and seek to improve the process. One chapter critically assesses the 'Sustainability Check' [in Dutch *Omgevingswijzer*] instrument and another develops a new tool: the Plan Review. These chapters suggest how CBA may be adapted as a learning tool, and how the results of spatial analysis may be merged with other evaluation tools to improve decision-making support. They also point out that planning evaluation needs to assume a stronger focus on its users.

Niels Heeres, Taede Tillema and Jos Arts develop their chapter in the context of current discussions in the Dutch Ministry of Infrastructure and Environment on how to improve the instruments that are used in the early stages of plan making. They show how these discussions lead to the development of new planning instruments that can assess the spatial effects of integrated infrastructure projects and can support a collaborative planning process with representatives of different government functions and different disciplinary backgrounds. Following a review of the planning instruments available, the authors critically assess the capability of the 'Sustainability Check' instrument with data derived from interviews with experts in the field. They find that, although, this instrument enables cooperation, learning and the finding of common ground for action, further refinement is needed to ensure that social values are embedded in the decision-making process.

Emile Dopheide's chapter aims to ensure that CBA is used effectively as a tool to exchange learning in the decision-making process rather than being a 'black box' where the end result is delivered by a consultant to the client. Accordingly, he focuses on the final users of CBA and what they understand about the content and scope of CBA. First he reviews the substantive and process bottlenecks in using CBA to assess the effects of infrastructure projects and then he argues that these bottlenecks should be made clear and transparent to the end user. He concludes that more research needs to be carried out to understand the relationship between CBA outcomes and the decision outcome and the extent to which the final users actually understand and can interpret the CBA results.

David Hamers, Like Bijlsma and Anton van Hoorn develop a new instrument – The Plan Review – to address some of the weaknesses of CBA. Their aim is to facilitate the consideration of multi-level policy goals in an increasingly dynamic planning practice where projects are often small-scale, and adaptive, and promoted by many different stakeholders. The Plan Review is a matrix that considers 16 different spatial conditions or spatial qualities (policy requirements) and enables the reviewers to reflect and consider how the project plan fits with higher-level objectives. The matrix structures a dialogue focusing on context sensitivity and the project plan's reasoning, to help the plan reviewers to compare alternatives, rank them and explore possible plan improvements.

Els Beukers clearly acknowledges some of the limitations of CBA and seeks to refine it as a tool to facilitate communication, learning and reflection. Her CBA-Dialogue tool is tested in two experiential case studies where it structures a two-way dialogue between the plan owners and the CBA evaluators. This works in her cases, marked by high levels of interpersonal skills and trust, to enable a valuable exchange of knowledge to help refine the integrated plan. She warns, though, that it is still difficult to see CBA as a standardized tool for integrated transport plan assessment, since the spatial context of each plan needs to be considered so that the spatial and synergistic effects of plan proposals can be carefully assessed.

Part IV: Evaluating Planning Intervention, Institutions and Spatial Change

This section reaffirms the importance of spatial awareness: evaluation in (and of) planning should include a clear understanding of the linkages between various spatial activities and land uses. One of these chapters' salient conclusions is the need for attention to institutional design – both of the concerned plans, projects and programs, and of the evaluation processes and contexts themselves. Different institutional arrangements are evaluated in a variety of contexts, from Swedish local planning through Budapest urban renewal and Italian land-use policy, to Scottish university communities. Evaluations apply diverse methods and innovative approaches, often integrating quantitative and qualitative measurement and analysis, to enhance contextual awareness in urban projects. These chapters demonstrate how evaluation instruments can be dynamic and provide timely evaluative information on institutions and spatial change.

Angela Hull's chapter applies institutional analysis to evaluate a project designed to encourage sustainability enhancing behaviour, based on research into individual and collective behaviour change. The case is a community project in a Scottish university to promote environmental projects such as bicycling, recycling, and communal gardening. The analysis found that bureaucratic obstacles prevented effective action, and concluded that the institutionalization of shared values is critical for achieving significant behaviour change.

Ann Åkerskog, Sylvia Dovlén and Abdul Khakee's contribution is an evaluation of planning: how well are sustainability factors integrated into Swedish local planning. Their qualitative analysis covers two case studies. One assesses the introduction of an environmental perspective in sampled localities before and after implementation of the EU-SEA Directive; the other evaluates the integration of energy efficiency in a set of municipal structure plans. The study found that communities' pre-existing sustainability orientation explains much of the differences between municipal plans.

Tom Kauko offers a place-based evaluation of Hungarian urban revitalization. An innovative quantitative-qualitative methodology is applied to assess the impacts of selected urban renewal projects in Budapest in an overall framework of evaluating their sustainability contributions. Self-Organizing-Mapping measures projects' impacts on property price stability; field survey evaluates their physical and social impacts through observation, plan-document analysis and interviews. Institutional analysis of project planning and financing yields insights on the effectiveness of alternative ways to organize and implement urban renewal.

Catarina De Lucia, Atif Kubursi and Dino Borri raise the issue of vulnerability in public policy analysis with a striking application of place-based evaluation This issue is important because adaptation to climate change demands consideration of possible catastrophic events: estimating the potential impacts (social and economic) of floods, landslides, earthquakes etc. This chapter presents a systematic method of vulnerability assessment, which is demonstrated by assessing the local impacts of landslides in Italy. The policy relevance of such analysis is illustrated in the chapter's conclusion tracing its implications for Italian planning, land-use and development control.

Key Conclusions

The central conclusions from the book are:

- Evaluation should be more clearly informed by local spatial characteristics. In particular, these include institutional capacities, local economic potential and benefits. The understanding is that evaluation work should be more area-based and should raise spatial awareness among the various public and private parties involved in land-use and infrastructure development.

- Evaluation tools should be developed and used to inform strategic planning. To be useful, they should be user friendly and function as an intermediary between public and private users. Evaluation instruments should be dynamic and provide consistent and timely information.
- Evaluation in (and of) planning should include a well-developed appreciation of the linkages between different spatial activities and/or land uses. This means expressing more indirect and longer-term impacts of planned interventions (such as strategic infrastructure projects and major facilities), and including co-costs and co-benefits, unobserved values, and transaction costs. The call here is for evaluators to think in terms of synergies, for example, between a road project and surrounding property.
- Attention to institutional design in evaluation activities is required. Current problems in the field of land-use and infrastructure development, such as the need to raise private funding and the need to better understand social impacts, make it necessary to articulate and evaluate the value of public spatial investment more explicitly. These evaluation demands imply the need to establish new organizational arrangements and rules.

References

Bertolini, L. 2012. Integrating mobility and urban development agendas: A manifesto. *DISP*, 48(1), pp. 16–26.

Black, W.R. 2010. *Sustainable Transportation: Problems and Solutions*. New York: The Guilford Press.

Busscher, T., Zuidema, C., Tillema, T. and Arts, J. 2014. Bridging gaps: Governing conflict between transport and environmental policies. *Environment and Planning A*, 46(3), pp. 666–681.

Geerlings, H., Shiftan, Y. and Stead, D. 2012. *Transition Towards Sustainable Mobility: The Role of Instruments, Individuals and Institutions*. Farnham: Ashgate.

Hijdra, A., Woltjer, J. and Arts, J. 2013. Value creation in capital waterway projects: Application of a transaction cost and transaction benefit framework for the Miami River and the New Orleans Inner Harbor Navigation Canal. *Land Use Policy*, 38, pp. 91–103.

Hull, A., Alexander, E., Khakee, A. and Woltjer, J. 2010. *Evaluation of Sustainability and Participation in Planning*. London: Routledge.

Khakee, A., Hull, A., Miller, D. and Woltjer, J. 2008. *New Principles in Planning and Evaluation*. Aldershot: Ashgate.

McCann, P., Barca, P. and Rodriguez-Pose, A. 2012. The Case for Regional Development Intervention: Place-Based versus Place-Neutral Approaches. *Journal of Regional Science*, 52(1), pp. 134–152.

Oliveira, V. and Pinho, P. 2010. Evaluation in Urban Planning: Advances and Prospects. *Journal of Planning Literature*, 24(4), pp. 343–361.

PART I:
Evaluating Value and Benefit in Land-Use and Infrastructure Development

Chapter 2

Evaluation in Institutional Design for Infrastructure Planning and Delivery

Ernest Alexander

Introduction

Institutional Design

Why do the planning and delivery of major strategic infrastructure involve institutional design? Because institutional design is needed to address some critical issues: how to plan the infrastructure project under consideration, how to ensure effective implementation of the planned project, and how to organize efficient delivery and life-cycle operation of the proposed infrastructure system and service.

Enlisting, creating or transforming organizational structures and processes for planning, procuring, constructing and operating infrastructure systems – networks and facilities and their related services – is an integral part of deployment and planning for such projects, involving organization and reorganization and possible changes in legislation, regulations, standard operating procedures and so on. These issues raise the question of appropriate institutions and institutional contexts for infrastructure planning and delivery.

Conventional approaches to this question reduce it to the common dispute for or against privatization. A more sophisticated approach disaggregates the infrastructure planning and delivery process, to explore the consequences of private vs. public provision of its relevant constituent elements (Cannadi and Dollery, 2005). Here the main issues are:

Procurement: public, private or a combination – public-private partnerships (PPPs) in various forms (Whittington, 2012). These can involve alternative infrastructure procurement structures and processes from design-build to design-build-finance-operate-maintain.

Financing: public vs. private; here again PPPs are possible. Public or private financing raises the question of value capture potential and alternative infrastructures for value capture, e.g. special assessments, special districts, or PPPs based on land sequestration (Alexander, 2012).

All these are essentially institutional design issues that are developed in more detail below.

Evaluation

How does evaluation come into institutional design? Institutional design demands three steps:

1. Analyse the relevant case/issue: identify critical elements (actors and transactions) and their relationships, assess strengths, weaknesses, problems and so on. In infrastructure procurement, for example, this might involve evaluating local contractors' value of time and their attitudes to risk. For infrastructure financing, value capture potential is critical. This demands place-related evaluation: assessing and evaluating the impacts of the subject (or similar) public project and their spatial distribution.
2. Design feasible alternatives; these are the subject of the evaluations that are the topic of this chapter. Such design is essential to follow the transaction-cost theory maxim of 'remediability', which demands comparative evaluation rather than prescribing some ideal standard or norm (Williamson, 1995). Alternative designs for infrastructure procurement can be drawn from the repertoire of various forms of PPPs that have been developed and tried, while alternative strategies for funding infrastructure can combine financing as a procurement element with the institutional forms of value capture discussed below.
3. Evaluate alternative institutional designs that address the relevant issue, problem or case. The base-line for this evaluation is the existing institutional configuration and context. This gives meaning to the maxim of 'remediability': if there is no real feasible alternative institutional design that proves superior to the *status-quo* the case in question is not a problem and can be effectively handled by the existing institutions.

This introduction spells out the logic generating this chapter's topic: evaluation in institutional design for infrastructure planning and delivery. The final product here is a review and assessment of various evaluation processes and methods applied in appraising alternative institutional designs for infrastructure planning and delivery. But first there is a brief general introduction to institutional design, followed by discussions of infrastructure planning and procurement, and of infrastructure financing and value capture.

Institutional Design[1]

What is Institutional Design

To understand institutional design, we have to define institutions. Institutions are:

> the rules of the game in society … the humanly devised constraints that shape human interaction … complexes of norms and technologies that persist over time by serving collectively valued purposes … some have an organizational form, others exist as pervasive influences on behavior. (North, 1990, p. 3)

Institutional design, then, means designing institutions: devising and realizing rules, procedures and organizational structures to enable and constrain behaviour and action and conform them to held values, achieve desired objectives or execute given tasks (Alexander, 2006, p. 4). When infrastructure planning and delivery demand new organization or reorganization, legislation, regulation or new routines and procedures, institutional design will be needed.

Institutional Design: Where and Who?

We can distinguish between three 'levels' that are associated with different types of institutional design. The highest level applies to whole societies or addresses significant macro-societal processes and institutions. This is sometimes called 'constitution writing' (e.g. the US Constitution and the EU and its institutions) but it also includes major national reorganizations and innovative strategic political-administrative programs (from the Justinian legal code to Roosevelt's New Deal) that often occur after major societal discontinuities.

The next level of institutional design is of interest to us, because it involves planning and implementation structures and processes. This includes establishing and operating interorganizational networks, creating new organizations and transforming existing ones, and devising and applying incentives and constraints in the form of laws, regulations and resource deployment to develop and implement policies, programs, projects and plans. This is the level associated with infrastructure planning and delivery.

At the lowest level we find intra-organizational institutional design, addressing organizational sub-units and small semi-formal or informal social units and processes, such as committees, teams, task forces, work groups etc. This occurs in every field, from the global corporation's 'matrix' organization to the weekly poker club in Brady's Bar, and is intended to ensure effective and timely task performance.

---

1 This section is condensed from Alexander (2012, pp. 164–167).

Institutional Design: Theory and Experience

Applying theoretical knowledge to institutional design is problematic, for several reasons. One is ignorance, another is the intrinsic nature of design, a third is the risk of the ecological fallacy: these make mechanical application of universal abstract knowledge problematic. Finally, there is the multi-party nature of institutional design itself: often the first challenge is to get all the involved actors to agree on their common purposes.

Nevertheless, there are some areas of knowledge that may help the practitioner. At the macro- and meso-levels institutional economics in general, and transaction cost theory in particular, can be useful tools for institutional analysis. At the meso-level, the Institutional Analysis and Design approach (Ostrom, 2005) and the concept of interorganizational coordination structures (Alexander, 1995) offer a repertoire of institutional design methods and solutions. At the intra-organizational micro-level, agency theory provides concepts and models that are essential for informed intra-organizational institutional design. At the same time, relevant case-studies can enable the practitioner to draw on the vast reservoir of institutional design experience (Alexander, 2006, pp. 12–24).

Infrastructure Planning and Delivery

Procurement

How to organize and structure project planning and procurement is one of the critical institutional design issues in infrastructure planning and delivery. The constituent elements/stages of infrastructure planning and delivery systems are: 1) (General) Planning; 2) Financing; 3) (Detailed) Design; 4) Construction; 5) Operation & Maintenance; and 6) Improvement. Conventionally a public agency is responsible for the whole process, though parts are often outsourced to other public organizations and/or bid out to private firms, e.g. a transportation planning consultant office might be commissioned to plan a highway, design is outsourced to an engineering firm, construction is bid out to contractors, and a transit operating corporation might be franchised for a mass transit system's O&M. In this approach such outsourcing is by conventional procurement practices and contracts.

Privatization of infrastructure planning and delivery joins the public and private sectors differently, forming public-private partnerships (PPPs) that combine these elements in a variety of possible ways. Thus privatizing 3) and 4) together produces the 'Design-Build' form of contract, 3) + 4) + 5) = 'Design-Build-Operate-Maintain', and 2) + 3) + 4) + 5) = 'Design-Build-Finance-Operate'. In initiating a strategic infrastructure project the adoption of an appropriate structure for the procurement and delivery process is an important institutional design challenge.

Different PPP structures may be more or less effective for different cases in different contexts,[2] with significant impacts on direct and indirect project costs. These include various transaction costs, which are often neglected in conventional *a priori* project cost estimates, such as preliminary administration and research, bid and contract administration and monitoring, and dispute arbitration and resolution. Another important factor is the direct and indirect cost or value of time to the various parties involved in or affected by the project: the cumulative costs of delays attributable to poor project management or execution can be large (Whittington, 2012).

Consequently, an important part of institutional design for infrastructure planning and delivery is designing the structure of the PPP that is being considered for running the project. Referring back to the institutional design process described above, this means designing alternative feasible PPP structures for the planned project and evaluating their relative effectiveness compared to the conventional public planning, procurement and delivery system. How such evaluation is and can be done is discussed in more detail below.

Financing and Value Capture[3]

An important aspect of infrastructure planning and delivery is financing. In conventional public sector infrastructure provision this was simply ignored: government approval of a proposed project included its appropriate budget funding. That approach has long been passé: endorsement of a major infrastructure undertaking by the relevant public agencies rarely implies their commitment to finance its costs with money allocated from regular revenues. The result of several decades' retreat from public financing of major infrastructure has been increased interest in value capture.

What is Value Capture (VC)?

The concept of VC developed to address a widespread problem in public finance: the inability to fund public investments in potentially beneficial projects entirely from government (central and or local) general or capital budgets. One common way to meet this shortfall in funding is to assess part or all of the costs to the direct beneficiaries, based on the economic principle: 'the user pays'. This is done for a wide variety of public investments ranging from toll highways through user fees for sewerage systems, and sometimes makes these projects feasible. But political, social and economic-functional considerations limit the efficacy of user fees, so that many valuable public projects end up costing more than the present and projected money available.

2 There is quite extensive evaluation literature that attests to this, partly reviewed in Whittington (2012).

3 This section is condensed from Alexander (2012, pp. 167–170).

The realization that the benefits of public investments are not limited to the direct users suggested the idea of VC. The activities supported by the project usually generate a variety of indirect benefits to the residents and property owners in affected areas, which are reflected in increased land values. Thus accessibility improvements attributable to transportation investments contribute significant value increments to adjacent commercial and residential properties, and raise land values in the community at large.

Value Capture makes the expected land value increments attributable to planned public projects available to fund part of their costs. A review of VC experience provides an institutional design repertoire of how VC can be implemented; this is summarised below.

Institutional Design for VC

VC mechanisms can be grouped under three headings, which indicate different basic approaches to VC. Compulsory capture uses mandated taxes or charges; voluntary capture involves the private sector in realizing the added value; land endowment realizes the added value by selling property rights.

Compulsory capture

A variety of mechanisms have been used to effect compulsory capture of added land values. Perhaps the oldest is the general *VC property tax*, which taxes property at a split rate – one (fixed) for improvements, and another (variable) for land, which is systematically adjusted to capture value increments attributable to public investments. In several countries local governments can capture added value through *betterment fees*, but their efficiency is limited.

The most widespread bounded VC mechanism is probably *impact fees*, which are charged to developers as a condition of approving their project proposals. There are also several ways in which countries have institutionalized VC by enabling a special property tax levy earmarked for funding particular public investments. One is by creating a functional *Special District* with taxing powers, e.g. school districts, library, police and public safety, firefighting, local and metropolitan park districts, metropolitan sewerage districts, and regional river basin management and flood control districts.

The *Benefit Assessment District* (BAD) is another bounded VC mechanism, which has been used for various purposes, e.g. BADs around planned mass transit stations to capture the added values produced by projected Transit Oriented Development (TOD). A specialized version of the BAD is the *Business Improvement District* (BID) e.g. for downtown development in the US and the UK. BIDs are partnerships between local government and area businesses with taxing authority enabling a special assessment subject to a majority vote of the district's businesses. *Tax Increment Financing* (TIF), based on the same principle as the BAD, enables local government to earmark the revenue stream anticipated from assessed property value increases in the TIF at area to finance specified public projects there.

Voluntary capture

The basic mechanism for voluntary VC is the *public-private partnership* (PPP), a tailor-made institutional arrangement that may include some of the other mechanisms reviewed here. Regulative incentives supporting voluntary VC include *zoning tools*: transferable development rights (TDRs), planned unit development (PUD), and incentive zoning.

Land endowment

This applies to VC on state owned land or publicly conferred development rights. Revenue from such property or development rights that are sold to private parties or assigned to a PPP is obviously an attractive source of funding to supplement public investments in potentially beneficial projects.

Table 2.1 Institutional design forms for Value Capture

Basic approach	Institutional design form *[existing agent]* [new agent]	Value Capture mechanism
Compulsory capture (general)	Property Tax *[Local government]*	Variable rate on land to capture value increment attributable to public investments.
	Betterment Fees *[Local government]*	Proportion of land value increment.
(bounded)	Impact Fees *[Local government]*	Implicitly related to land value increment.
	[Special District] (e.g. Mass-Transit, Metro-Parks)	Tax on beneficiaries related to value increment and/OR levy on local governments.
	[Benefit Assessment District] (BAD) (e.g. for Transit-Oriented Development area)	Tax on beneficiaries related to value increment.
	[Business Improvement District] (BID) (e.g. for downtown/ neighbourhood centre)	Tax on beneficiaries related to value increment.
	Tax Increment Financing (TIF) *[Local government]*	Earmark value increment-related increment of property tax in proclaimed area to fund public investments.
Voluntary capture	Public-Private Partnership (PPP)	Land endowment (below). Public contribution: compulsory capture mechanisms (above), regulatory incentives: zoning.
Land endowment	PPP – State-owned land and/OR development rights	Share of profit from land development and/OR realization of conferred development rights.

Source: Alexander (2012).

Evaluation in Infrastructure Institutional Design

The institutional design of infrastructure planning and delivery is recognized as important, but the complexity of planning and delivering infrastructure projects makes empirical analysis daunting. To evaluate private sector provision of local public infrastructure in Australia, Cannadi and Dollery (2005) proposed five performance dimensions: 1) Economic performance (including economic efficiency, financial performance, aggregate welfare and economic development); 2) Social performance (including service price, quality and value, employment, and equity considerations); 3) Democratic performance (accountability, due process, participation and transparency); 4) Political, including power and political process impacts; and 5) Legal: regulatory and competition policy issues. While this is a valuable conceptual framework its operational feasibility is questionable,[4] and I have chosen to focus on two aspects of infrastructure planning and delivery.

Infrastructure Planning and Procurement

Today, conventional evaluation of public infrastructure investments includes some institutional design considerations.[5] Alternative forms of PPPs are identified: design-bid-build, design-build, design-build-operate, design-build-finance-operate, and pure Operation & Maintenance. Evaluating feasible alternatives is included in overall project evaluation as cash-flow analysis of discounted costs and revenues, applying maximum NPV as the public investment decision criterion (Miller, 2000). However, in this evaluation approach transaction costs receive little attention, besides identifying some under the heading: 'Transaction Costs Incurred in the Private Sector' (Miller, 2000, pp. 184–185).[6]

A recent meta-analysis of public infrastructure projects suggests higher awareness of institutional design with the growing importance of PPPs, recommending governance as a topic for more research. Build-operate-transfer is the most frequently observed among the diverse procurement approaches that have emerged during the authors' period of reference: 1998–2003, but there is no systematic comparison or evaluation of alternative forms of PPPs (Ke et al., 2009). The same applies to another such meta-analysis, which reviewed over 20 sources to identify 'success factors/barriers' in effectuating PPPs. These included the role of government, PPP forms and public procurement practices, but again recommendations are based on a qualitative analysis of experience and outcomes rather than systematic evaluation (Kwak, Chih and Ibbs, 2009).

4 Indeed, the authors did not apply it systematically in their own analysis, and prescribe case by case evaluation.

5 Of course, these are not consciously identified as such.

6 The transaction costs of delays under alternative PPP forms may be included in the discounted cash flow analysis method, as they may affect the project's estimated NPV if they are significant.

Aziz (2007) proposes the analysis of PPPs' success in infrastructure development from two perspectives. One is 'finance-based' – see below. The other is 'services-based', which defines success as optimizing time and cost efficiency in service delivery. Several projects in the US and Canada are compared, relating them specifically to the US context. An important criterion for success is meeting 'value-for-money' objectives, which suggests a form of benefit-cost analysis[7] for empirical application.

Whittington (2008) pioneered transaction cost based evaluation of alternative forms of public infrastructure procurement, in *ex post* assessment of the efficiency of US state highway projects. The conceptual framework for this approach estimates all the transaction costs associated with project planning, procurement and delivery under different forms of PPP, explicitly distinguishing them from the direct infrastructure production costs that are usually taken into account.

This framework was applied to analyse and evaluate design-build procurement in several State highway programs (Dowall and Whittington, 2006) and for a systematic empirical evaluation comparing two freeway interchange projects in the state of Washington (Whittington, 2012). These studies identified three types of transaction costs. The first are project management and administration: preliminary administration and research, procurement, monitoring and control[8] which can vary depending on many factors; here the critical factor was the procurement form. Whittington found interesting cost variations here between the two projects, but overall they seem to cancel out in light of the small difference between their total project costs: ~2.2 per cent savings of design-build over design-bid-build.

The second kind of transaction costs relate to differences in the execution of specific tasks under the different forms of procurement. Thus engineering design cost ~13 per cent less under design-build. Classic economics might attribute this to the efficiency of competitive private execution, but detailed case study suggests that public sector inefficiency here is the result of intermittent work due to interruptions in project funding. Design-build demands front-end funding, so work on the project can be continuous and more effective.

Another task, construction, cost ~15 per cent more under design-build This difference is really another transaction cost, attributable to contractors' strategic bidding to anticipate their higher risk under this form of procurement. Under the heading 'The public value of time' the study diagnoses the reason for the 5 years' difference in project duration (from inception to completion), which is not design-build's superior efficiency but again the other project's intermittent funding

This study was a pilot project, so the third kind of transaction costs are only implied or mentioned as items for future research. These are actors' and affected

7 This is not explicitly spelled out, and it is unclear whether transaction costs are considered, but transaction cost based analysis could also be indicated here.

8 In Whittington (2012) these are: preliminary administration/ancillary studies, bid administration, contract administration, outside agreements/damages/change orders and disputes.

parties' indirect costs related to project procurement and execution. A more complete evaluation should include as time-related transaction costs actors' costs and benefits attributable to project scheduling and duration under the different forms of PPP. Costs might include contractors' increased capital costs due to delays, and affected parties' indirect costs attributable to postponement of project completion or commencement of service. Benefits might include government savings by delaying funding vs. the front-end funding required by design-build. Other items discussed are impacts on labour organization and contract agreements, and on environmental and regulatory compliance (where different procurement forms may raise difficulties synchronizing project planning and decision making with regulatory deadlines, e.g. environmental review and delivery of EIS.[9]

Infrastructure Financing

Evaluating alternative infrastructure for project financing is usually limited to the forms of privatization discussed above, without considering the potential of value capture. This applies to the prescribed simple cash-flow analysis approach with max.NPV as the optimization criterion (Miller, 2000), and variations such as Ye and Tiong's (2000) combination of NPV/weighted average cost of capital with 'dual risk-return' methods. It is even true for Cannadi and Dollery's (2005) comprehensive evaluation framework.

More complex evaluation frameworks are proposed, which could be applied to assessing alternative institutional design for value capture as well as the forms of PPP for which they were used in their case demonstrations. This is implied in Aziz's (2007) criteria for his 'finance-based' approach, which include meeting financial objectives,[10] risk allocation consequences, and value-for-money objectives. Even more promising is a proposed theoretical framework based on multi-criteria analysis and applying the Analytical Hierarchy Process, which is contrasted with conventional benefit-cost analysis based approaches (Thomopoulos, Grant-Miller and Tight, 2009).

The only case I know of explicit evaluation of alternative institutional design for infrastructure planning and delivery is the analysis of alternative forms of organization for the Trans-Israel Highway. The consultants here developed a multi-criterion decision analysis framework (with weights and scores) and applied it to three forms of organization for project planning and delivery: 1) Government (through the Transportation Ministry and the DPW) planning and delivering the project as a regular public highway; 2) Authority: creating a new special statutory state agency – a Highway Authority – for planning, construction and O&M of the highway as a toll-highway; and 3) Privatization: design-finance-build-O&M

9 These do not seem amenable to empirical operationalization as estimated transaction costs.

10 Aziz refers to 'private financial objectives', but there is no reason why public financial objectives could not be used as well or instead.

the highway as a toll-highway. The criteria included NPV, construction duration/efficiency, operational efficiency, funding continuity, financing sources, user considerations: cost and service, and governmental considerations: risk, control and impacts on the national economy. Though this framework was adequate for its professed purpose, its application was fatally flawed[11] (Alexander, 1998).

Summary

Evaluation in infrastructure planning and delivery usually applies to project plans or programmes (alternative technologies, locations, modes, routes or ROWs) and designs. The institutional design of major strategic infrastructure projects has rarely been subjected to deliberate systematic evaluation. More recently, concerns about public sector efficiency and limitations of public funding have heightened incentives for privatization and awareness of institutional design.

Institutional design options for infrastructure projects focus on two aspects of infrastructure planning and delivery: alternative forms of PPPs for public infrastructure procurement, and alternative models of project institutionalization to supplement public funding by value capture. The available options in both these areas are presented above. Then we ask: how can these be evaluated? reviewing the limited literature and experience.

Most of the evaluation of PPPs is discursive: discussion of appropriate performance criteria and speculation on the possible impacts of various forms of public infrastructure procurement. Few concrete empirical assessment methods have been proposed, and even fewer applied. The most prevalent approach is to subsume this factor under the project's overall financial performance, applying conventional measures such as NPV or more sophisticated variations. This clearly has limitations: the evaluation targets direct costs and savings, but other possibly important factors are ignored, risking conclusions that may be biased or wrong.

More promising methods apply an expanded evaluation framework, where financial performance is only one of a larger set of criteria. A possible form for such a framework is a modified benefit-cost analysis. This could include direct and indirect costs and benefits that can be monetised, expressed as NPV or a B/C ratio, and embrace other (more intangible or difficult to measure or estimate) impacts in a kind of impact analysis.[12]

Another useful evaluation approach is transaction cost analysis, focusing on the marginal differences in identified transaction costs between PPP alternatives.

11 Analysis revealed several methodological omissions and errors that have a simple explanation. The whole evaluation exercise was undertaken to produce a predetermined result, so as to rationalize a decision that had already been taken: choosing privatisation among the offered organizational alternatives.

12 An adaptation of the Community Impact Evaluation model (Lichfield, 1996) comes to mind.

While limited so far in its empirical application, this method has potential. Though the quantification of expected transaction costs is difficult and demanding, identifying and estimating transaction cost differences (even on a scale of orders of magnitude) can yield revealing findings and insights.

For evaluating alternative infrastructure for value capture for a particular project, a well-developed multi-criterion decision analysis should be appropriate.[13] Such a framework also integrates the evaluation of alternative infrastructure for a particular infrastructure project as a whole. This evaluation could feature PPP options and value capture alternatives under the various feasible configurations of project planning, delivery and operation, and include estimated transaction costs and savings among its decision criteria. While it would be a significant expansion of conventional infrastructure project alternatives' evaluation, the ultimate difference that institutional design might make to the project's effectiveness is likely to make the added effort worthwhile.

References

Alexander, E.R. 1995. *How Organizations Act Together: Interorganizational Coordination in Theory and Practice.* Amsterdam: Gordon & Breach.

Alexander, E.R. 1998. 'Evaluation in Israeli Spatial Planning: Planning Theory vs. Planning Practice'. In N. Lichfield, A. Barbanente, D. Borri and A. Prat (eds) 1998. *Evaluation in Planning*. Dordrecht: Kluwer.

Alexander, E.R. 2006. Institutional design for sustainable development. *Town Planning Review* 77 (1), pp. 1–27.

Alexander, E.R. 2012. Institutional design for value capture and a case: The Tel-Aviv metropolitan park. *International Planning Studies* 17 (2), pp. 162–177.

Aziz, A.A. 2007. Successful delivery of public private partnerships for infrastructure development. *Journal of Construction Engineering and Management* 133 (2), pp. 918–931.

Cannadi, J. and Dollery, B. 2005. An evaluation of private sector provision of public infrastructure in Australian local government. *Australian Journal of Public Administration* 64 (3), pp. 112–118. Available at: http://unpan1.un.org/intradoc/groups/public/documents/APCITY/UNPAN030163.pdf [accessed 5 January 2013].

Dowall, D. and Whittington, J. 2006. *Transaction-Cost Economic Analysis of Institutional Change Toward Design-Build Contracts for Public Transportation.* Berkeley, CA: University of California Transportation Center. Available at: http://escholarship.org/uc/item/97k3c3hs [accessed 5 January 2013].

Ke, Y., Wang, S.Q., Chang A.P. and Cheung, E. 2009. Research trend of public-private partnership in construction journals. *Journal of Construction Engineering and Management* 135 (10), pp. 1076–1086.

13 Programmed methods are available for this, such as AHP and EVAMIX.

Kwak, Y.H., Chih, Y.Y. and Ibbs, C.W. 2009. Towards a comprehensive understanding of public private partnerships for infrastructure development. *California Management Review* 51 (2), pp. 51–78.

Lichfield, N. 1996. *Community Impact Evaluation*. London: UCL Press.

Miller, J.B. 2000. *Principles of Public and Private Infrastructure Delivery*. Dordrecht: Kluwer.

North, D.C. 1990. *Institutions, Institutional Change and Economic Performance*. Cambridge, UK: Cambridge University Press.

Ostrom, E. 2005. *Understanding Institutional Diversity*. Princeton: Princeton University Press.

Thomopoulos, N., Grant-Muller, S. and Tight, M.R. 2009. Incorporating equity considerations in transportation infrastructure evaluation: Current practice and a proposed methodology. *Evaluation and Program Planning* 32 (4), pp. 331–359.

Whittington, J. 2008. 'The Transaction-Cost Economics of Highway Project Delivery: Design-Build Contracting in Three States'. DCRP, University of California-Berkeley: PhD Dissertation.

Whittington, J. 2012. When to partner for public infrastructure? Transaction cost evaluation of design-build delivery. *Journal of the American Planning Association* 78 (3), pp. 269–285.

Williamson, O.E. 1995. 'Transaction cost economics and organization theory'. In O.E. Williamson (ed.) 1995. *Organization Theory: From Chester Barnard to the Present and Beyond*. New York: Oxford University Press.

Ye, S. and Tiong, R. 2000. NPV-at-risk method in infrastructure project investment evaluation. *Journal of Construction Engineering and Management* 126 (3), pp. 227–233.

Chapter 3

Creating and Evaluating Co-Benefits and Co-Costs of Environmental Policies and Investments in Urban Areas

Matthias Ruth, Junming Zhu, Nancy S. Lee and Sahar Mirzaee

Introduction

Environmental policies are typically designed and investment decisions are implemented to address specific, well-described issues – be it to reduce, for example, traffic congestion or cut emissions of greenhouse gases into the atmosphere. This chapter, however, does not focus on these primary, targeted outcomes, but instead concentrates on two related aspects that can have far-reaching implications for policy and investment making – the ancillary cost and benefit impacts of any action, and the indeterminacy of causal relationships between system interventions and outcomes.

It is the rule rather than the exception that policies and investments generate co-benefits and co-costs, and since their magnitude may ultimately sway a decision, they must be reflected in policy and investment analysis (Ng, 1983). The generation of environmental and health benefits that accrue, for example by reducing traffic congestion or curbing emissions from power generation may in the end be the dominant outcome compared to the stated goal of improving the flow of transportation or of stabilizing the climate. Total ancillary benefits may exceed primary benefits (Pearce, 2002), and even where they do not, their inclusion may help magnify the primary benefits (Ekins, 1996a) and can alter the rank-ordering of preferred policy choices (Aunan et al., 2004). Inclusion of co-benefits and co-costs thus adds complexity to any cost-benefit comparison and poses challenges for decision making and implementation, as well as institutional arrangements for them both.

Furthermore, existing or proposed plans, policies, investments and technologies not related to climate change, for example – such as the protection of a wetland for habitat of an endangered species – may influence the benefits of climate action, because the wetland may enhance much needed carbon sequestration and flood control services (cf. Rübbelke, 2003; Markandya and Rübbelke, 2004). Similarly, tightening of air quality standards may reduce the benefits that can be ascribed to a greenhouse gas emissions reduction policy that comes on top of air quality standards (Burtraw and Toman, 1997). Correcting for the effects from other air emissions related policies, Burtraw and Toman (1997) find co-benefits of carbon emissions mitigation in the US to be on average $12–$18 per ton of carbon, and $20–$30 at the margin.

To define and capture co-benefits and co-costs requires systematic treatment of system boundaries in space and time, and commensurability of these costs and benefits with other performance measures. For example, the costs of improved insulation of residential structures and the associated carbon emissions reductions can, in principle, be readily calculated. Similarly, monetary values may be imputed for public health costs of declining indoor air quality that results from reduced ventilation of living spaces. Changes in comfort, for example, are more difficult to quantify in monetary terms, as are the increases in national energy security that result from reductions in domestic heating and cooling requirements. As the scope of analysis increases, the ancillary effects of policies and investments become more diffuse, and the ranges of estimates for costs and benefits broaden (Figure 3.1).

In an effort to ultimately guide the use of co-benefit and co-cost measures for environmental policy and investment decision making, the next two sections of this chapter present selected highlights from a growing literature on co-benefit and co-cost assessment by first focusing on broader conceptual issues and then on specific applications. In both sections, we draw on studies in energy and climate economics and policy to illustrate current practices and provide an overview of the state of the art. The section on network analysis uses a select set of 35 scientific studies and reports on greenhouse gas emissions management to illustrate the relationships that exist between climate action and actual or perceived outcomes. Here, we use network analysis to structure the wealth of information in the literature and to offer information on the potential ancillary benefits and costs of investment and policy choices.

Much of the literature on which we draw for our analysis addresses national-level issues. However, co-benefits and co-costs can vary geographically quite significantly due to, for example, differences in affected populations, infrastructures and ecosystems, as well as physical environmental conditions. Many of the investments, in turn, will be targeted towards regional and local emissions. Thus, the next section of the chapter focuses on urban areas instead of larger regional, national or global outcomes because urban areas have recently been given heightened attention in policy and investment decision making to curb greenhouse gas emissions and adapt to residual climate change (Ruth and Baklanov, 2012), and because these areas may be more conducive to community-based approaches to the identification and capture of co-benefits and minimization of co-costs.

While policy interventions are usually geared towards meeting well-described objectives, the actual outcomes, in contrast, are often rather diffuse, and causal linkages between the original plan or action and the outcome are hard to establish. Such difficulties in establishing causal relations are well known for complex systems in which, for example, technologies, behaviours and environmental conditions interact in non-linear, time-lagged fashion. The last section of this

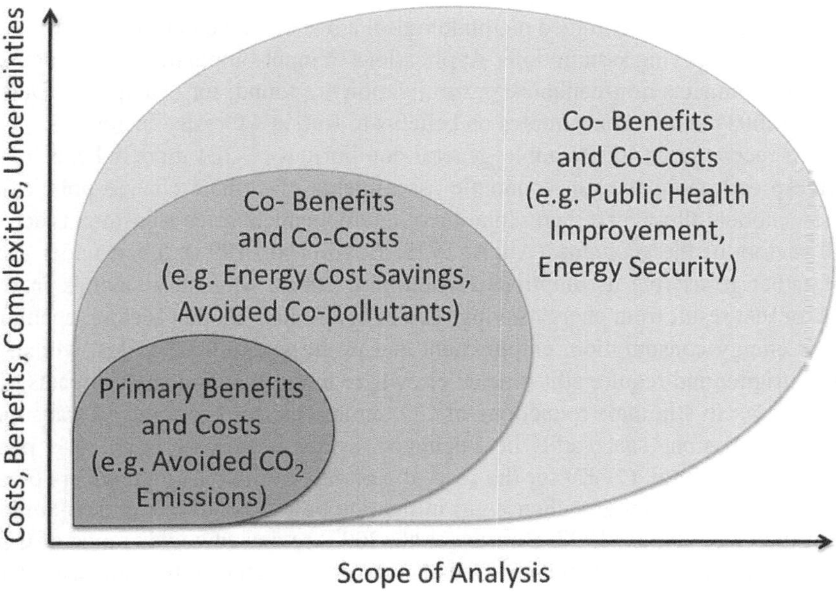

Figure 3.1 Changes in measurements with increasing scope of analysis

chapter, therefore, explores how co-benefit and co-cost analysis may be used to help shape policy and investment making at the urban scale, and expands on the insights generated by the illustrations from the preceding parts of this chapter to highlight the institutional innovations needed for effective capture of co-benefits, and minimization of co-costs. The chapter closes with a brief summary and a note on the challenges for research, investment and policy making.

Capturing Co-Benefits and Co-Costs for Quantitative Analysis

The notion of ancillary or co-benefits and co-costs has long shaped, in different ways, the assessment of policy and investment outcomes, and is by no means new. Input-output analysis, for example, is widely used to quantify the multiplier effects associated with a change in expenditures in one sector of the economy as that change ripples through the economy to affect purchases and sales of other sectors (Leontief, 1966; Converse, 1971; Johnson and Bennett, 1981). All else equal, the policies with the highest income or employment multipliers, which capture all the direct and indirect effects, are preferred. Specifications of input-output models benefit from the richness of official statistics collected on firms and households at national and subnational levels, typically rest on assumptions of linear relationships among sectors of the economy, and are based on the notion that the demand and supply system equilibrates after an exogenous intervention took place, though continued methodological advances have helped to relax some of these simplifying assumptions. Applications of input-output models to calculate ancillary impacts of greenhouse gas mitigation are found, for example, in Dudek et al. (2003), which concentrates on benefits to Russia's forestry sector.

In recent years, computable general equilibrium (CGE) models have found widespread application in economic assessments of climate change policies as these models allow for a representation of interdependent price adjustments across all sectors of the economy (Allais, 1978; Boyd et al., 1995), albeit under a set of rather restrictive assumptions (Ackerman, 2002). Monetary savings in one sector that result from energy savings can be reinvested in other sectors, changing their energy consumption, employment and profit generation profiles, which, in turn, trigger and require adjustments elsewhere in the economy. The impacts of a carbon tax to stimulate reductions of CO_2 emissions, for example, on transport-related emissions can readily be calculated, as has been done in an early piece by Glomsrød et al. (1992) for the case of Norway. In their seminal papers on air quality benefits from greenhouse gas mitigation policies, Burtraw and co-authors (Burtraw and Toman, 1997; Burtraw et al., 2007) reviewed a wide range of CGE models and pointed out their various shortcomings, especially with respect to: (a) the limited recognition in these models of the role that non-greenhouse gas (GHG) policies can have in driving future emissions and thus ancillary benefits; and (b) the role of geographic variability of co-benefits that may result, for example, from differences in population densities. Little attention, however, was given to the fact that standard CGE practice requires rather restrictive assumptions about conditions for market equilibration, and the fact that important co-benefits and co-costs may not be part of a market-based calculus at all, such as un-priced environmental goods and services.

In contrast to standard input-output and CGE modelling, systems dynamics models have been developed to explicitly capture the time-lagged and often non-linear feedback relationships that may dampen or amplify the effects of a stimulus exerted by one part of the system on others, and back onto the source of the original stimulus (Meadows, 1999), for example, when changes in affluence result in resource degradation that undermines affluence (Meadows et al., 2004; Turner, 2008). These models have found widespread applications, for example, in linking demographic, economic, and environmental dynamics with each other (Ruth et al., 2011) and thus to explore the costs and benefits of systems interventions across multiple subsystems, across space, and across time. Data limitations to establish all relevant functional relationships among subsystem components often relegate model use to scoping tools and scenario generators for the exploration of possible outcomes under altered initial conditions or subsystem interactions. Yet given the non-equilibrium nature of many of these models and the diversity of subsystems these models typically consider, the notion of costs and benefits usually includes much more than monetary outcomes.

More recently, a series of empirically driven assessments of policy and investments have, on the one hand, broadened the concepts of co-benefits and co-costs to include longer-term and larger-scale social and political issues, for example, when national security implications of increased energy efficiency are considered. On the other hand, in these studies, the set of such by-products of policy and investments has been far from systematic and tended to concentrate on co-benefits without comparable attention to potential co-costs. Other empirical extensions have broadened the analysis to encompass environmental cost accounting (Gale and Stoke, 2001; Jasch, 2006), explored methods to arrive at aggregate social cost of carbon estimates (Ackerman and Stanton, 2012), and combined financial and environmental life cycle costing to capture a suite of emissions and their business and environmental impacts across supply chains, production process and end uses (Carlsson Reich, 2005). As the scope of these assessments broadened, new data quality issues emerged and awareness surrounding uncertainties of the results increased (Fahlén and Ahlgren, 2012).

Since an increasing number of publications in recent years have focused on the ancillary outcomes of climate change policy, and the investments in infrastructures and technologies to curb greenhouse gas emissions, the remainder of this chapter focuses on the co-benefits and co-costs associated with those policies and investments. The absence of national strategies and policies has prompted much planning and action at the regional and local scales, which in many cases has presented data and methodological challenges for the application of traditional approaches of co-benefit and co-cost quantification, such as input-output, general equilibrium and systems dynamics models. As a consequence, what has and has not been counted as a co-benefit or co-cost is often decided on an *ad hoc* basis, rather than on a rigorous methodological foundation. Furthermore, the mechanisms to guide decisions towards maximization of total net benefits have also lacked structure.

Quantifying the Co-Benefits from Climate Action

Reducing emissions of greenhouse gases has become a goal of firms, communities, regions and countries around the world. The dominant strategies to achieve this goal include improvements in energy conversion efficiencies in the power and end use sectors, such as transportation and buildings; switching to fuels of lower carbon content, such as moving from coal to natural gas; limiting the direct use of fossil fuels, such as with nuclear power; expanding renewable energy supplies, such as through solar, wind and biomass generation; and carbon capture and sequestration, either through natural processes such as reforestation and soil conservation, or by technical means of capturing carbon at the point of fuel combustion or removing it from the atmosphere after release occurred. Since the processes that lead to carbon emissions typically also entail the generation of other pollutants, climate mitigation can help cut the emissions of those pollutants as well. Notable examples are reductions in SO_X, NO_X and particulate matter (Dessus and O'Connor, 1999;

Wang and Smith, 1999), but also water pollution and land use impacts that come from extraction, processing and combustion of fossil fuels.

In the case of transportation, for example, achievements in emissions reductions may not only be had by switching among fuels and increasing fuel use efficiencies, but also by altering infrastructures and behaviors so that transportation volumes decline. Co-effects of these strategies may include a reduction of congestion, noise and road damage (see e.g. Barker et al., 1993; Barker and Rosendahl, 2000), declines in traffic accidents and fatalities, increased physical activity and thus better public health (Woodcock, et al. 2009), but also changes in land and property values. In the building sector, improved insulation may reduce unwanted heat loss or cooling requirements, resulting in direct monetary savings from lower fuel consumption, but that may compromise indoor air quality, potentially leading to adverse health effects or requiring investments to counteract them.

The ramifications of emissions reduction on human health have found considerable attention in the literature. Aunan, et al. (2004), for example, quantify the welfare impacts from SO_2 and PM_{10} emissions reductions in Shanxi, China, that are caused by CO_2 emissions abatement. Starting with population-weighted exposure levels to air emissions from coal combustion, and exposure-response functions from epidemiology, they estimate health outcomes in terms of changes in life expectancy. Their results suggest unit prices of health impacts that warrant emissions reductions through a range of technological options available to the industrial, power generation and household sector. Crawford-Brown et al. (2012) estimate the economic benefits from reductions in mortality and non-fatal disease that would result if Mexico were to cut its CO_2 emissions by 77 per cent below a baseline growth scenario and compared the outcomes to those of other human health risk management strategies. On the basis of that comparison they argue for climate policy to be an important element of a larger set of environmental policies targeted towards improved human health outcomes. And since benefits from improvements in health may be larger for the poor, climate change policy induced reductions of local air emissions may bring with them improvements in social justice and equity, as has been argued by Garg (2007) for the case of New Delhi, India. However, despite strong and growing evidence of the various positive health impacts from climate mitigation (Bell et al., 2008; Nemet et al., 2010), co-benefits estimates have so far had limited impact on policy and practice (Jack and Kinney, 2010).

In contrast to human health impact assessments, the quantification of co-benefits from CO_2 mitigation that result in reduced harm to ecosystems remains sparse. Early work by Ayres and Walter (1991), for example, considered the impacts of reduced air pollution on forest values in the Federal Republic of Germany and in the US. Their analysis, carried out for data from the 1980s, calculates co-benefits associated with human health impacts, corrosion of materials, and reduced degradation of soils and vegetation from reductions in particulates, SO_X, NO_X, volatile organic compounds (VOC), CO and hydrocarbons from mobile sources and power plants. These co-benefits are shown to be on the order of \$20–\$60

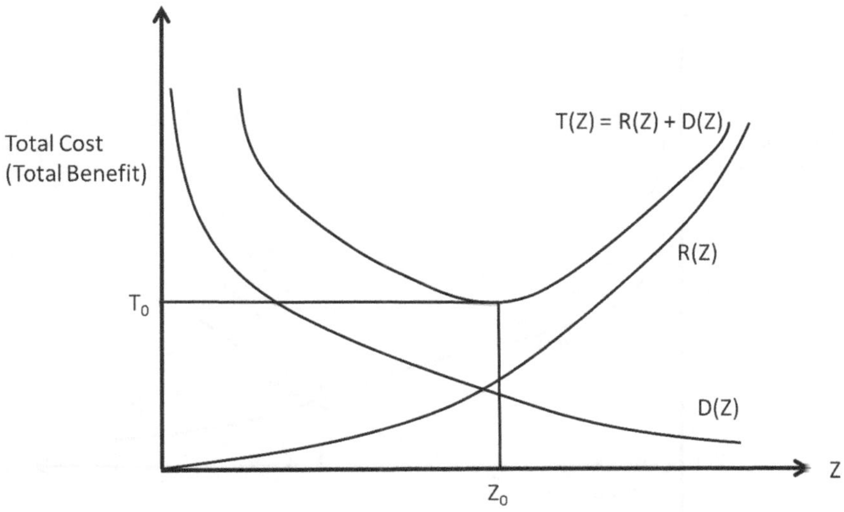

Figure 3.2 Standard damage cost and abatement cost profiles

per ton of CO_2 avoided. Alfsen et al. (1992) expanded on the analysis by Ayres and Walter (1991) to also include recreational values from Norway's forests and reductions in acidification of freshwater lakes as a result of implementing a carbon tax. More recently attention has also been given to the improvements in livelihoods and changes in adaptive capacity of local communities from reduced emissions associated with deforestation (Chhatre et al., 2012), though such co-benefits have not been quantified.

Conceptually, the economics of carbon (and other pollution) abatement requires the identification of cost and benefit profiles across different abatement schedules (Ayres and Walter, 1991; Ekins, 1996a; 1996b). Figure 3.2 shows the standard case of damage costs, D(Z), declining for increasing levels of emissions reduction, Z. The costs of carrying out those reductions, R(Z) are usually assumed to rise both monotonically and exponentially because avoiding the next unit of pollution will be ever harder to do. Economic use of resources would dictate identifying those emissions reductions levels at which total costs T(Z) are minimized. At that point (T_0), the slope of the total cost curve is zero and the marginal costs of abatement and damages are equal to each other. In other words, at the optimal abatement level Z_0, the benefits from reducing emissions by the next unit just cover the costs of avoiding that next unit from being released into the environment.

In the presence of co-benefits, however, the costs of emissions reductions may be negative, implying that reductions of some initial emissions have benefits that outweigh the costs, at least for some levels of effort (Figure 3.3). Examples include the implementation of energy efficiency and fuel switching programs that have high economic returns and also yield, for example, environmental health

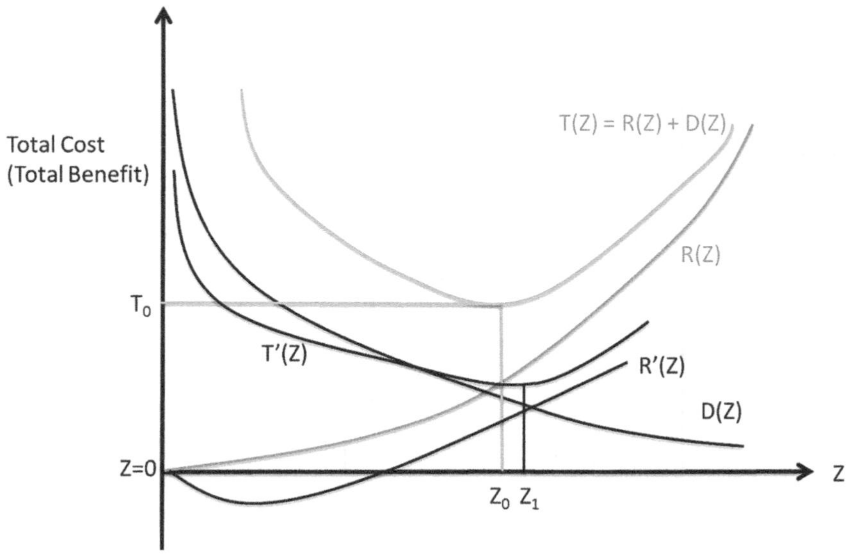

Figure 3.3 Damage cost and abatement cost profiles with co-benefits of abatement

benefits from lower ambient pollution concentrations. The result is an equilibrium abatement level Z_1 that is higher than the original one (Z_0). If, however, emissions abatement results in only co-costs (Figure 3.4), then the new abatement cost curve (R''(Z)) lies above the original one (R(Z)) and the new equilibrium occurs at the lower level, Z_2. An example of potentially high co-costs of emissions reductions is the rapidly declining indoor air quality that may come with increased insulation. The emergence of mold spores and rising radon concentrations are two prominent examples (Shim et al., 2011). Of course, the potential for both co-benefits and co-costs may arise, resulting in optimal abatement levels somewhere between Z_0 and Z_2.

The arguments contained in figures 3.2 to 3.4 can similarly be made for climate adaptation actions. The standard assumptions would posit that costs of adaptation rise monotonically and damages from increased adaptation decline. When adaptation generates co-benefits, optimal adaptation would be higher than in the absence of the co-benefits. For example, if wetlands are protected to provide flood control services, property values may rise as a result of increased recreational opportunities, scenic beauty, and species diversity. In contrast, when adaptation generates co-costs, optimal adaptation levels will be lower than in the absence of those co-costs. If, for example, expansion of wetlands for flood control provides habitat for vector-borne diseases, considerable dis-amenities and adverse public health outcomes may be generated by the adaptation action.

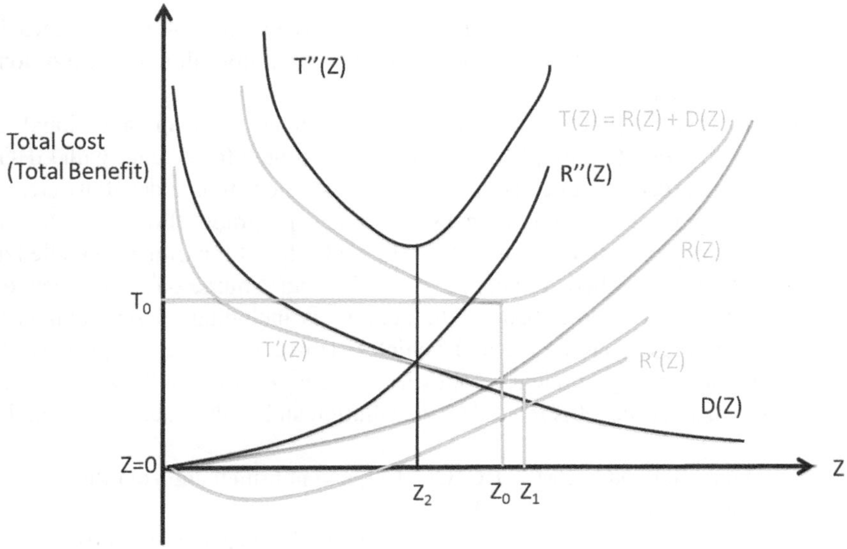

Figure 3.4 Damage cost and abatement cost profiles with co-costs of abatement

Network Analysis of Co-Benefits from Climate Action

A network demonstration is a convenient way both to represent the detailed coupling of climate actions and consequences, and to capture in systematic and aggregated form the emergent feature of actions and consequences. To illustrate potential relations between actions and (beneficial) consequences, we have established an action-consequence network on the basis of 35 research papers and reports that are related to climate change mitigation. The list of papers and reports is provided in Appendix A – collectively, they provide an initial glimpse into the way co-benefits are conceptualized and portrayed in the literature, and as a consequence offer insights into the relationships that have been selected and explored by the more than 100 researchers who have worked on these papers and reports.

The network includes 62 actions, represented as black nodes, and 78 consequences, represented as white nodes (Figure 3.5). The size of a node is scaled according to the number of links it has, i.e. the number of consequences or actions to which the node relates. Because the reduction of GHGs is the primary policy target, the node for GHGs emission is in the centre of the network and links to most of the actions. Salient co-benefits that can be associated with actions that reduce GHGs include air pollution reduction – PM_{10}, NO_x, SO_2 – and general health benefits. An example of a set of action-consequence is the replacement of coal-fired power in China with wind energy. The replacement led to co-benefits in the form of reduced GHGs and avoided cases of premature mortality, of chronic bronchitis, and of hospitalizations for respiratory or cardio-pulmonary conditions (Partridge and Gamkhar, 2012).

Comprehensive climate policy programs, such as those on a national or regional level, promote multiple approaches (i.e. actions) for dealing with GHGs and thus cause broad sets of consequences. Some of the action nodes illustrated in Figure 3.5 represent such comprehensive climate programs. For example, the energy efficiency programs discussed by He et al. (2010) for China – labelled as EEPCN in the upper right quadrant of Figure 3.5 – lead to numerous consequences. The EEPCN action is described by the authors as including energy savings in industry, buildings, households, and vehicles. However, the consequences are attributed to the energy efficiency policy as a whole, resulting in the reduction of GHGs and many co-benefits related to air pollution and health outcomes, namely:

- reduced SO_2, NO_x, $PM_{2.5}$, BC, OC, CO, and non-methane volatile organic compounds (NMVOCs); and
- avoided cases of cardiopulmonary mortality, lung cancer mortality, other mortality, respiratory morbidity, cardiovascular morbidity, and outpatient visits.

Other actions, and the associated acronyms representing actions and consequences, are listed in Appendix B.

For the respective subset of the literature that generates this network, the majority of actions are associated with a reduction of PM_{10}, NO_x, and SO_2, followed by actions associated with the co-benefits of reduced eutrophication, smog, and acidification. Therefore, our subsequent discussion of the network analysis focuses on co-benefits. To focus on the co-benefits, the network in Figure 3.5 is re-conceptualized by removing the node representing GHGs emissions and the nodes representing actions that only cause GHGs emission reduction without other co-benefits or co-costs being addressed in the respective papers and reports. This change creates a smaller network with 51 actions and 77 consequences, as shown in Figure 3.6. Because all climate actions are no longer connected together by the consequence of GHGs emission reduction, the network splits into a main component with dense interconnections between 43 actions and 60 consequences, and seven small isolated components. The density of the connections indicates relationships between and among the actions and resultant co-benefits. The more pronounced the clustering the more connectivity there is between and among actions and ancillary consequences. In contrast, small, isolated groupings like those shown on the upper left, outer edge of Figure 3.6 indicate fewer connections and therefore fewer ancillary consequences. As shown in the main component, the major actions still constitute the group of comprehensive policy programs, while the major consequences continue to include air pollution reduction and health benefits.

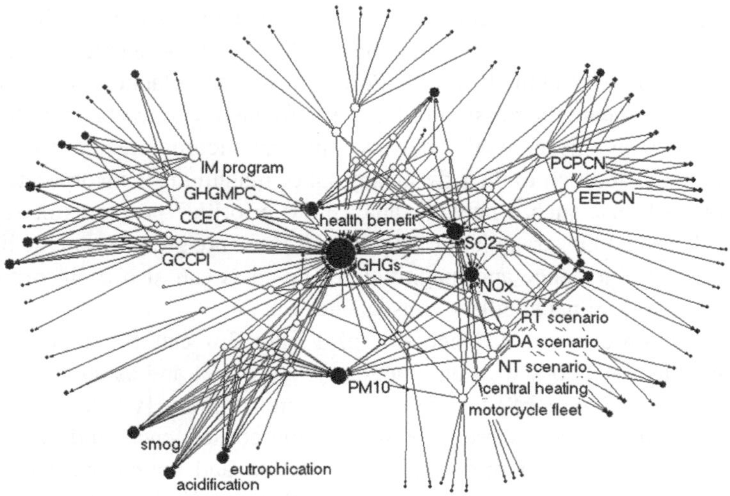

Figure 3.5 Interconnected network of climate actions (black nodes) and consequences (white)

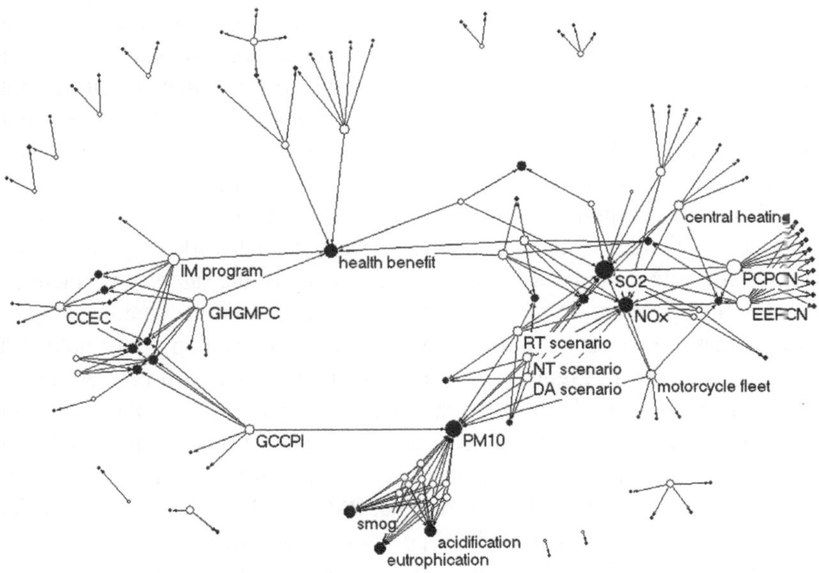

Figure 3.6 Network of the climate actions and co-benefits

By removing the GHG consequence node, the network diagram in Figure 3.6 highlights the relationships between those actions and co-benefits that have been explicitly dealt with in the literature. Policy makers and planners could use such a diagram to compare the possibility for, if not frequency of, co-benefit incidence.

Building on this network analysis, one can incorporate the magnitude of co-benefits in addition to their frequency. Once magnitude is incorporated, policy makers could use such a network analysis to prioritize programs or regulatory actions based on the strength of co-benefits as those are grounded in the literature. Similarly, the incidence and magnitude of co-costs can also be incorporated to yield a network of net ancillary consequences.

The action-consequence network of Figure 3.6 can be re-organized and displayed as a network with only actions (Figure 3.7) and as another with only consequences (Figure 3.8) for the ease of demonstration. Two actions are linked with each other if they share at least the same consequence, and vice versa. In Figure 3.7, the IM Program (vehicle inspection and maintenance program in Bangkok) discussed by Li and Crawford-Brown (2011) and the GHGMPC (GHG mitigation policies in Chile) presented by Dessus and O'Connor (1999) share many of the same consequences i.e. co-benefits in this example as the GHG benefit has been removed from the network. A thicker and wider link indicates that more consequences are shared by the two connected actions. The actions are generally clustered into four groups: the largest cluster in the upper middle includes mainly comprehensive energy or environmental policy programs and technology improvements on the national or regional scale; a cluster on the right includes 10 actions that comprise a set of climate strategies that are implemented over time and in varying degrees of stringency; the cluster in the lower middle includes four national climate policies, two about renewable energy development, and one vehicle maintenance program; finally the standalone actions along the left edge of the network indicate actions that create unique consequences, according to this set of literature. Presented in this way, the network diagram suggests actions that create synergy because they share many of the same consequences or a set of actions that are, in principle, able to stimulate climate change mitigation actions across a wide range of arenas by fostering actions along finer links. Each of the three clusters in Figure 3.7 corresponds to a cluster with both actions and consequences at the similar position in Figure 3.6. By referring to their position in Figure 3.6, it is clear that the three clusters of climate actions all share the benefits in PM_{10} emission reduction and human health, without which the three clusters will be isolated from each other. This shows that PM_{10} and health issues are two focal points addressed by different kinds of policies and strategies. In comparison, the reductions of SO_2 and NO_x emissions are addressed by the same kind of policies and strategies, although the absolute number of climate actions covering them is similarly large, indicated by the node sizes of SO_2 and NO_x in Figure 3.6.

In a similar manner to Figure 3.7, Figure 3.8 shows a network with 77 consequences. The middle and left portions of the diagram include the largest cluster of benefits that largely stem from reduced air pollution and energy use,

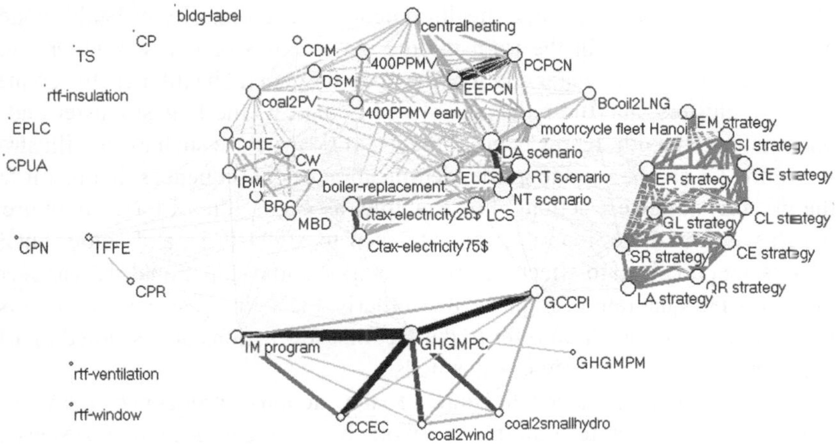

Figure 3.7 Network of climate actions

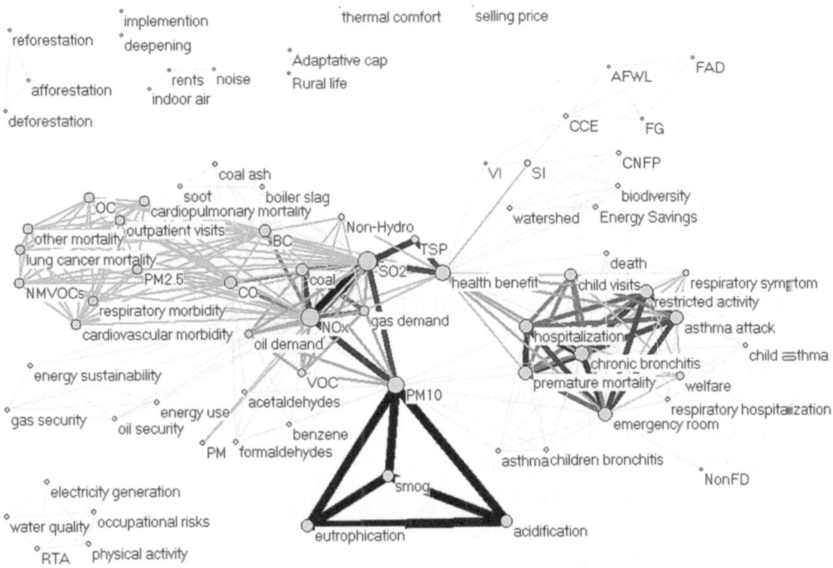

Figure 3.8 Network of co-benefits

alongside with several positive health outcomes. The majority of health benefits are clustered densely in the right middle portion. Above it is a smaller cluster related to forestry and water quality. The node for general health benefits connects all three clusters, and the reduction of PM_{10} connects the largest cluster and the one related to health. Knowledge of the network structure such as that illustrated in Figure 3.8 can be used to guide identification of consequences that result from the greatest or fewest actions. For example, as exemplified by the literature on which we draw here, many actions are seen to affect PM_{10} and smog, but few actions are perceived to affect electric generation and occupational risks as seen in the lower left quadrant of Figure 3.8. Similarly, the actions associated with nodes in the upper right quadrant refer to the fact that few actions are assumed to affect biodiversity and protection of watersheds.

In a final re-conceptualization of the action-consequence network in Figure 3.5 we generate a simpler network by aggregating similar actions and similar consequences into more general groups (Figure 3.9). The actions are aggregated into the following nine groups:

1. national climate policies,
2. other energy or environmental policies,
3. technology improvement,
4. energy or CO_2 taxes,
5. building retrofit and optimization,
6. renewable energy development,
7. fuel substitution,
8. vehicle maintenance, and
9. local participation.

The consequences are also aggregated into nine groups:

1. GHGs emission reduction,
2. reduction of other air pollutants,
3. reduction of energy use,
4. promotion of health benefits,
5. air quality improvement,
6. water quality improvement,
7. improvement of forest health,
8. improvement of living quality, and
9. local development.

The most influential groups of actions that appear in this network are national climate policies, other energy or environmental policies, and technology improvements. Absent from this network are, for example, behavioural changes at the level of individual households or firms, which lays open the main thrust of the existing literature.

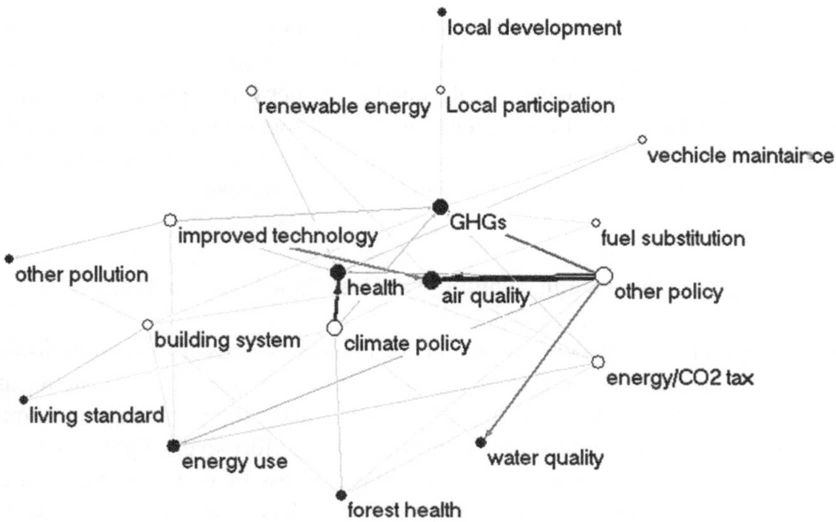

Figure 3.9 A network with grouped actions and consequences

The most frequently caused consequences are GHGs emission reduction, air quality improvements, and health benefits. While all the action groups contribute to GHGs emission reduction, it is worth noting that at least the three most influential action groups more frequently lead to co-benefits in air quality or health – indicated by the thickest and darkest links – rather than GHGs emission reduction.

From a policy and planning perspective, Figure 3.9 indicates that based on the literature included in this network analysis, there are relatively fewer consequences for improved forest health, water quality, local development, and living standards than for GHG emissions, air quality, and health benefits. If for example, improving water quality and living standards are important objectives and similar policies and actions as the ones reviewed for this network are being considered, then the policy maker or planner may want to rebalance the scope of actions yielding co-benefits to water quality and living standards to the extent it is feasible to do so.

The network diagrams illustrated in figures 3.5 to 3.9 can be used to demonstrate the frequency of the occurrence of actions and consequences – at least to the extent that those are perceived as relevant in the 35 studies and programs reviewed here. This information may guide the design of policies and regulations or combine actions to increase the chances of a desired outcome. Another interpretation is that GHGs, air quality, and health effects have received more attention than other possible outcomes. More generally, the promotion of sustainability goals requires healthy and resilient environmental, economic, and social sub-systems. Given the literature reviewed for this network analysis, the emphasis has been on the environmental systems, which are a critical element for the evaluation of co-benefits and co-costs either because environmental systems are directly affected or serve as the conduit through which the policy and plan implementation impact economic and social system. However, further attention may need to turn to a richer set of economic outcomes (subsumed in the network representations above as 'living standard' and 'local development') and social outcomes (subsumed in 'local participation'). Such attention, in turn, will likely make more complex the setting within which co-benefits and co-costs are realized and can be managed.

Capturing Co-Benefits and Avoiding Co-Costs in a Complex World

The discussion in the previous three sections points towards at least the following six key challenges when attempting to identify and quantify co-benefits and co-costs, and when trying to guide investment and policy making on climate mitigation and adaptation towards cost-minimizing effort levels. First, the diffuse nature of co-benefits and co-costs across space, across time, and across different segments of the population makes it impossible to fully capture them in the analysis in all but the most trivial cases. Yet, the fact that we know that co-benefits and co-costs exist means that one must properly reflect them, unless one is prepared to carry out an economic analysis to find optimal resource allocation for climate action that is, by definition, not able to pinpoint that optimum.

Second, many of the co-benefits and co-costs of climate action affect, or are mediated by, social and environmental processes that have no representation in the market place. The effectiveness of a warning system to alert people of an impending flooding event or heat wave, for example, is fundamentally shaped by the social fabric of the populations at risk (Lange et al., 2009), as recent experiences in Europe and elsewhere have clearly shown (see, for example, Stone, 2012).

Third, the abilities of people – ordinary citizens and investment and policy makers alike – to gauge risk and to make consistent decisions across time, however, are rather limited even in simple experimental settings where probabilities of the occurrence of particular outcomes are unambiguously assigned to the solution space (see, for example, Kahneman et al., 1991; Kahneman, 2003; Thaler and Sunstein, 2009). Yet, being able to overcome such limitations in real-world settings is not just critical to the ability to rank-order alternative climate actions but particularly crucial when the ramifications of climate change at the local places where decisions need to be made are, themselves, not well understood either.

Unclear thinking meets the challenges of solving yet unknown problems. So far, the approach largely concentrated on the latter – trying to increase knowledge about local climate impacts by generating ever more sophisticated forecasts. Downscaling of global climate models to resolutions that are commensurate with the resolution at which, for example, infrastructure or land use planning take place (Hewitson and Crane, 1996) is a booming cottage industry of the climate science community. In such exercises attempts are made, for example, to infer from global models what the changes in local temperatures, precipitation events or wind speeds, and thus wind-based power generation potentials or impacts on buildings and other infrastructures may be (Pryor et al., 2005; Pryor and Barthelmie, 2011).

However, the spatial and temporal resolution of the rather coarse social and economic information rarely matches with the resolution of downscaled climate models. Population forecasts are driven by census-based information and demographic models that often disregard essential feedback between the human population and its environment. Economic forecasts are available, at best, at coarse spatial and temporal scales for only a few years to come. Energy use data, which closely track population and economic activity, and also climatic conditions, often lump several fuel types and end use sectors together. These data on the human side of the climate change story rarely match up with information coming from climate change models. Given the shortcomings in demographic, economic and other essential data to identify and manage bi-directional human-environment relations, one may ask: What good will it do to provide ever more disaggregated data on climate change without commensurate efforts to refine the understanding, at local and regional scales, of the socioeconomic processes to which localized climate models and data are supposed to be linked?

While better information – if it can be properly managed and interpreted – is an important requisite for informed decision making, not all information has equal value. Reminiscent of Liebig's Law of the Minimum (Liebig, 1890), ignorance in some areas cannot be overcome by knowledge in others.

Fourth, the economics of climate action portrays both the problem and the solution as being perfectly amenable to marginal analysis (Bosello et al., 2010). For example, economists ask and try to answer questions of the following kind: 'What is the marginal contribution to risk reduction of preserving another acre of land or spending another dollar on early warning systems or the height of a sea wall?' Instead of marginal adjustments, though, it is often the discrete nature of impacts and the lumpiness of the impacted systems that matter. From some critical level of impacts onward, ecosystems, social systems or infrastructure systems either work or they don't. These systems are, in some sense, binary. Treating them as such, and later learning that there was room for marginal changes may be a safer approach to identifying 'optimal' climate action then the one dominating economics today. Similarly, real-world climate action is largely driven by action plans that contain portfolio-like bundles of investment and policy actions. These are usually arrived at through a process of citizen and expert dialog, and constrained by the budget available for action, as it is the case with most climate action plans at the regional and community level. An example of an analysis of co-benefits generated by such an action plan is found in Chae (2010), which reports on the co-benefits associated with SO_X, NO_X, PM_{10} and odor control from implementing the various measures proposed by the air quality management plan and CO_2 emissions reductions of the Seoul metropolitan area, such as fuel switching for its vehicle fleet and landfill gas reuse. Rather than asking, for example, what the co-benefits of the next avoided ton of CO_2 would be, the study presents the discrete items of a portfolio of management options and rank orders them with respect to their ability to generate primary and ancillary outcomes. By doing so, it avoids the challenges of marginal benefit-cost analysis and is closer to the actual decision-making environment within which the policy and investment processes unfold.

Fifth, the notion is prevalent that optimal climate action – as identified in the network analysis section above – requires balancing the costs of action against the cost of inaction. While such thinking is valuable at the aggregate societal level, it does beg the question how that balance is actually achieved in the real world, where those who benefit from the consequences of reduced climate change are not necessarily the ones who pay for emissions reductions or adaptation. What are the institutional mechanisms for welfare transfer between those who bear the costs and those who enjoy the benefits? To the extent that the implementation of a policy includes welfare transfers – such as with tradable permit schemes or implementation of the Clean Development Mechanism (Ellis and Kamel, 2007) – the presence of, and ability to capture, co-benefits and co-costs may affect the magnitude of those transfers (Rübbelke, 2006). At local scales, the winners and losers from policy and investment making may be more readily identifiable, and potentially more vocal than at larger regional and national scales.

Sixth, climate change is only one of many key impacts that will shape society's future. Infrastructures deteriorate, local communities and ecosystems change. In some sense, climate change is a multiplier of existing threats or magnifier of existing needs. For example, the performance of physical infrastructure systems typically does not get better with climate change, nor does the need to care for the elderly and the poor (Ruth and Ibarraran, 2009). What is often framed as a climate challenge, in essence, has root causes of which climate change and the impacts it creates are mere symptoms. Unfettered economic growth, social and economic inequality, myopic decision making and a host of other drivers shape greenhouse gas emissions profiles and climate vulnerabilities.

As a corollary of the thread multiplier and need magnifier effects of climate change, decision makers in urban areas, as well as at national levels, may find that effective ways of dealing with climate impacts will come from making existing institutions – such as those that deal with infrastructure development and management or with social and ecological issues – more effective and making climate issues part of the rationale under which they operate. Rather than generating co-benefits and minimizing co-costs of climate action, greenhouse gas emissions reductions and adaptation action may become the co-benefits of 'mainstreaming climate policy' (Huq et al., 2003). This may be particularly relevant for urban areas where the relationship between urban resource allocation towards climate mitigation has little to no noticeable effects on global climate change itself.

Reframing Policy Objectives and Performance Measurement for Urban Climate Action

With the struggles to find a successor arrangement for the Kyoto Protocol attention has turned to the role of cities in generating carbon emissions and reducing the risks of climate change to their inhabitants, their infrastructures and ecosystems (IEA, 2008; ICLEI, 2012). At the urban scale, the implications of local climate mitigation actions for global atmospheric concentrations of greenhouse gases are tenuous at best. In contrast, the cause-effect relationships are more easily identified for adaptation actions undertaken by cities.

The possibility to generate through climate mitigation and adaptation co-benefits in the form of improved quality of life and lower risks in urban areas further lends support to local climate action. While conceptually appealing and potentially of strategic interest in guiding select investment options, the presence of co-benefits and co-costs has not yet been rigorously handled in decision making. The fact that co-benefits and co-costs can be found across the economic, social and environmental realm, and only be expressed in monetary terms in the most simple of cases, suggests however a need for richer approaches to the evaluation of alternative policy and investment options than are found, for example, in traditional economic analysis. Similarly, the fact that different parts of society differently enjoy, or suffer from, implementation of those options requires a broader social science based approach to analysis and decision making. Multi-criteria analysis, consensus conferences (Grundahl, 1995; Guston, 1999; Nelson et al., 2009) and development of local narratives (Elwood, 2006) are just a few of the strategies to arrive at more holistic assessments of policy and investment options. But since the full range of co-costs and co-benefits of climate action only emerges through time in a changing economic, social and climate context, mechanisms need to be in place through which society fine-tunes and continuously adjusts the social and economic parameters that shape both mitigation and adaptation strategies. Such an approach fundamentally differs from the more traditional economic optimization of a single objective function, given a small number of control variables and known constraints.

But where is the national and international climate policy that recognizes that climate change is the global symptom of local problems? Where is the economics that recognizes the limits of marginal analysis of impacts and response strategies? Where is the social science research that shows how human behaviour, technology choice and environmental impact relate to each other in the urban setting, where the majority of people live and the bulk of economic activity takes place? Where are the governance models that harmonize local with global interests? After decades of discourse on climate change, the answers are still few and far between to help planners and policy makers meet the existing and impending challenges.

Appendix A: Articles and Reports Included in the Network Models

Categorized by the Actions

1. Fuel substitution co-benefits:
   - Chae, 2010
   - Minjares et al., 2013
   - Partridge and Gamkhar, 2012
   - Rusu, 2012
   - Zhang et al., 2012

2. Proper Maintenance of equipment co-benefits:
   - Li and Crawford-Brown, 2011

3. Improved Technology co-benefits (new technology, retrofit):
   - Aunan et al., 2004
   - Fahlén and Ahlgren, 2012
   - Li and Crawford-Brown, 2011
   - Lv Yongtao et al., 2011
   - Minjares et al., 2013
   - Oanh et al., 2012

4. Building Systems and Operations co-benefits:
   - Eckelman et al., 2008
   - Jakob, 2006

5. Tax Policy co-benefits:
   - Ayres and Walter, 1991
   - Boyd et al., 1995
   - Burtraw et al., 2003
   - Ekins, 1996b
   - Mao et al., 2012

6. Policy Implementation and Scenarios co-benefits:
   - He et al., 2010
   - Kuosmanen, et al., 2009
   - Mao et al., 2012
   - Ren et al., 2012
   - Rusu, 2012
   - Selvakkkumaran and Limmeechokchai, 2013
   - Takeshita, 2012
   - van Vuuren et al., 2006
   - Zhang and Wang, 2011

7. Climate Policy co-benefits:
   - Alfsen et al., 1992
   - Burtraw and Toman, 1997
   - Crawford-Brown et al., 2012
   - Dessus and O'Connor, 1999
   - Dudek et al., 2003
   - Garg, 2007
   - Jack and Kinney, 2010
   - Markandya and Rübbelke, 2003
   - OECD, 2000
   - Pittel and Rübbelke, 2008

8. Indigenous/Local Participation co-benefits:
   - Chhatre et al., 2012

Appendix B: List of Acronyms and Abbreviations in Illustrations of Network Diagram

For Climate Actions and Consequences

Table 3.1 Acronyms and abbreviations for climate actions

Acronym/abbreviation	Action
	Fuel substitution
BCoil2LNG	Switch from oil to liquefied natural gas
Coal2PV	Replace coal-fired power with large-scale photovoltaic
Coal2smallhydro	Replace coal-fired power with small hydro
Coal2wind	Replace coal-fired power with wind
	Proper maintenance of equipment
IM program	Implement vehicle inspection and maintenance program
	Improved technology
Boiler-replacement	Replace inefficient boilers with state-of-the-art boilers
BRQ	Bind coal through briquetting
Central heating	Retrofit centralized heating with energy-efficient systems
CoHE	Implement co-generation of heat and electricity
CW	Remove dust by coal washing
IBM	Improve boiler management
MBD	Modify boiler design
Motorcycle fleet	Change technology for motorcycle fleet

Acronym/abbreviation	Action
	Building systems and operations
Bldg-label	Use energy-efficiency labeling
Rtf-insulation	Retrofit wall insulation
Rtf-ventilation	Retrofit ventilation and filtration systems
Rtf-window	Retrofit window replacement
	Implement tax policy
Ctax-electricity25$	US$25 per ton carbon tax in electricity sector
Ctax-electricity75$	US$75 per ton carbon tax in electricity sector
TFFE	Taxes on fossil fuel extraction
	Phased policy implementation and tiered scenarios
400PPMV	400 PPMV CO_2 stabilization scenario
400PPMV early	400 PPMV CO_2 stabilization scenario with early CO_2 reduction
CDM	Clean development mechanism
CE strategy	Convex early stabilization strategy
CL strategy	Convex late stabilization strategy
DA scenario	Domestic action: meet Kyoto targets allowing only for internal emission trading
DSM	Energy efficiency in demand side
EEPCN	Energy-efficiency policies in China (e.g. industry energy savings, household energy savings, vehicle energy efficiency improvements)
ELCS	Enhanced low carbon scenario
EM strategy	Early mid-intensive reduction strategy
ER strategy	Equal reduction strategy
Euro forest growth	European forest growth policy
GE strategy	Gradual early start strategy
GL strategy	Gradual late start strategy
LA strategy	Late action strategy
LCS	Low carbon scenario
NT scenario	Normal trade: full use of Kyoto mechanism but allows trading of surplus allowance
PCPCN	Pollution control policies in China (e.g. particulate matter control in industry, improvement in rural cooking condition, vehicle emission standards)
QR strategy	Quick and release strategy
RT scenario	Restricted Trade: full use of Kyoto mechanism but without allowing surplus allowance
SI strategy	Step-wise increase strategy
SR strategy	S-curve development of reduction strategy
	Climate policy
CCEC	Controlling carbon emissions in China
CP	Climate policies
CPN	Climate policies in Norway

Acronym/abbreviation	Action
	Climate policy
CPR	Climate policies in Russia
CPUA	Climate policies in urban areas
GCCPI	Global climate change policies in India
GHGMPC	GHG mitigation policies in Chile
GHGMPM	GHG mitigation policies in Mexico
	Indigenous/local participation
EPLC	Effective participation of local communities
TS	Tenure security

Table 3.2 Acronyms and abbreviations for climate consequences

Acronym/abbreviation	Consequence
	Greenhouse gases
GHGs	Greenhouse gas emissions
	Frequency of health impacts
Asthma	Asthma symptoms
Asthma attack	Acute asthma attacks
Cardiopulmonary mortality	Cardiopulmonary mortality
Cardiovascular morbidity	Cardiovascular morbidity
Child asthma	Child asthma
Child visits	Child medical visits
Children bronchitis	Acute bronchitis in children
Chronic bronchitis	Chronic bronchitis
Death	Avoided deaths
Emergency room	Emergency room visits
Health Benefit	General health benefits
Hospitalization	Hospitalization for respiratory and cardiovascular disease
Lung cancer mortality	Lung cancer mortality
NonFD	Non-fatal diseases
Occupational risks	Occupational risks
Other mortality	Mortality not caused by lung cancer or cardiopulmonary disease
Outpatient visits	Outpatient visits
	Frequency of health impacts
Physical activity	Physical activity
Premature mortality	Premature mortality
Respiratory hospitalization	Respiratory related hospital admissions
Respiratory morbidity	Respiratory morbidity

Acronym/abbreviation	Consequence
	Frequency of health impacts
Respiratory symptom	Acute respiratory symptom days
Restricted activity	Restricted activity days
RTA	Transportation accidents
	Air quality constituents (C, NO_x, SO_x, VO_x, PM)
BC	Basic carbon emission
CCE	Corrosion of capital equipment
CO	CO emission
NMVOCs	Non-methane volatile organic compound emission
NO_x	NO_x emission
PM	PM emissions
PM_{10}	PM_{10} emission
$PM_{2.5}$	$PM_{2.5}$ emission
Smog	Smog
SO_2	SO_2 emission
TSP	Total suspended particulates
VOC	Volatile organic compounds
	Physical and chemical constituents
Acetaldehydes	Acetaldehyde emission
Benzene	Benzene emission
Boiler slag	Boiler slag
Coal ash	Coal ash
Formaldehydes	Formaldehydes emission
Soot	Soot emission
	Fuel and energy demand and production
Coal	Coal demand
Electricity generation	Electricity generation
Energy savings	Energy savings
Energy sustainability	Sustainability of secure energy supply
Energy use	Total energy demand
Gas demand	Gas demand
Gas security	Gas supply security
Non-hydro	Non-hydro renewables generation demand
Oil	Oil demand
Oil security	Oil supply security
	Indigenous/local participation
Adaptive cap	Adaptive capacity
Deepening	Democratic deepening
Implementation	Implementation costs
Rural life	Improved rural livelihoods

Acronym/abbreviation	Consequence
	Forest health
Afforestation	Afforestation
Biodiversity	Biodiversity
CNFP	Creating non-timber forest products
Deforestation	Avoided deforestation
Reforestation	Reforestation
SI	Soil improvement (agriculture)
VI	Vegetation improvement (agriculture)
	Water quality
AFWL	Acidification of fresh water lakes
Acidification	Acidification
Eutrophication	Eutrophication
Water quality	Water quality
Watershed	Watershed protection
	Living standard
Indoor air	Indoor air quality
Noise	Noise
Rents	Rental income
Selling price	Housing sales price
Thermal comfort	Thermal comfort
Welfare	Welfare changes

References

Ackerman, F. 2002. Still dead after all these years: Interpreting the failure of general equilibrium theory. *Journal of Economic Methodology*, 9(2), pp. 119–139.

Ackerman, F. and Stanton, E.A. 2012. Climate risks and carbon prices: Revising the social cost of carbon. *Economics*, 6(10), pp. 1–25.

Alfsen, K., Brendemoen, A. and Glomsrod, S. 1992. *Benefits of Climate Policies: Some Tentative Calculations*, Discussion Paper No. 69. Norway: Central Bureau of Statistics.

Allais, M. 1978. 'Theories of general economic equilibrium and maximum efficiency'. In G. Schwödiauer (ed.) 1978. *Equilibrium and Disequilibrium in Economic Theory*. Dordrecht: D. Reidel Publishing Company, pp. 129–201.

Aunan, K., Fang, J., Vennemo, H., et al. 2004. Co-benefits of climate policy – lessons learned from a study in Shanxi, China. *Energy Policy*, 32(4), pp. 567–581.

Ayres, R.U. and Walter, J. 1991. The greenhouse effect: Damages, costs and abatement. *Environmental and Resource Economics*, 1(3), pp. 237–270.

Barker, T. and Rosendahl, K.E. 2000. 'Ancillary benefits of GHG mitigation in Europe: SO_2, NO_X and PM_{10} reductions from policies to meet Kyoto targets using the E3ME model and extern evaluations'. In Intergovernmental Panel on Climate Change (ed.) 2000. *Assessing the Ancillary Benefits and Costs of Greenhouse Gas Mitigation.* Paris: Organisation for Economic Co-Operation and Development, pp. 413–450.

Barker, T., Johnstone, N. and O'Shea, T. 1993. *The CEC Carbon/Energy Tax and Secondary Transport-Related Benefits. Energy–Environment–Economy Modelling*, Discussion Paper No. 5. Cambridge: University of Cambridge.

Bell, M.L., Davis, D.L., Cifuentes, L.A., et al. 2008. Ancillary human health benefits of improved air quality resulting from climate change mitigation. *Environmental Health*, 7(41), pp. 1–18.

Bosello, F., Carraro, C. and Cian, E.D. 2010. Climate policy and the optimal balance between mitigation, adaptation and unavoidable damage. *Climate Change Economics*, 1(2), pp. 71–92.

Boyd, R., Krutilla, K. and Viscusi, W.K. 1995. Energy taxation as a policy instrument to reduce CO_2 emissions: A net benefit analysis. *Journal of Environmental Economics and Management*, 29(1), pp. 1–24.

Burtraw, D. and Toman, M. 1997. *The Benefits of Reduced Air Pollutants in the U.S. from Greenhouse Gas Mitigation Policies, Resources for the Future*, Discussion Paper No. 98-01-REV. Washington, DC: Resources for the Future.

Burtraw, D., Krupnick, A., Palmer, K., et al. 2003. Ancillary benefits of reduced air pollution in the US from moderate greenhouse gas mitigation policies in the electricity sector. *Journal of Environmental Economics and Management*, 45(2003), pp. 650–673.

Carlsson Reich, M. 2005. Economic assessment of municipal waste management systems – case studies using a combination of life cycle assessment (LCA) and life cycle costing (LCC). *Journal of Cleaner Production*, 13(3), pp. 253–263.

Chae, Y. 2010. Co-benefit analysis of an air quality management plan and greenhouse gas reduction strategies in the Seoul metropolitan area. *Environmental Science and Policy*, 13(3), pp. 105–216.

Chhatre, A., Lakhanpal, S., Larson, A.M., et al. 2012. Social safeguards and co-benefits in REDD+: A review of the adjacent possible. *Current Opinion in Environmental Sustainability*, 4(6), pp. 654–660.

Converse, A.O. 1971. On the Extension of Input-Output Analysis to Account for Environmental Externalities. *American Economic Review*, 61(1), pp. 197–198.

Crawford-Brown, D., Barker, T., Anger, A. and Dessens, O. 2012. Ozone and PM related health co-benefits of climate change policies in Mexico. *Environmental Science and Policy*, 17(March), pp. 33–40.

Dessus, S. and O'Connor, D. 1999. *Climate Policy without Tears: CGE-Based Ancillary Benefits Estimates for Chile*, OECD Development Centre Working Paper No. 156 (formerly Technical Paper No. 156). OECD/OCDE.

Dudek, D., Golub, A. and Strukova, E. 2003. Ancillary benefits of reducing greenhouse gas emissions in transitional economies. *World Development*, 3(10), pp. 1759–1769.

Eckelman, M.J., Anastas, P.T. and Zimmerman, J.B. 2008. Spatial assessment of net mercury emissions from the use of fluorescent bulbs. *Environmental Science & Technology*, 42(22), pp. 8564–8570.

Ekins, P. 1996a. How Large A Carbon Tax is Justified by the Secondary Benefits of CO_2 Abatement? *Resource and Energy Economics*, 18(2), pp. 161–187.

Ekins, P. 1996b. The Secondary Benefits of CO_2 Abatement: How Much Emission Reduction Do they Justify? *Ecological Economics*, 16(1), pp. 13–24.

Ellis, J. and Kamel, S. 2007. *Overcoming Barriers to Clean Development Mechanism Projects*. Paris: Organisation for Economic Co-operation and Development, International Energy Agency.

Elwood, S. 2006. Beyond cooptation or resistance: Urban spatial politics, community organizations, and GIS-based spatial narratives. *Annals of the Association of American Geographers*, 96(2), pp. 323–341.

Fahlén, E. and Ahlgren, E.O. 2012. Accounting for external environmental costs in a study of a Swedish district-heating system – an assessment of simplified approaches. *Journal of Cleaner Production*, 27(May), pp. 165–176.

Gale, R.J.P. and Stokoe, P.K. 2001. 'Environmental cost accounting and business strategy'. In C. Madu (ed.) 2001. *Handbook of Environmentally Conscious Manufacturing*. Norwell, MA: Kluwer Academic Publishers, ch. 6.

Garg, A. 2011. Pro-equity effects of ancillary benefits of climate change policies: A case study of human health impacts of outdoor air pollution in New Delhi. *World Development*, 39(6), pp. 1002–1025.

Glomsrød, S., Vennemo, H. and Johnsen, T. 1992. Stabilization of emissions of CO_2: A computable general equilibrium assessment. *Scandinavian Journal of Economics*, 94(1), pp. 53–69.

Grundahl, J. 1995. 'The Danish consensus conference model'. In S. Joss and J. Durant (eds) 1995. *Public Participation in Science: The Role of Consensus Conferences in Europe.* London: Science Museum, pp. 31–40.

Guston, D. 1999. Evaluating the first U.S. consensus conference: The impact of the citizens' panel on telecommunications and the future of democracy. *Science, Technology and Human Values*, (24)4, pp. 451–482.

He, K., Lei, Y., Pan, X., et al. 2010. Co-benefits from energy policies in China. *Energy*, 35(11), pp. 4265–4272.

Hewitson, B.C. and Crane, R.G. 1996. Climate downscaling: Techniques and application. *Climate Research*, vol. 7, pp. 85–95.

Huq, S., Rahman, A., Konate, M., et al. 2003. *Mainstreaming Adaptation to Climate Change in Least Developed Countries (LDCS)*. London: International Institute for Environment and Development.

ICLEI. 2012. *Global Protocol for Community-Scale Greenhouse Gas Emissions (GPC).* Bonn: International Council for Local Environmental Initiatives

IEA. 2008. *World Energy Outlook, Energy Use in Cities*. Paris: International Energy Agency, ch. 8.

Jack, D.W. and Kinney, P.L. 2010. Health co-benefits of climate mitigation in urban areas. *Current Opinion in Environmental Sustainability*, 2(3), pp. 172–177.

Jakob, M. 2006. Marginal costs and co-benefits of energy efficiency investments the case of the Swiss residential sector. *Energy Policy*, 34(2), pp. 172–187.

Jasch, C. 2006. How to perform an environmental management cost assessment in one day. *Journal of Cleaner Production*, 14(14), pp. 1194–1213.

Johnson, M.H. and Bennett, J.T. 1981. Regional environmental and economic impact evaluation: An input-output approach. *Regional Science and Urban Economics*, 11(2), pp. 215–230.

Kahneman, D. 2003. Maps of bounded rationality: Psychology for behavioral economics. *The American Economic Review*, 93(5), pp. 1449–1475.

Kahneman, D., Knetsch, J.L. and Thaler, R.H. 1991. The endowment effect, loss aversion, and status quo bias. *The Journal of Economic Perspectives*, 5(1), pp. 193–206.

Kuosmanen, T., Bijsterbosch, N. and Dellink, R. 2009. Environmental cost-benefit analysis of alternative timing strategies in greenhouse gas abatement: A data envelopment analysis approach. *Ecological Economics*, 68(6), pp. 1633–1642.

Lange, H., Garrelts, H., Osthorst, W. and Selmi, F. 2009. 'Distributional effects and change of risk management regimes: Explaining different types of adaptation in Germany and Indonesia'. In M. Ruth and M. Ibarraran (eds) 2009. *Distributional Impacts of Climate Change: Social and Economic Implications*. Cheltenham, UK: Edward Elgar Publishers, pp. 183–207.

Leontief, W.W. 1966. *The Structure of American Economy: An Empirical Application of Equilibrium Analysis*. New York: Oxford University Press.

Li, Y. and Crawford-Brown, D.J. 2011. Assessing the co-benefits of greenhouse gas reduction: Health benefits of particulate matter related inspection and maintenance programs in Bangkok, Thailand. *Science of the Total Environment*, 409(10), pp. 1774–1785.

Machol, B. and Rizk, S. 2013. Economic value of U.S. fossil fuel electricity health impacts. *Environmental International*, 52(February), pp. 75–80.

Mao, X., Yang, S., Liu, Q., et al. 2012. Achieving CO_2 emission reduction and the co-benefits of local air pollution abatement in the transportation sector of China. *Environmental Science & Policy*, 21(August), pp. 1–13.

Markandya, A. and Rübbelke, D.T.G. 2004. Ancillary benefits of climate policy. *Jahrbücher für Nationalökonomie und Statistik*, 224, pp. 488–503.

Meadows, D. 1999. *Leverage Points*. Hartland, VT: The Sustainability Institute.

Meadows, D.H., Randers, J. and Meadows, D.L. 2004. *Limits to Growth: The 30-year Update*. Post Mills, VT: Chelsea Green Publishing Company.

Minjares, R., Blumberg, Ķ. and Sanchez, F.P. 2012. Alignment of policies to maximize the climate benefits of diesel vehicles through control of particulate matter and black carbon emissions. *Energy Policy*, 54(March), pp. 54–61.

Nelson, J.W., Scammell, M.K., Atlman, R.G., et al. 2009. A new spin on research translation: The Boston consensus conference on human biomonitoring. *Environmental Health Perspectives*, 117(4), pp. 495–499.

Nemet, G.F., Holloway, T. and Meier, P. 2010. Implications of incorporating air-quality co-benefits into climate change policymaking. *Environmental Research Letters*, 5(1), pp. 1–9.

Ng, Y.-K. 1983. *Welfare Economics: Introduction and Development of Basic Concepts* (revised edition). Basingstoke: Macmillan.

Oanh, N.T.K., Phuong, M.T.T. and Permadi, D.A. 2012. Analysis of motorcycle fleet in Hanoi for estimation of air pollution emission and climate mitigation co-benefit of technology implementation. *Atmospheric Environment*, 59 (November), pp. 438–448.

OECD. 2000. *Ancillary Benefits and Costs of Greenhouse Gas Mitigation, Proceedings of an IPCC Co-Sponsored Workshop held on 27–29 March 2000, in Washington, D.C.* Paris: Organisation for Economic Co-operation and Development.

Partridge, I. and Gamkhar, S. 2012. A methodology for estimating health benefits of electricity generation using renewable technologies. *Environmental International*, 39(1), pp. 103–110.

Pearce, D. 2000. 'Policy frameworks for the ancillary benefits of climate change policies'. In OECD (ed.) *Assessing the Ancillary Benefits and Costs of Greenhouse Gas Mitigation*. Paris: Organisation for Economic Co-operation and Development, pp. 517–560.

Pittel, K. and Rubbelke, D.T.G. 2008. Climate policy and ancillary benefits: A survey and integration into the modelling of international negotiations on climate change. *Ecological Economics*, 68(1–2), pp. 210–220.

Pryor S.C., Schoof, J.T. and Barthelmie, R.J. 2005. Climate change impacts on wind speeds and wind energy density in northern Europe: Empirical downscaling of multiple AOGCMs. *Climate Research*, 29, pp. 183–198.

Pryor, S.C. and Barthelmie, R.J. 2011. Assessing climate change impacts on the near-term stability of the wind energy resource over the USA [online]. Proceedings of the National Academy of Sciences. Available at: doi: 10.1073/pnas.1019388108 [accessed 11 March 2014].

Ren, W., Geng, Y., Xue, B., et al. 2012. Pursuing co-benefits in China's old industrial base: A case of Shenyang. *Urban Climate*, 1(November), pp. 54–64.

Rübbelke, D.T.G. 2003. An analysis of differing abatement incentives. *Resource and Energy Economics*, 25(3), pp. 269–294.

Rusu, M. 2012. Social cost of carbon: Opportunities and environmental solutions. *Procedia Economics and Finance*, 3(2012), pp. 690–697.

Ruth, M. and Baklanov, A. 2012. Urban climate science, planning, policy and investment challenges. *Urban Climate*, 1(1), pp. 1–3.

Ruth, M. and Ibarraran, M. (eds) 2009. *Distributional Impacts of Climate Change: Social and Economic Implications*. Cheltenham, UK: Edward Elgar Publishers.

Ruth, M., Kalnay, E., Zeng, N., et al. 2011. Sustainable prosperity and societal transitions: Long-term modeling for anticipatory management. *Environmental Innovation and Societal Transition*, 1(1), pp. 160–165.

Selvakkumaran, S. and Limmeechokchai, B. 2013. Energy security and co-benefits of energy efficiency improvement in three Asian countries. *Renewable and Sustainable Energy Reviews*, 20(April), pp. 491–503.

Shim, Y., Hobbs, B.F., Ruth, M. and Williamson, S. 2011. *Meeting Maryland's Greenhouse Gas Reduction Goals: Electrical Reliability Impacts from Maryland's Climate Action Plan.* College Park, Maryland: University of Maryland, Center for Integrative Environmental Research.

Shrestha, R.M. and Pradhan, S. 2010. Co-benefits of CO_2 emission reduction in a developing country. *Energy Policy*, 38(5), pp. 2586–2597.

Stone, B. 2012. *The City and the Changing Climate: Climate Change in the Places We Live.* Cambridge: Cambridge University Press.

Takeshita, T. 2012. Assessing the co-benefits of CO_2 mitigation on air pollutants emissions from road vehicles. *Applied Energy*, 97 (September), pp. 225–237.

Thaler, R.H. and Sunstein, C.R. 2009. *Nudge: Improving Decisions About Health, Wealth, and Happiness*. New York: Penguin Books.

Tol, R.S.J. and Verheyen, R. 2004. State responsibility and compensation for climate change damages – a legal and economic assessment. *Energy Policy*, 32(9), pp. 1109–1130.

Turner, G. 2008. A comparison of 'The Limits to Growth' with thirty years of reality. *Global Environmental Change*, 18(3), pp. 397–411.

Van Vuuren, D.P., Cofala, J., Eerensa, H.E., et al. 2006. Exploring the ancillary benefits of the Kyoto Protocol for air pollution in Europe. *Energy Policy*, 34(4), pp. 444–460.

von Liebig, G.F. 1890. Nekrolog: Justus von Liebig: Eigenhändige biographische Aufzeichnungen [online]. *Berichte der deutschen chemischen Gesellschaft*, 23(3), pp. 817–828. Available at: doi: 10.1002/cber.18900230391 [accessed 11 March 2014].

Woodcock, J., Edwards, P., Tonne, C., et al. 2009. Public health benefits of strategies to reduce greenhouse-gas emissions: Urban land transport. *Lancet*, 374(9705), pp. 1930–1943.

Yongtao, L., Chao, H., Xiuqin, M. and Fankun, W. 2011. Co-benefit research on energy efficiency reforming project of district heating system. *Energy Procedia*, 5, pp. 598–603.

Zhang, D., Tang, S., Lin, B., et al. 2012. Co-benefit of polycrystalline large-scale photovoltaic power in China. *Energy*, 41(1), pp. 436–442.

Zhang, J. and Wang, C. 2011. Co-benefits and additionality of the clean development mechanism: An empirical analysis. *Journal of Environmental Economics and Management*, 62(2), pp. 140–154.

Chapter 4
Social, Economic and Ecological Benefits of Landscape Park Projects: Using Benefit Transfer to Assess Green Infrastructure Projects

Karsten Rusche and Jost Wilker

Introduction

Urban planning of public investments is often confronted with several decision options for policy makers, framed by their budgets. Based on this, they carefully need to decide, which of the planning alternatives they choose to be realized. Looking at recommendations from economic theory, a major source of information on that issue is a benefit cost analysis (Brent, 2006). By looking at those projects that offer the highest net benefits for society, a ranking of relevant investments can be done and compared with the annual budget.

This general principle is also valid in the case of public good investments, but there is one major drawback in the benefit cost analysis of those goods: they are not sold on markets, so there are no prices easily available for the calculation of societal benefits. This is especially relevant for green infrastructure investments. In administration, they are mainly viewed as investments that cost money but do not generate revenue. This schematic view often hinders an optimal supply with public green infrastructure.

This chapter aims at showcasing two important issues. On the one hand, it will develop a benefit cost based benefit transfer approach which can be easily adopted by city departments. On the other hand it will demonstrate that economic valuation is an important tool in supporting decision-making processes.

In the next sections a case study for the German city Stuttgart is described in its structure, methods and results. The benefit transfer approach presented here is focused on several possible investments in green infrastructure, which is in this article understood as a network of parks, open spaces, playing fields, woodlands, allotments and private gardens. These places are all valuable parts of the urban economy (Natural England, 2009, p. 7). So, green infrastructure delivers economic, social and environmental benefits such as tourism, recreation and biodiversity for the inhabitants of city regions. Nevertheless, due to shortages in public finances and the competition with other land uses, investments in green infrastructure need to be justified economically against politicians and the public, as mentioned above. By means of this study on 18 Landscape Park Neckar projects in Stuttgart, this issue is investigated further by exploring the possibilities and limits of benefit transfer and cost-benefit analysis to guide policy analysis and local decision making in urban green infrastructure planning. The projects cover a wide range of different investment types that are linked to the urban green infrastructure of Stuttgart, such as river renaturations, improvements of parks and development of large urban regeneration projects.

The next section briefly outlines the methodological approach of benefit transfer. Further, we explain the role of planning evaluation, and in particular how benefit transfer practice may be embedded in a cost-benefit analysis. Based on these results, the chapter makes a series of recommendations for further research on valuation, evaluation and a better use of benefit transfer in green infrastructure policy and planning. The final section summarises the main outcomes of our research and discusses its application and relevance in practice.

Benefit Transfer

Benefit transfer belongs to the field of cost-benefit analysis (Johnston and Rosenberger, 2010). The term describes the usage of economic research data gathered at a certain place and time, called 'study site', to make predictions of welfare estimates for another site, called 'policy site', for which primary information is not available (Rosenberger and Loomis, 2000; Ready and Navrud, 2005; Wilson and Hoehn, 2006; Johnston and Rosenberger, 2010). That means, benefit transfer is only applicable 'if a study already exists that valued a good similar to the good in question' (Ready and Navrud, 2005, p. 196).

Recent research distinguishes commonly between two types of benefit transfer: unit value and function transfer (Johnston and Rosenberger, 2010). The first approach is suitable when the imagined impacts of the project site can be measured in independent, non-overlapping units which are identified through review of previous valuation studies for that good (Ready and Navrud, 2005). To guarantee high validity study site and policy site should be similar and at close range. So the benefit values from the study site can be projected unit by unit to the policy site. In many cases the values taken from a study site will have to be adjusted for differences in source and policy contexts, i.e. currency, time or culture (Bateman et al., 2011). When the context of the study site is different to the one of the policy site function transfer may increase the reliability of the benefit transfer. Here, a value function predicts the benefit values of the project site on the basis of the measurable characteristics of the study site. By knowing the significant factors affecting WTP for the study site (e.g. income, age, opinions, etc.) this information can be used as a function to assess the benefits of the policy site (Pearce et al., 2006; Ready and Navrud, 2005).

Function transfer is dependent on several conditions regarding study site attributes and uses data of a meta-analysis which combines value estimates from different studies to provide more flexible and better applicable value functions. The general conclusion in literature is that function transfers typically outperform unit value transfers (Johnston and Rosenberger, 2010; Loomis, 2014).

Both types of benefit transfer share common positive and negative aspects. On the downside there are two main important errors that easily accrue when benefit transfer is not conducted thoroughly: measurement and also transfer errors (Plummer, 2009). The first are related to imprecise empirical designs in the original studies. Due to those differences between a true underlying value and a primary study value will occur. Contrarily, the transfer errors occur within the transfer process from a study site to a policy site. They are also labelled 'generalization errors' as they can be caused by 'the correspondence between sites and populations, the commensurability of non-market goods and policy contexts, and the benefit transfer method applied' (Johnston and Rosenberger, 2010, p. 486). In addition to those errors there is always a methodological problem in aggregating single benefit values from different study sites into one monetary value for a policy site (TEEB, 2010). In this process there is a high risk of double-counting of values, because there is often no clear-cut differentiation on what exactly has been measured on-site or has been asked in questionnaires. In summary, those downside of benefit transfer leads to the conclusion that transferred values are to a certain extend overestimated. International studies on that specific error in benefit transfer set the bias rate at a typical range from 4 up to 40 per cent, but in extreme cases it can be 100–200 per cent (Loomis, 2014).

In contrast to the issues discussed above, benefit transfer also offers several positive characteristic that lead to its wide spread acceptance in public policy. Due to its relative simplicity, benefit transfer is commonly conducted worldwide by agencies and institutions to assess economic benefits and costs of development projects to justify investments (Wilson and Hoehn, 2006). Benefit transfer offers a time-saving and inexpensive option to support the policy progress instead of conducting original studies when benefit estimates are required. Therefore, benefit transfer is often the only feasible alternative to estimate monetary benefits of environmental goods and services (Loomis, 2014), although 'the use of primary research to estimate values is generally preferred' (Johnston and Rosenberger 2010, p. 479). Those characteristics also underline the adequacy of the benefit transfer method to be a relevant tool to support decision-making processes.

Despite its increasing importance for policy making, practitioners are still looking for best practice standard on how to apply benefit transfer (Wilson and Hoehn, 2006). Here, the scientific discussion on the one hand and the practical use of benefit transfer for policy analysis on the other hand are not perfectly linked with each other and show divergence (Ready and Navrud, 2005). While academic researchers have concentrated on the use of function transfer and other approaches to reduce transfer errors and improve validity, practitioners are interested in less complex but more easily accessible concepts and guidance (Rosenberger and

Johnston, 2009). However, despite the challenges outlined here, benefit transfer is an unavoidable part of benefit cost analysis and crucial for planning policy analysis in North America (Plummer, 2009). In Germany, benefit transfer is part of the scientific discussion but rarely used (Ahlheim and Lehr, 2002; Thiele and Wronka, 2001). This chapter, therefore, will make an assessment of a case in Stuttgart, thereby focusing on the role of benefits and how they may be evaluated in this context.

Case Study

The case study for the benefit transfer approach in Stuttgart deals with a set of investments which are part of the strategic masterplan 'Landscape Park Neckar in Stuttgart'. Here, 18 different green infrastructure projects are planned (and are already partly built) in the proximity to the river Neckar that flows through the Eastern parts of Stuttgart (see Figure 4.1).

The main political aims for the City of Stuttgart regarding these investments are:

- to reactivate the river as an urban open space that can be used by inhabitants,
- to attract more tourists to the recreational sites and
- to save urban habitats to enrich Stuttgart's biodiversity.

In sum, green infrastructure projects shall raise quality of life in Stuttgart. To achieve this, the accessibility of the river was and will be improved, along with more and extended possibilities for recreation and leisure and also ecological improvements along the river Neckar will be undertaken. The range of projects has different emphases. Recreationally focused investments in the heart of Stuttgart's East are planned together with plainly ecological projects in the more remote areas.

To convince policy makers in Stuttgart to support the planned network of green infrastructure measures – by enacting budgets to start investment works – the City administration decided to commission an economic valuation study, which should put money values on the variety of possible benefits that are created by the Landscape Park projects. So, the study is used as a decision-making support tool, because its main purpose is not just the pure ex-post justification of public actions but rather an instrument that is used for negotiations between different resorts of city administrations. The main motivation of the department of city planning was to highlight the benefits from green infrastructure investments that are not within the standard way of approaching budgetary and strategic internal discussions.

An assessment of the question which benefit transfer approach may fit best requires a specification of the policy context relevant to the case (Johnston and Rosenberger, 2010). Two major policy issues have framed the way in which valuation was conducted. Due to a considerably low budget and time constraints, it was not possible to do primary data collections through local investigations, experiments

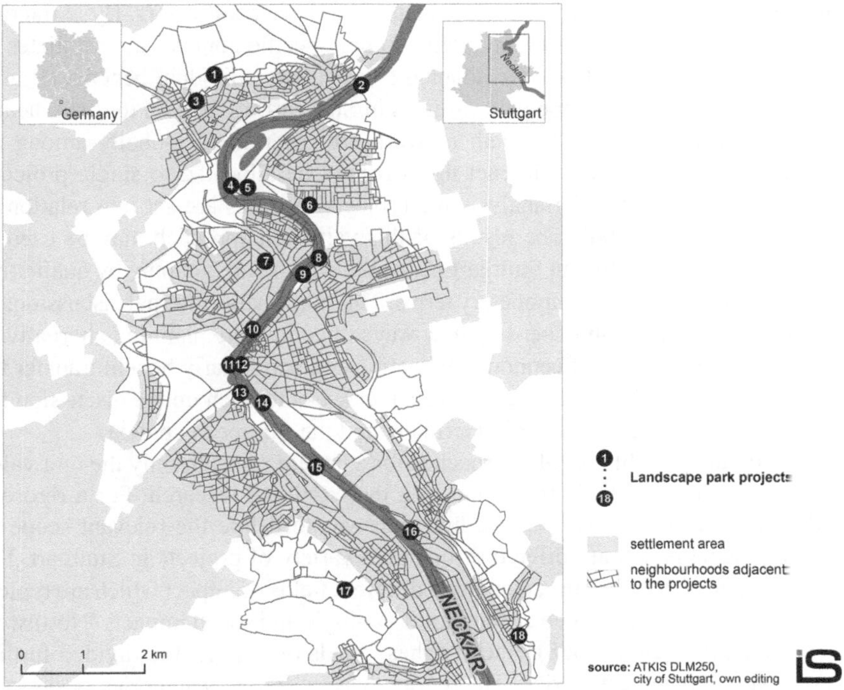

Figure 4.1 Location of the investments
Source: City of Stuttgart, ATKIS DLM 250, own editing 2014.

or surveys within Stuttgart or to concentrate on meta-analysis regression within a value function transfer. Furthermore, the project was highly policy oriented and progress was regularly fed back with the client. The administration of the City of Stuttgart wanted to have an indication on the benefit cost relations that can be expected from the set of green infrastructure investments, no exact calculation of compensatory amounts. For the political administration this was an innovative approach to underpin green infrastructure planning decisions and policies with impact assessments based on economic valuations.

Economic Valuation Approach for the Case Study

To capture the projects′ benefits in economy, society and environment (Niemelä et al., 2010) this section develops a system of relevant impact fields as a basis for benefit transfer. For instance, ecological benefits are measured in the categories 'landscape and biodiversity', 'water management and flood protection' and 'climate mitigation and adaptation'. Within each impact field the benefit transfer method was used to identify suitable quantity and price indicators. By extracting available information from other studies to our specific context, the estimates of benefits from these studies were adapted to calculate individual benefits (Pearce et al., 2006) of the landscape park projects in Stuttgart. Effects on tourism, health and well-being and from the path network are determined globally among all projects as effects on these impact fields cannot be allocated to single projects. By means of a benefit-cost analysis the projects′ benefits are set into relation to their costs (Brent, 2006). The results of this valuation approach may be used to give reasons for investing in Stuttgart's green infrastructure not only by qualitative indicators, but expressed in monetary terms. This may enable internal discussions in public administration and opens up new ways of comparing planning alternatives. Against the importance of economic valuation and particularly benefit transfer for policy analysis in North America, this chapter seeks to open up the discussion for the potential of benefit transfer in Germany and the European context.

Within the possible set of approaches for a benefit transfer, only the unit value transfer remained suitable for conducting the empirical research. For a rigorous and structured benefit transfer the first step was to define the relevant scope of different benefits that are offered through the variety of projects in Stuttgart. For this, the approach concentrated on three different fields of impact which are typical for classical green investments: economic, ecologic and social impacts (Northwest Development Agency, 2008). Each of these fields of impact was divided further into seven impact categories. Based on these different categories a desktop research was started to find a set of original studies that offer information on relevant indicators that can be used for benefit transfer within the given categories (see Appendix for an overview of used literature values).

Within benefit transfer literature it is well known that it is crucial to try to find the best suitable study sites that mirror the policy sites (Ready and Navrud, 2005). In addition, this study site information should be taken from case studies that are situated within the same country as the policy sites to avoid generalization errors. In line with these academic recommendations the search for relevant study sites followed the rule: Whenever possible, the site should be within Germany, then within German speaking countries in Europe, then within Europe. In the case there were no appropriate studies that be found within this spatial scope, we had to consider study sites from abroad.

Impact categories that deal with social benefits could be supplemented with information from studies within Germany or Switzerland. Many of these studies deal with the value of urban parks, recreational areas, cycle tracks or river renaturation by working with stated preference approaches (contingent valuation/choice modelling). In contrast, studies on the ecological benefits of urban green infrastructure are rare in the German and Europe. They are much more spread across other continents like Asia, Africa or the Americans. According to the similarity of study and policy sites, here is the highest risk of causing generalization errors. The economic aspects were captured by using regional survey based statistics for the expenditures of tourists. Figure 4.2 lists the relevant indicators that were used for

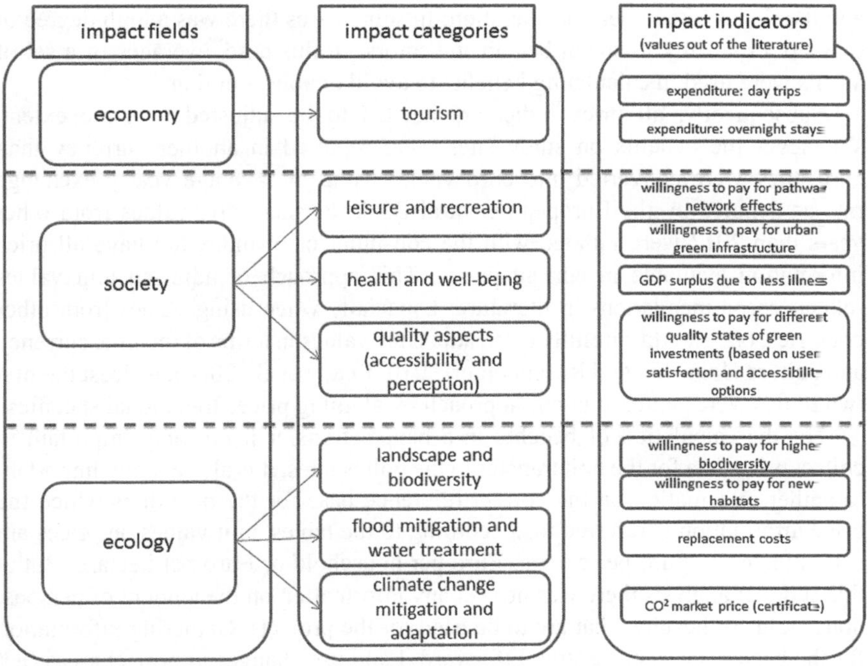

Figure 4.2 Methodology of benefit transfer

each project in terms of individual benefit transfers. For each of these indicators publications were used to gather benefit value information for the transfer to Stuttgart's 18 landscape park projects (see Appendix).

One important extension to the project-based approach needs to be highlighted. In the process of recognizing relevant impact categories, some of them needed to be measured altogether for all 18 projects instead of being individually calculated: network effects on the pathway system, impulse on tourism and health effects. In these cases it is crucial to follow a certain rationale that was made up in discussions with local stakeholders. Each project on its own is too small to have a significant impact on one of these categories. Only if the whole bundle of projects from the Masterplan Landscape Park Neckar is implemented together, they will create benefits that are big enough to be measured in these three categories.

The calculation of benefits follows a stepwise approach. In the first stage, relevant impact categories were identified, on which each of the project is suggested to have an impact. This was discussed together with stakeholders from the City of Stuttgart to enrich this process with local expert knowledge. Based on this step unit values of the indicators for the benefits were chosen from the list as in Figure 4.2. Again, decision was based on the character of each investment to ensure a high similarity between study site from which the unit value was taken and the policy site under consideration. In some cases there was a high degree of overlap, especially in social benefit indicators. In this case, averages of a set of WTPs were used in calculating benefits to avoid double counting.

Subsequently, all price indicators needed to be adjusted to some extent. Whenever the benefits on study sites were reported in another currency than euro, they were converted into euro values using the average yearly exchange rate as denoted by the European Central Bank. Furthermore, values from other years than 2012 were inflated with the consumer price index too have all price information available in current prices. This approach of adjusted unit values reflects recommendations in literature: Especially when using values from other countries, one should carefully try to adjust the values in terms of income, currency and culture (Johnston and Rosenberger, 2010; Ready et al., 2004). At least the first two topics were captured in our approach by aligning prices to national specifics.

For the calculation of benefits in a benefit transfer it not only important to collect price data for the unit transfer to the policy sites, it is also equally important to gather information on the correct reference base for the quantities which the price information is referred to. According to the tables, unit values on prices are denominated in Euro per person, Euro per household or Euro per hectare. But at this stage of analysis there was not yet any information on the amount of persons, households or hectares that are to be used for the projects. Gathering information for the hectares was quite straightforward. Land use changes in project areas (net surface acreage of development) were easy to obtain, because the projects were already modelled in a GIS. Collecting data on the number of people that would possibly use the policy sites for recreation and leisure, sports or other activities needed a more sophisticated approach.

In order to approximate the number of beneficiaries of the project, a buffer approach that used GIS data on urban population was developed. Based on the amount of impact fields the projects would have effect on, three different spatial buffer zones were constructed around each project. They range from 300 meter for projects with little impact on the surrounding environment up to a buffer of 1200 meter for projects that influence on ecology, economy and society altogether (see Figure 4.3). All adolescent inhabitants that reside within these buffer zones were defined as beneficiaries of the projects. For each project the individual buffer zones were used to calculate the reference base for the unit value transfer. In this case, overlapping buffers (and populations) were allowed because people can use (and therefore value) different projects in their neighbourhood. For the analysis of global effects (network and health), only the non-overlapping sum of inhabitants was used to calculate the benefits. About 90,000 inhabitants were defined as being affected by the benefits that are and would be created by the bundle of landscape park projects in Stuttgart. In addition, the estimated effect on local tourism was drawn up by the assumption that daily visits as well as overnight stays will increase by 0.5 per cent in relation to Stuttgart's annual amount of tourists. This number was confirmed by local experts from Stuttgart's tourist office.

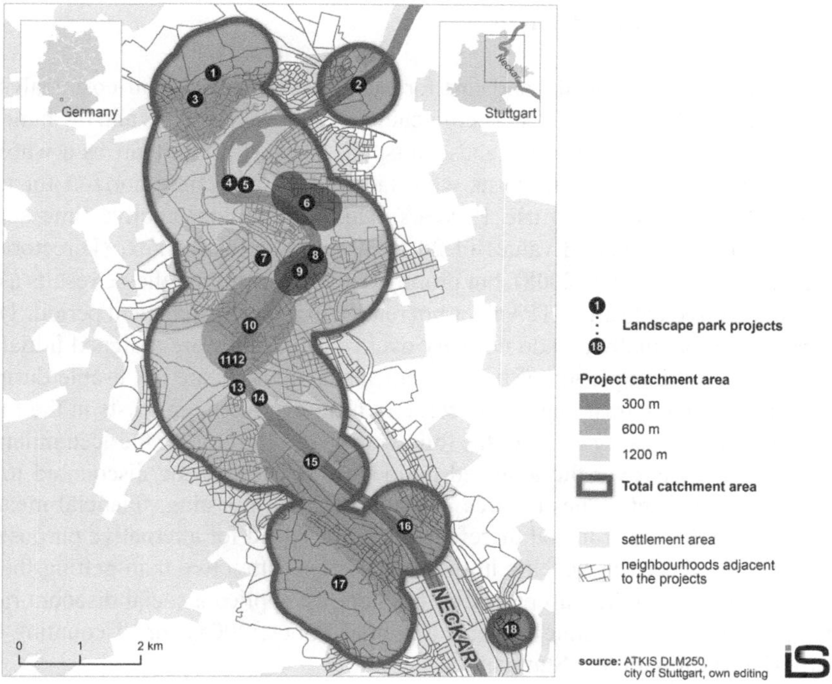

Figure 4.3 Impact catchment area of the investments
Source: Stadt Stuttgart, ATKIS DLM 250, own editing 2014.

For all global benefits the sum of impacts was added to the individual calculations for each project by using an allocation key (paths lengths, relative attractiveness as defined by stakeholders).This step was essential to enable the benefit cost analysis to deliver results for individual projects. For example, the network effect for cycling paths was calculated using the number of people that will use the new tracks (inhabitants as well as some number of cyclists from the city region) multiplied by the willingness-to-pay value for cycling tracks from a study site within the Stuttgart region. The share for each of the 18 projects was then divided by looking at the percentage share of total length of pathways of all projects for each individual project. This way of benefit allocation ensured that only projects were given extra benefits from 'global' effects if they had at least little pieces of pathways in the project plans. Thereby, bigger parks and paths improvement were assigned the highest shares.

The last step that was done before conducting the benefit cost analyses was the collection of investment and maintenance costs for all investments. These data was provided by the City of Stuttgart.

Results of Benefit-Cost Analysis

After calculating the benefits and cost for each investment a benefit-cost analysis was conducted to estimate the cost efficiency of single investments and the bundle of projects of the 'Masterplan Landschaftspark Neckar' in Stuttgart as a whole. For this reason, benefits and costs were faced. As a base time horizon for the calculations we decided to use 15 years. This a quite conservative time span in comparison to standard values of 20 to 30 years in this context (Directorate General Regional Policy, 2008), but promises to provide more reliable results as a lot of transferred data are WTP values not related to a specific payment period. The benefits occur annually and do not vary over time, but they are assigned initially in the year after completion of the investments as the sites are not usable during construction. In contrast, the costs accrue in form of investment costs in the first year and as maintenance costs for the following years. To compare the accumulated benefits and costs over the assigned period, values need to be discounted to a gain the net present values (Pearce et al., 2006). By investing, financial means are bound in the long run and therefore cannot be spend for alternative purposes. Furthermore, receiving benefits in the future is less attractive than getting them today. According to the European Commission, we applied a social discount rate of 3.5 per cent (Directorate General Regional Policy, 2008) for discounting of study sites information in Stuttgart.

Table 4.1 displays the benefit-cost ratios for different time periods and the scenario of a 50 per cent benefit cut over 15 years. A ratio of 1 expresses the break-even point, while a lower value means that the costs exceed the benefits and a value higher than 1 reveals benefit gains. Accordingly, it has to be mentioned that the projects` benefit-cost ratios vary a lot due to their different amounts of benefits and costs. Focusing on the 15 year ratios it becomes obvious that investments related to mainly ecological impacts such as project 18, 9 and 2 would not be cost covering after 15 years and therefore may cause higher costs than creating benefits. This is due to the fact that these kinds of investments would affect particularly nature itself and would provide only indirect benefits for people which are difficult to capture. Project 13 is so expensive that even high benefits within all impact fields would not result in a benefit-cost ratio of 1 or better. All other investments would create higher benefits than causing costs. Especially the benefit gain through recreation and leisure, quality of space through improved river access, and health and well-being would be high. The investment number 17 could be an outlier as its cost-benefit ratio is by far the highest, but this can be explained by very low investment and maintenance costs in relation to investments with a high amount of green spaces which would provide higher absolute benefits. The results of the benefit-cost analysis reveal that the 18 landscape park projects would generate estimated benefits of about 300 million euros in total, which would exceed five times their costs.

Table 4.1 Benefit-cost ratios

Projects	15 year ratio	10 year ratio	20 year ratio	15 year ratio (50% benefit cut)
1	5.03	3.83	1.31	2.52
2	0.04	0.03	1.37	0.02
3	11.40	8.40	1.36	5.70
4	16.09	11.76	1.37	8.04
5	6.78	4.74	1.43	3.39
6	17.47	12.80	1.36	8.73
7	25.64	20.20	1.27	12.82
8	18.41	13.61	1.35	9.21
9	0.05	0.04	1.36	0.03
10	8.09	5.88	1.37	4.04
11	13.34	9.94	1.34	6.67
12	1.66	1.22	1.36	0.83
13	0.97	0.70	1.38	0.49
14	6.23	4.58	1.36	3.12
15	1.70	1.25	1.35	0.85
16	4.87	3.63	1.34	2.44
17	50.43	41.80	1.21	25.21
18	0.05	0.04	1.30	0.03
Total	5.39	3.94	6.58	2.69

To classify the mentioned results for the time horizon of 15 years we did also extrapolations for time frames of 10 and 20 years as well as compiled a scenario for a 50 per cent budget cut over 15 years (see Table 4.1). Comparing the benefit-cost ratios shows that a longer time horizon would result in higher benefits while a shorter time frame would affect the benefits negatively. The influencing factor is the annual maintenance costs which would be lower as the yearly monetary benefits among all projects. The overall picture of the benefit-costs ratios of the single investments would not change a lot due to a change in the time horizon. Certainly, a benefit cut by 50 per cent decreases the benefit-cost ratios by the same degree. Still, only 6 out of 18 projects would not be break-even after 15 years and the total benefit-cost ratio over all investments would be 2.69 (see Table 4.1).

Nevertheless, not all projects would break-even after 20 years under normal benefit conditions; even 50 years would not be enough (see Figure 4.4). However, most investments would reach the break-even point at an early stage and only three projects are estimated to be break-even later than the fifth years. Combined with the previous results this means that especially investments which would be shortly break-even and would provide high absolute benefits would be suitable for contemporary implementation as they theoretically generate high benefits in a rather short period of time.

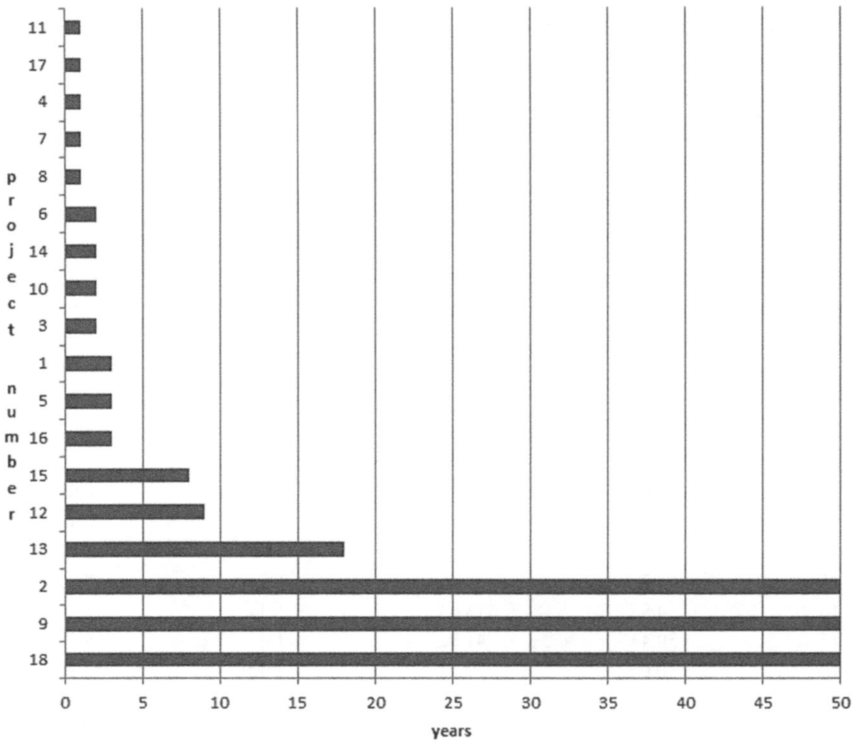

Figure 4.4 Break-even points

The Potential Use of 'Benefit Transfer' in 'Green Infrastructure Planning'

Looking at the approach of unit value transfer in the case study from an academic point of view, it is worth to discuss what are the main drawbacks and which research approaches can improve the efficiency of such analyses.

As stated in the description of the case study motivation, the aim of the City of Stuttgart was to get acquainted with the economic valuation of benefits of public goods. In times of austerity this becomes more and more important ensure an efficient allocation of public money. So, using the rather simplistic approach of a unit value transfer was constructive, because it helped practitioners to get these ideas applicable in urban planning. Nevertheless, the impact of the theoretical discussion on the optimal form of benefit transfer on practical planning is very limited. This is because with an increasing degree of complexity (as with meta-analysis regressions) the capability of stakeholders to grasp the idea of what is done is decreasing.

The results of our approach for this case study are good for their purpose of giving an indication about the benefit-cost relations for public investments, but there is much to improve when decision supporting elements shall be modelled in more detail. For this purpose it is essential to rely on primary, place-based data wherever possible.

On the basis of the approach in this chapter and suggestions in literature, an optimal place-based approach will combine both, survey/primary data and benefit transfer components. For all locally measurable information more experience is needed in Europe and especially in Germany to widen the basis of study site analyses. For this, not only stated preference data should be collected more eagerly, but also primary data on changes in ecosystem services due to project investment activities is needed. As discussed in the third section, these are the issues on which there is a huge lack of case studies which would help to give more guidance in which approaches are best to assess urban planning issues. These data could then be used to collect a set of study sites that can be used in more complex approaches like benefit function transfer. By using a wide set of local data, these approaches could more easily be explained to practitioners because they have more access to the topics that are dealt with in the study sites. In a plain unit value transfer approach as in this chapter, much time is needed to discuss issues of why one publication could be appropriate for the analyses and how the values can be set in relation to others that are in other currencies and from other countries.

Conclusions

This chapter investigates the use and suitability of benefits transfer as component of benefit-cost analysis for green infrastructure planning and decision making. Benefit transfer is a popular discussed topic among researchers and practitioners are also interested in its employment. Nowadays, economic arguments outweigh all other arguments and therefore are seriously needed to support decision making in green infrastructure planning to highlight the benefits of urban green. Unfortunately, the divergence between academic literature and current practice as well as the lack of guidance and protocols creates uncertainty and mistrust of benefit transfer and complicates is practical application in green infrastructure planning.

In our study, we used unit value transfer within a benefit-cost approach to assess the impacts of 18 landscape park projects in Stuttgart. This approach was mainly context and policy oriented as the administration of the City of Stuttgart wanted to have an indication on the benefit-cost relations to be expected from the set of green infrastructure investments and no exact calculation of compensatory amounts. We worked closely together with the municipal planning experts and developed a system of impact fields, categories and indicators to transfer most suitable unit benefit values from original studies to the policy sites in Stuttgart. The amount of affected population by each investment was measured through the introduction of different buffer zones around the sites. By this means, the projects′ economic, social and ecologic benefits were estimated and allocated to contrast them with the investment and maintenance costs in a benefit-cost analysis.

The results of the benefit-cost analysis reveal that the 18 projects generate benefits of about 300 million euros, which exceed five times the costs. Especially the benefit gains through recreation and leisure, quality of space through improved river access, and health and well-being would be high. Within a time horizon of 15 years, only four projects would not break-even; but most of them would already create net benefits after a few years. The overall picture of the benefit-costs ratios of the single investments would not even change a lot due to a change in the time horizon or a 50 per cent benefit cut. That means that investment in green infrastructure in Stuttgart would be worth as the benefits outbalance the costs. This proves that a comprehensible, adjusted and detailed unit value transfer within a benefit-cost approach is well convenient for practical application in green infrastructure planning as it delivers clearly the needed information within the funding and time limitations characteristic of the policy process. In addition, the approach presented relies heavily on intense interaction with policy experts from the Department for Town planning and regeneration of the City of Stuttgart. Linking the academic approach to the logic of local spatial planning increased political acceptance and scientific accuracy altogether. So, using a more integrated approach enhances a place-based understanding of local benefits. Furthermore, the results of this study strengthen the position of urban green within the competition of other land uses for limited public means by providing hard economic evidence. Nevertheless, there is further research needed to achieve two things. At first, the database on economic valuation studies in Europe and especially Germany is comparably low and needs to be extended by a variety of primary date generating case studies. Secondly, this increased access and usage of economic valuation in spatial planning would help to get practitioners familiar with concepts and opportunities that are offered through this research area.

Appendix

Table 4.2 Impact categories and indicators

Impact category: tourism		
Indicator: average expenditure for daily trips		
Value in literature	*Unit*	*References*
31.10	EUR / person & day	Deutscher Tourismusverband e.V. (DTV), 2010
Indicator: average expenditure for overnight stays		
Value in literature	*Unit*	*References*
169.30	EUR / person & night	Maschke, 2005

Impact category: health and well-being		
Indicator: average GDP		
Value in literature	*Unit*	*References*
187.06	EUR / person & day	Official statistics
187.05	EUR / person & day	Official statistics
Impact category: leisure and recreation		
Indicator: willingness to pay for green infrastructure		
Value in literature	*Unit*	*References*
492.00	USD / ha & year	Economics for the Environment Consultancy, 2005
14.95	EUR / person & year	Matz, 2006
9.09	EUR / person & year	Wilker, 2009
10.50	EUR / person & year	Klaphake and Meyerhoff, 2003
Indicator: added value through path network effects		
Value in literature	*Unit*	*References*
11.85	EUR / person & year	Wilker and Rusche, 2013
Impact category: quality aspects (accessibility and perception)		
Indicator: willingness to pay for higher levels of quality		
Value in literature	*Unit*	*References*
148.00	USD / year & household	Collins et al., 2005
492.00	USD / ha	Collins et al., 2005
16.91	GBP / person & year	Hanley et al., 2006
14.95	EUR / person & year	Matz, 2006
10.50	EUR / person & year	Klaphake and Meyerhoff, 2003
Impact category: quality aspects (accessibility and perception)		
Indicator: willingness to pay for better/easier access to green infrastructure		
Value in literature	*Unit*	*References*
124.50	CHF / person	Schwarzwälder et al., 2010
153.50	CHF / person	Schwarzwälder et al., 2010
114.00	CHF / person	Schwarzwälder et al., 2010
Impact category: landscape and biodiversity		
Indicator: added value through increased biodiversity		
Value in literature	*Unit*	*References*
906.00	GBP / ha	Northwest Regional Development Agency, 2008

178.00	EUR / household & year	Liebe and Meyerhoff, 2005
11.90	willingness to pay per household	Meyerhoff and Dehnhardt, 2004
Indicator: added value through habitat conservation		
Value in literature	*Unit*	*References*
48.00	GBH / household	Collins et al., 2005
201.00	USD / ha	Economics for the Environment Consultancy, 2005
20.17	willingness to pay in GBP	Hanley et al., 2006
Impact category: flood mitigation and water treatment		
Indicator: replacement/avoidance costs		
Value in literature	*Unit*	*References*
500.00	EUR / ha & year	Meyerhoff and Dehnhardt, 2004
464.00	USD / ha	Economics for the Environment Consultancy, 2005
288.00	USD / ha	Economics for the Environment Consultancy, 2005
45.00	USD / ha	Economics for the Environment Consultancy, 2005
Impact category: climate change mitigation and adaptation		
Indicator: CO_2 market prices		
Value in literature	*Unit*	*References*
10.50	USD / t CO_2	Jiao Xi, 2009
20.00	EUR / t CO_2	Verspecht et al., 2010

Acknowledgements

This chapter has been produced out of a study for the City Stuttgart in 2012. The authors want to thank the City of Stuttgart, Department for Town planning and regeneration for many interesting discussions and the close cooperation related to the BCA.

References

Ahlheim, M. and Lehr, U. 2002. Das Sparmodell der Umweltbewegung. *Perspektiven der Wirtschaftspolitik*, 3 (1), pp. 85–104.

Bateman, I.J., Mace, G.M., Fezzi, C., et al. 2011. Economic Analysis for Ecosystem Service Assessments. *Environmental and Resource Economics*, 48, pp. 177–218.

Brent, J.B. 2006. *Applied Cost-Benefit Analysis*. Cheltenham: Edward Elgar.

Collins, A., Rosenberger, R. and Fletcher, J. 2005. The economic value of stream restoration. *Water Resource Research*, 41 (2), W02017, doi: 10.1029/2004WR 003353.

Deutscher Tourismusverband e.V. (DTV). 2010. *Toursimus in Deutschland 2010.* Bonn: DTV.

Directorate General Regional Policy (Hg.). 2008. *Guide to Cost-Analysis of Investment Projects.* Structural Funds, Cohesion Fund and Instrument for Pre-Accession: European Commission.

Economics for the Environment Consultancy. 2005. *The Economic, Social and Ecological Value of Ecosystem Services – A Literature Review.* London: EFTEC.

Hanley, N., Wright, R.E. and Alvarez-Farizo, B. 2006. 'Estimating the economic value of improvements in river ecology using choice experiments – An application to the water framework directive'. In Navrud, S. and Ready, R. (eds) 2006. *Environmental Value Transfer – Issues and Methods. The Economics of Non-Market Goods and Resources*, 9. Dordrecht: Springer Netherlands, pp. 111–130.

Johnston, R.J. and Rosenberger, R.S. 2010. Methods, trends and controversies in contemporary benefit transfer. *Journal of Economic Surveys*, 24 (3), pp. 479–510.

Jost Wilker and Karsten Rusche. 2013. Economic valuation as a tool to support decision-making in strategic green infrastructure planning. *Local Environment*, 11(2013), pp. 1–13.

Klaphake, A. and Meyerhoff, J. 2003. 'Der ökonomische Wert städtischer Freiräume. Eine Anwendung der Kontingenten Bewertung auf eine städtische Parkanlage in Berlin'. Working Paper on Management in Environmental Planning, 08/2003. Berlin: TU Berlin.

Liebe, U. and Meyerhoff, J. 2005. 'Die monetäre Bewertung kollektiver Umweltgüter. Theoretische Grundlagen, Methoden und Probleme'. Working Paper on Management in Environmental Planning, 13/2005. Berlin: TU Berlin.

Loomis, J.B. 2014. 'Economic Valuation: Concepts and Empirical Methods'. In Fischer, M.M. and Nijkamp, J. (eds) 2014. *Handbook of Regional Science.* Berlin, Heidelberg: Springer, pp. 973–992.

Maschke, J. 2005. *Tagesreisen der Deutschen.* Schriftreihe 50. München: DWIF.

Matz, K. 2006. 'Was ist ein Stadtpark wert? Ökonomische Bewertung des Görlitzer Parks in Berlin mit einer Zahlungsbereitschaftsanalyse'. Schriftreihe der Fakultät Architektur Umwelt Gesellschaft, Bd. 128. Berlin: TU Berlin.

Meyerhoff, J. and Dehnhardt, A. 2004. 'The European Water Framework Directive and Economic Valuation of Wetlands. The Restoration of Floodplains along the River Elbe'. Working Paper on Management in Environmental Planning, 11/2004. Berlin: TU Berlin.

Natural England and Land Use Consultants. 2009. *Green Infrastructure Guidance.* Natural England, NE176.

Niemelä, J., Saarela, S., Söderman, T., et al. 2010. Using the ecosystem services approach for better planning and conservation of urban green spaces: A Finland case study. *Biodiversity Conservation*, 19, pp. 3225–3243.

Northwest Regional Development Agency (Hg.). 2008. *The Economic Benefits of Green Infrastructure – An Assessment Framework for the NWDA*. Northwest Regional Development Agency.

Pearce, D., Atkinson, G. and Mourato, S. 2006. *Cost-Benefit Analysis and the Environment. Recent Developments*. Organisation for economic Co-operation and Development: OECD Publishing.

Plummer, M.L. 2009. Assessing benefit transfer for the valuation of ecosystem services. *Frontiers in Ecology and the Environment*, 7, pp. 38–45.

Ready, R. and Navrud, S. 2005. Benefit Transfer – The Quick, the Dirty, and the Ugly? *Choices – The Magazine of Food, Farm and Resource Issues*, 20 (3), pp. 195–199.

Rosenberger, R.S. and Johnston, R.J. 2009. Selection effects in meta-analysis and benefit transfer: Avoiding unintended consequences. *Land Economics*, 85, pp. 410–428.

Rosenberger, R.S. and Loomis, J.B. 2000. Using meta-analysis for benefit transfer: In sample convergent validity tests of an outdoor recreation database. *Water Resources Research*, 36, pp. 1097–1107.

Rusche, K. 2012. Ökonomischer Nutzen grüner Infrastruktur – mehr Lebensqualität durch Stadtgrün? *Berichte zur deutschen Landeskunde*, 86 (3), pp. 255–268.

Schwarzwälder, B., Arnold, M., Beer-Toth, K. and Baumgart, K. 2010. Mehrwert naturnaher Wasserläufe – Zahlungsbereitschaft für Revitalisierungsprojekte in der Schweiz. *Natur und Landschaft*, 85 (8), pp. 329–333.

The Economics of Ecosystems and Biodiversity (TEEB). 2010. *The Ecological and Economic Foundations*. London: Earthscan.

Thiele, H.D. and Wronka, T.C. 2001. 'Umweltgüter und ihre Bewertung: Möglichkeiten und Grenzen des Benefit Transfers'. Working Paper EWP 0106, Department of Food Economics and Consumption Studies, University of Kiel.

Verspecht, A., Vanermeulen, V., Vermeire, B., et al. 2010. *Development and Application of an Instrument to Measure the Economic Impact of Green Investments in the Urban Fringe around Bruges within the Framework of the Land Use Plan Field Zone Bruges*. Ghent: University of Ghent.

Wilker, J. 2009. 'GrünAnlage – Was ist Stadtgrün wert? Ökonomische Bewertung des Westparks in Dortmund'. Diploma thesis, TU Dortmund.

Wilson, M.A. and Hoehn, J.P. 2006. Valuing environmental goods and services using benefit transfer: The state-of-the art and science. *Ecological Economics*, 60, pp. 335–342.

Xi, J. 2009. *Valuation of Ecosystem Services in Xishuangbanna Biodiversity Conservation Corridors Initiative Pilot Site, China*. Greater Mekong Subregion Core Environment Program.

Chapter 5
A Multi-Attribute Comparative Evaluation of Value Capture Financing Mechanisms: A Case Study

Anastasia Roukouni, Francesca Medda,
Maria Giannopoulou and Athanasios Vavatsikos

Introduction

The anticipation that transport investment can contribute to sustainable economic growth has become a motivating factor for public investment in large-scale infrastructure. Land Value Capture has been used increasingly as a tool for funding high cost public transport systems such as rail transit. Crossrail is a new rail project that will develop high frequency and quality connections across London. The expected cost of the whole project is £14.8 billion, £4.1 billion of which will be generated by the Greater London Authority (GLA) through the Business Rate Supplement (BRS), a fiscal method based on Land Value Capture Finance. The objective of this chapter is to examine and evaluate the GLA's BRS and an alternative financing solution that consists of a combination of distance-based taxes that could be used instead of it. The task of selecting the most appropriate financial scheme is achieved using Multi-attribute Decision Models and the final decisions are taken based on a variety of quantitative and qualitative criteria.

The aspects of value and distance are generally considered very important for the urban planning process. The present chapter emphasises the vital role of the aforementioned parameters in planning evaluation. Moreover, by clarifying the existence of strong linkages between property and infrastructure, it highlights the advantages of a place – based approach to infrastructure development.

The London Crossrail project is a modern railway currently under construction in the United Kingdom. Its route will pass through 37 stations and run 118 km, from Maidenhead and Heathrow in the west, through new twin-bore 21 km tunnels below central London to Shenfield and Abbey Wood in the east (Figure 5.1). The railway is expected to transport 1.5 million more commuters in less than one hour to London's strategic business locations and it is also anticipated to increase London's rail capacity by approximately 10 per cent and reduce congestion on the Tube, Docklands Light Railway (DLR) and National Rail services. At many locations Crossrail will be influential in supporting and accelerating new

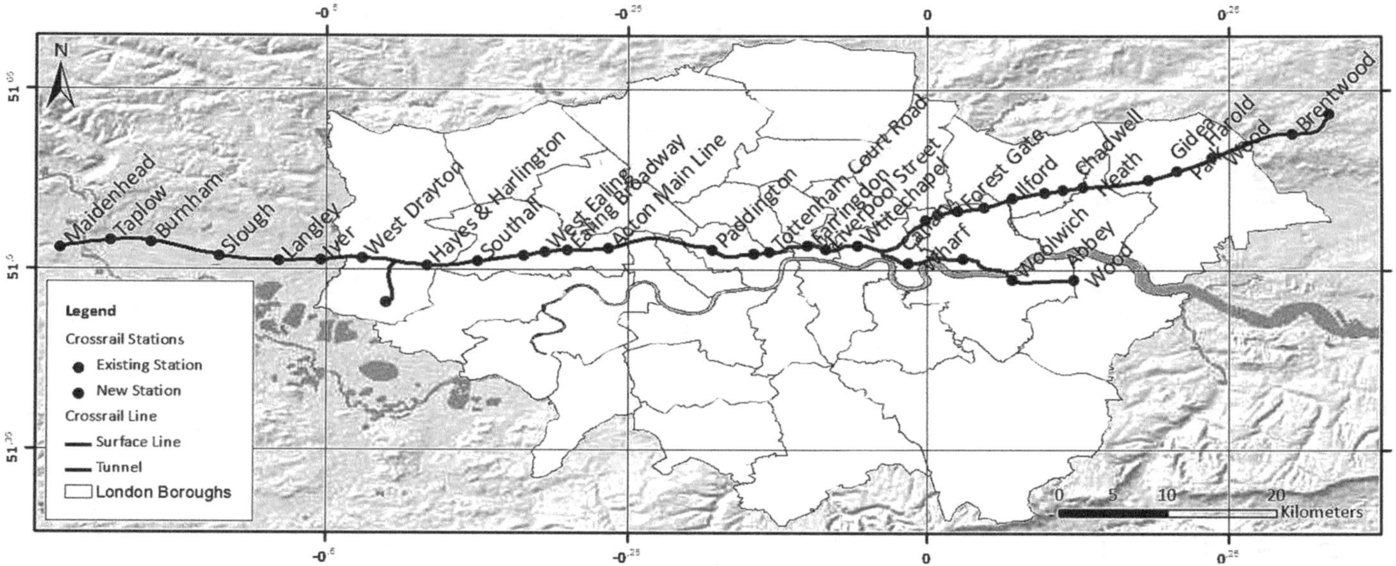

Figure 5.1 Crossrail route and stations map

development; it is already having an impact on investment decisions (Crossrail Property Impact Study, 2012).

The expected cost of the project is £14.8 billion, £4.1 billion of which will be covered by the Greater London Authority (GLA) through the Business Rate Supplement (BRS), a fiscal method based on Land Value Capture Finance. Crossrail is presumed to generate £1.24 billion annually for the London economy by the year 2026. With its contribution to ongoing development of key sectors and locations, it will impact widely on London in socioeconomic terms (Greater London Authority, 2010).

This chapter extends the line of research opened by Roukouni and Medda (2012), where, the effectiveness of the Business Rate Supplement scheme (BRS) proposed by Greater London as one of the repayment methods for the Crossrail project, was investigated. After having analysed the financial consequences of the BRS scheme and identified its weaknesses, an alternative financial mechanism was proposed, combining two distance-based taxes: a modified version of the existing BRS scheme and a Stamp Duty Levy.

Against this background, the present chapter conducts an evaluation both of the existing BRS scheme and of this alternative scheme, as two different potential funding mechanisms for the Crossrail project. The task of selecting the most appropriate financial scenario is achieved using Multi-Attribute Decision Models and the final decisions are taken based on a range of suitable quantitative and qualitative criteria and subcriteria. Significant effort is placed on the development of a decision model that combines Analytic Hierarchy Process (AHP) and Ideal Point Methods (IPM). In particular, AHP is used in order to provide criterion weights while IPMs are providing the decision rules that allow the generation of a variety of norms for combining decision makers' per criterion preferential system.

The chapter is structured as follows. In the second section, the literature review on the existing relationship between transport infrastructure and land value is presented and the concept of Value Capture Finance is discussed. The presentation of the two examined financing strategies follows in the third section. The fourth section, includes the methodology of the analysis, describing Multi-attribute Decision Analysis Models, followed by problem formation and evaluation criteria in the fifth section. The discussion of the results takes place in the sixth section. The chapter concludes in the seventh section with a critical analysis of the results.

Transport Accessibility and Land Value: The Concept of Value Capture Finance

Transport development can have a crucial role in achieving economic growth and moreover stimulate land development; this expectation has become an incentive for investing in transportation infrastructure of an area both for public and private parties (Salon and Shewmake, 2011; Ahlfeld and Feddersen, 2010; Banister and Berechma, 2002; OECD, 2002). There is a high level of interaction between transportation systems and land markets which is reflected on land use patterns. This interaction is described through the concept of accessibility (Iacono et al. 2009; Litman, 2011). Geurs and van Wee (2004, p. 128), define accessibility as 'the extent to which land-use and transport systems enable (groups of) individuals to reach activities or destinations by means of a (combination of) transport mode(s)'.

Most of the research done on the relationship between transit and property values indicated a noteworthy positive association between improved transport facilities and nearby property values (Roukouni et al., 2012). The effect of transit however has proved not to be the same for residential and commercial properties, as the impact on the latter appears to be weaker than the corresponding impact on the former. The studies of commercial properties find the positive impact limited to adjacent areas; nevertheless in those areas the effect is more intense than for residential properties (Drennan and Brecher, 2012). Smith et al. (2012) in their paper present an overview of 95 studies worldwide on the causality between transport development and land values.

Focusing on London, Gibbons and Machin (2004) stated that property prices increase at least 1.5 per cent for every 1 km closer to a train station. London's Jubilee Line Extension Impact Study's findings are in line with the aforementioned observation; on average, a 32 per cent raise on commercial property values and a slightly lower (+30 per cent) impact on residential property was reported, with the degree of impact being less visible as distance from stations increases (Chesterton Property Market Study, 2002). Moreover, Don Riley (2001) in his book includes evidence that London's Jubilee line extension caused an increase of £1.3 billion in the rent value of nearby properties. Crossrail Property Impact Study, published in October 2012, strengthens and confirms the results of the previous studies, as it asserts that Crossrail could help create additional residential and commercial value of approximately £5.5 billion along the route between the years 2012 and 2021. The study also projects that commercial office values around Crossrail stations in central London will experience an increase due to Crossrail over the next decade, with an uplift of 10 per cent in capital value above a rising baseline projection (Crossrail Property Impact Study, 2012).

The capital costs of urban transport investment could be recovered by capturing (partially or totally) the increase in land value that results from the investment; this may be accomplished through a fiscal mechanism such as Value Capture Finance (Medda and Modelewska, 2010). The term Value Capture Finance (VCF – also known as Land Value Capture) refers to a family of public finance mechanisms that raise funds in proportion to the increase in land value associated with new or improved public infrastructure. Land value capture has been used increasingly as a tool for funding high cost public transport systems such as rail transit (Levinson and Istrate, 2011; Doherty, 2008; Huxley, 2009).

A key point for the comprehension as well as the implementation of Value Capture mechanisms is examining the correlation between property and infrastructure. Hence, in planning evaluation it is essential to decide who benefits from the infrastructure and moreover, to clearly recognize the impacts of infrastructure development in terms of market and distance. Value capture applications are based on the so – called 'benefit principle' which states that the beneficiaries of a particular infrastructure or service should also pay for it. Provision of new public transit systems or improvements to existing services increases accessibility-related benefits to the adjacent properties, which are then positively capitalised into higher land values (Mathur, 2012, Kemp et al., 2012, Mejia and Muñoz, 2009).

In the context of the mass transit system investment, three main methods of Land Value Capture can be identified: Betterment tax, Tax increment finance and Joint Development mechanism. Betterment tax – in which category the BRS scheme that will be described in the following section belongs – is directed towards the beneficiaries of increased accessibility, of reduced congestion and pollution, and of lower transport costs achieved due to a transport investment. In a mass transit system such as light rail, a significant part of the capital cost is constituted by the land acquisition (Medda, 2012).

In case when population or economic growth or/and other factors which influence the demand for residential or commercial development fail to reach the projections, there is a high risk that the costs of infrastructure will not be recovered, which would lead to a financial loss (Iacono et al., 2009). Therefore, the successful application of a value capture strategy depends heavily on the factor 'timing'. The majority of such strategies need strong market conditions in order to motivate property owner interest due to the fact that value capture strategies are based on development; opportunities to capture value are influenced by the same factors as the real estate market (Reconnecting America's Center for Transit Oriented Development, 2008).

Presentation of the Examined Scenarios

As mentioned in the introductory section, Crossrail will cost £14.8 billion. Within this portfolio, the contribution of Greater London Authority (GLA) and Transport for London (TfL) to the project costs will be approximately £7.15 billion and the revenue will be generated through three different mechanisms: Business Rate Supplement (£4.1 billion), Community Infrastructure Levy (CIL) (up to £0.3 billion) and Section 106 Contributions (£0.3 billion) (Modern Railways special report, 2011).

The existing BRS scenario is examined towards an Alternative financing scenario which uses a place – based approach. The distance – based nature of the Alternative scenario was chosen due to the fact that accessibility has been proven to have a high impact on land values, as discussed in detail in the previous section. Therefore, it was attempted to evaluate the suitability of such a financing mechanism for urban transportation infrastructure projects.

The presentation of the existing BRS scenario and the proposed Alternative scenario, which from now on will be referred to as Plan A and Plan B respectively, follow.

The GLA's Business Rate Supplement Scenario: Plan A

According to the White Paper for Business Rate Supplements (2007), 'National non-domestic rates (NNDR), or Business Rates, is a tax levied on occupation of non-domestic properties as part of the wider framework of business taxation in the UK. Occupiers' liability is a function of the assessed rateable value of their property, known as a hereditament, and a multiplier. The multiplier is a figure usually specified as a number of pence in the pound'. Rateable value represents the open market annual rental value of a business/ non-domestic property. The rating lists represent all the rateable values in England and Wales. New rating lists are published every five years in a process called revaluation (Valuation Office Agency, 2012).

The BRS is applied to all 32 London Boroughs and the Common Council of the City of London, to all non-domestic properties with rateable values above £55,000. The estimated total number of hereditaments liable to the Crossrail BRS in London is 46,468 properties. Its implementation commenced in April 2010 and is estimated to last until 2038. The multiplier was set to 2p for each year of the Crossrail BRS, with the intention to be reviewed every five years after each rating revaluation. It is assumed that the rateable values, and hence the revenue, will increase 15 per cent in every reevaluation, except from the first of 2015, for which a lower 6 per cent is projected. The BRS is a flat levy, which means that the rate is exactly the same for every London Borough (Greater London Authority, 2010). The estimated total revenue collected during the plan's lifecycle is £8.315 billion.

The two substantial weaknesses of this plan that were identified include equity issues and extra – optimistic assumptions of growth rates. As mentioned previously, the tax is absolutely the same for all London Boroughs, a fact that contradicts to the benefit principle which implies that the cost of transportation for a contributor should be proportional to the benefits received, as the Boroughs situated closer to the Crossrail route will benefit more due to the increased accessibility. Moreover, it was stated that the success or failure of the plan depends profoundly on assuming a noteworthy future increase of rateable values; the long – lasting effects of the economic crisis and the resultant recession have not been considered thoroughly (Roukouni and Medda, 2012).

The Alternative Scenario: Plan B

The alternative plan comprises two taxes: a modified version of the BRS scheme and a Stamp Duty Land Tax,[1] imposed on the sale of residential properties, because they will also benefit from increased accessibility, as corroborated in the literature review above, and the implementation of the tax aimed to capture a proportion

1 'Stamp Duty Land Tax (SDLT) is a tax charged on land and property transactions in the UK. The tax is charged at different rates and has different thresholds for different types of property and different values of transaction' (HM Revenue and Customs, 2012).

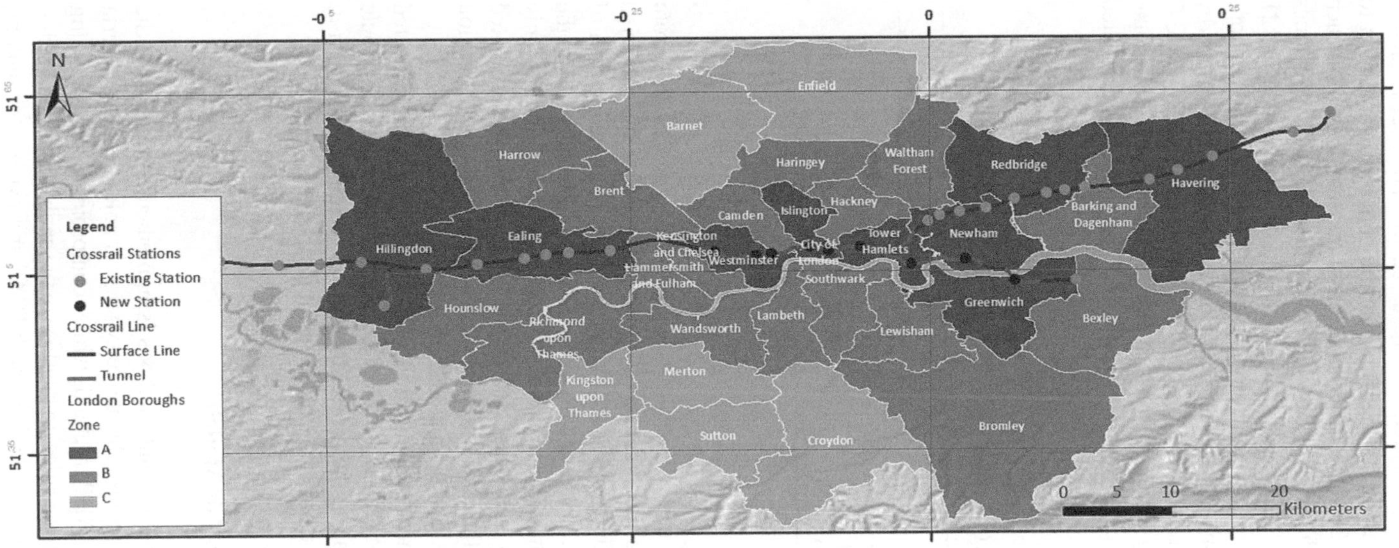

Figure 5.2 **The three Zones of London Boroughs according to Plan B**

of their increased value. Its implementation is assumed to start in 2013 and is estimated to last until 2040. Taking into account that the betterment tax ought to be an equitable levy (Medda, 2009), a differentiation in tax rates based on each Borough's location was introduced, and Boroughs were divided into three zones on the basis of their proximity to Crossrail stations (Roukouni and Medda, 2012).

Zone A includes the nine London Boroughs that have Crossrail stations located within them, while Zone B comprises the Boroughs adjacent to or nearby Boroughs in Zone A. This group is the largest with 17 Boroughs. Finally, Zone C is formed by the six remaining Boroughs that are distant from the Crossrail route (Figure 5.2). The different levels of tax rates applied to each zone are presented in Table 5.1.

Table 5.1 Differentiation of tax rates in Plan B

	Modified BRS	Stamp duty levy
Zone A	2.00p	1.50p
Zone B	1.50p	0.75p
Zone C	1.00p	0.25p

Source: Roukouni and Medda (2012).

The assumption of 15 per cent growth of rateable values for the BRS from 2015 onwards is reduced to a more conservative 12 per cent for the modified BRS. The Stamp Duty tax estimation of future revenue was made using data of monthly property prices for each Borough between 1995 and 2012 to calculate the historic annual growth rates. In order to provide a sufficient safety margin, only one-third of these growth rates have been used for the calculations. The estimated total revenue collected during the plan's lifecycle is £8.364 billion (£6.921 billion from the modified BRS and the rest from the Stamp Duty Tax).

As shown in Figure 5.3, which presents the spatial distribution of the expected gained revenue by each plan, in Plan B, the Boroughs closer to the Crossrail route contribute more to the financing of the project than the Boroughs distant from it, in contrast with Plan A, where a rather confused result is revealed.

Multi-Attribute Decision Analysis Models

A combined use of AHP and ideal point methods is implemented to support the evaluation of value capture strategies. The above synergy enables decision criteria weights elicitation using the AHP method while compromise programming methods are performed, as decision rules, to synthesize the analysis factors. In that manner a variety of operators are available to both DMs' and experts for ranking

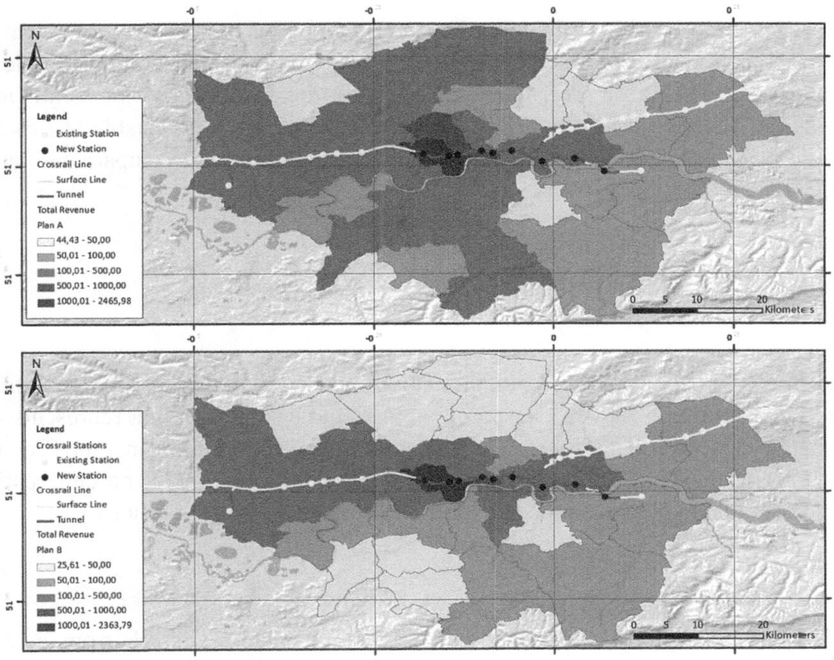

Figure 5.3 Spatial distribution of the revenue gained by Plan A and Plan B respectively

alternatives according to their remoteness from the ideal solution. The proposed synergy is expected to provide flexibility to the performed analysis given that the implementation of a variety of distance metrics provides significant information with respect to the stability of the alternatives final rankings.

Analytic Hierarchy Process (AHP)

Developed by T.L. Saaty (1977), AHP is a multicriteria method for addressing complex decision-making problems in which many competing alternatives (projects, actions, scenarios) exist. Since its release, the method has been widely used because it elicits DMs' preferences in a friendly and easily understood manner. As a procedure, AHP belongs to the family of methods that use pairwise comparisons in order to estimate relative preferences among decision analysis parameters in semi-structured decision problems. Alternatives are ranked according to their contribution to the overall goal of the analysis. Technically, this is achieved by forming pairwise comparison matrices to obtain the relative priorities among decision analysis parameters. The decision problem in AHP is modelled using hierarchies of objectives and sub-objectives incorporating the knowledge, experience and intuition of the DM for the specific problem (Banai-Kashani, 1989).

Relative priorities of the hierarchy are obtained by comparing every element in a branch against its peers with respect to the parent node using a nine-level fundamental scale of preferences to establish measurement of the dominance relation. Relative priorities Wj of the row elements are then established as the right principal eigenvector λ_{max} (Equation 1) of the pairwise comparison matrix (Saaty, 1996).

$$(A - \lambda_{max} I) \times w = 0 \tag{1}$$

Ideal Point Methods

Ideal point methods rank a set of alternatives A according to their separation from an ideal solution. They are based on an aggregating function that represents the relative closeness that originates in the compromise programming method (Zeleny, 1982, Hwang et al., 1997). The principle of compromise is that the best alternative is closer to a hypothetical ideal solution (PIS: Positive Ideal Solution):

$$A^{PIS} = \left\{x_{i1}^{+}, x_{i2}^{+}, \ldots, x_{im}^{+}\right\}$$

that maximizes the DM's preferential system and as far as possible from an anti-ideal solution (NIS: Negative Ideal Solution):

$$A^{NIS} = \left\{x_{i1}^{-}, x_{i2}^{-}, \ldots, x_{im}^{-}\right\}$$

that minimizes it for each one of the analysis criteria (Hwang et al., 1993). The latter is achieved using Minkowski L_p measures (Equation 3). According to the physical property of the distance metrics, when p increases, the total is formed with greater emphasis given to the largest deviation (Lai et al. 1994).

$$f_j^{+} = \left\{f_1\left(x_{i1}^{+}\right), f_2\left(x_{i2}^{+}\right), \ldots f_m\left(x_{im}^{+}\right)\right\} \qquad f_j^{-} = \left\{f_1\left(x_{i1}^{-}\right), f_2\left(x_{i2}^{-}\right), \ldots f_m\left(x_{im}^{-}\right)\right\} \tag{2}$$

$$L_p = \left(\sum_{j=1}^{m}\left(\left|f_j^{+} - f_j(x)\right|\right)^p\right)^{1/p} \quad \text{where } p \geq 1 \tag{3}$$

where $x \in A$

In practice, particular interest presents the distance measures for p values equal to 1, 2 and ∞, corresponding to the well-known Manhattan (L_1), Euclidean (L_2) and Chebychev (L_∞) distances, respectively.

Manhattan distance:
$$L_1 = \sum_{j=1}^{m}\left(\left|f_j^{+} - f_j(x)\right|\right) \tag{4}$$

Euclidean distance: $$L_2 = \left(\sum_{j=1}^{m} \left(\left| f_j^+ - f_j(x) \right| \right)^2 \right)^{1/2} \quad (5)$$

Chebychev distance: $$L_\infty = \max_j \left\{ \left| f_j^+ - f_j(x) \right| \right\} \quad (6)$$

where $x \in A$

Given that the final ranking of the candidate locations is achieved by performing the addition operation of the DM's preferential system per criterion, criterion maps need to be standardized. Standardization (also known as normalization) enables criterion maps to be combined on the basis of a common reference scale, which most commonly ranges between 0 and 1. When ideal point methods are considered as the decision rule, the standardization process is usually linearly accomplished using the maximum score or the score range approaches (Zeleny, 1982; Hwang et al., 1993). As derived above, ideal point methods implementation demands that the DMs not only define the analysis criteria magnitude *Wj* but also the ideal solution's attraction to the evaluation criteria through the parameter *p* (Zeleny, 1982).

Evaluation Criteria and Decision Hierarchy

By applying scientifically documented specifications and methods, the evaluation framework was formed using four primary assessment criteria: efficiency, equity, sustainability and feasibility (Iacono et al., 2009) and various subcriteria, with the aim to determine the effectiveness of the two different value capture strategies, Plan A and Plan B. The criteria and subcriteria used are described below.

Criterion C1: Efficiency

As an evaluation criterion, efficiency refers to a scheme's operational and economic efficiency, including its impact on economic development. The evaluation of each plan's efficiency relies on the benefit principle, which states that the cost to contributors should be proportional to the benefit they receive. Furthermore, it is examined to what extend each plan may hinder economic development and if it provides incentives to influence traveller's behaviour.

Subcriterion C1.1: Estimated revenue from Zone A/Total estimated annual transport and economic benefits in Zone A (criterion type: ascend) This subcriterion examines the ratio of Crossrail tax revenue that is projected to be generated from Boroughs that have stations located in them (Zone A) – and therefore are likely to benefit more – with respect to the total annual transport and economic benefits estimated for the same Zone, for each scenario, in order to investigate the potential of hindering or enhancing the economic development in this Zone.

Subcriterion C1.2: Estimated revenue from Zone A/Total Revenue (criterion type: ascend) The use of this subcriterion measures the contribution of the most beneficiary group of Boroughs (Zone A) to the total revenue in each plan, in order to investigate whether the cost to contributors is related to the received benefit.

Criterion C2: Equity

Equity as an evaluation criterion comprises benefit equity and users' capacity – to – pay equity. In terms of benefit equity, the fairness of each plan is examined, attempting to identify the equity issues with respect to different geographical areas. It is also attempted to identify the existing relationship, if any, between each plan and the personal capacity to pay of London's residents.

Subcriterion C2.1: Personal income reduction per Zone % (criterion type: descend) Concerning the capacity to pay equity, a lower rate corresponds to a more affordable tax burden; thus, it is likely to be paid under any circumstance, even during economic recession. This subcriterion, divided to the three following subcriteria, examines the impact each tax plan would have to a Londoner's net personal income, whether she/he is a resident of Zone A, B or C respectively.

- Subcriterion C2.1.1: *Personal Income reduction in Zone A %*
- Subcriterion C2.1.2: *Personal income reduction in Zone B %*
- Subcriterion C2.1.2: *Personal income reduction in Zone C %*
- Subcriterion C2.2: *Increase in total taxation due to the Crossrail tax in the poorest Boroughs %* (criterion type: descend)

The degree of fairness incorporated in the two plans is evaluated using this subcriterion. The poverty indicator which is used for the purposes of this analysis is the percentage of low – paid residents by borough (%). Boroughs which have over 18% low – paid residents were included in the calculation process.

Subcriterion C2.3: Distance weighted income (criterion type: descend) This subcriterion uses the straight-line distance between Crossrail route and each Borough's centroid, calculated in the ArcGIS 10 software package, with the aid of which, all maps included in this chapter were created. The spatial distribution of C2.3 for each one of the two plans is illustrated in Figure 5.4.

Criterion C3: Sustainability

Sustainability involves evaluating the two plans in terms of their adequacy, potential of growth, stability and elasticity as funding sources.

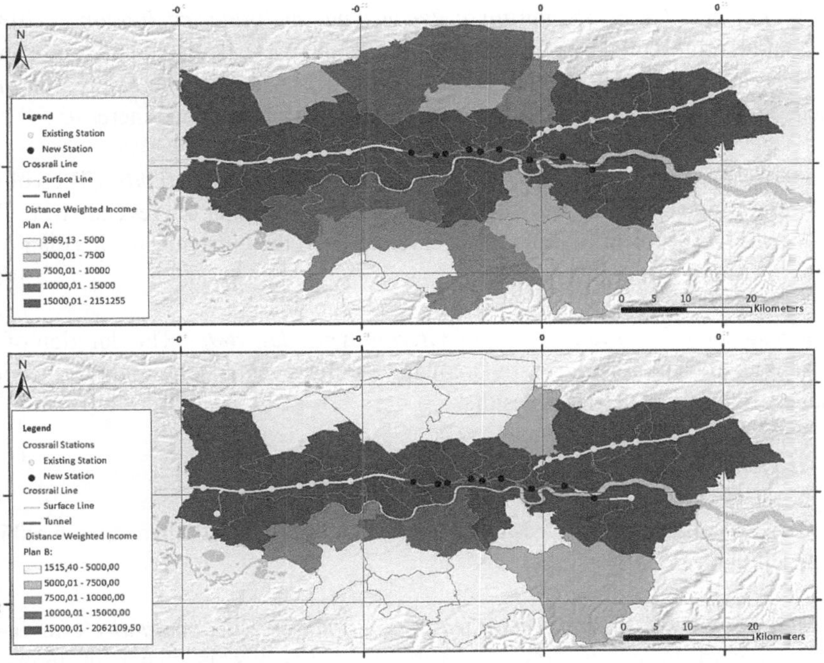

Figure 5.4 Spatial distribution of subcriterion 2.3 to the London boroughs in Plan A and Plan B respectively

Subcriterion C3.1: Estimated total revenue from non-domestic ratepayers (criterion type: descend) The estimated total revenue projected to come from non-residential sources from each plan was compared at this stage. It is assumed that a lighter tax burden imposed to the commercial ratepayers enhances the potential of growth of this group of ratepayers.

Subcriterion C3.2: Estimated revenue from Zone C/Total income of Zone C (criterion type: descend) This subcriterion examines to what extend can the revenue catch up with income growth, in Zone C, which includes the Boroughs that are situated far away from the Crossrail route and stations.

Subcriterion C3.3: Tax base of the scenario (criterion type: ascend) The tax base in the case of Plan A is defined by the number of hereditaments liable to the BRS, whereas in the case of Plan B, the tax base consists of the aforementioned figure plus the number of contracts referring to the Stamp Duty Land Tax. The range of the tax base can be used as an indicator for the volatility of revenue of each plan, and thus, its general stability.

Subcriterion C3.4: Safety margin (criterion type: qualitative evaluation) Growth potential, represented by growth rates of revenue, was calculated based on a more conservative approach in Plan B, as discussed previously. The safety margin is hence notably higher to account for potential inflation and need increases.

Subcriterion C3.5: Estimated total revenue (criterion type: ascend) The volume of the revenue each plan can raise, which indicates whether the revenue from Plan B would be sufficient to replace the revenue from the current scheme (Plan A) is an indicator of the plan's adequacy.

Subcriterion C3.6: Plan duration (criterion type: descend) The duration of the plan is an indicator of how easily predictable is the revenue; long – term projections have a higher risk of biased results.

Criterion C4: Feasibility

As an evaluation criterion, feasibility refers to political and administrative feasibility of each plan.

Subcriterion C4.1: Administration feasibility (criterion type: qualitative evaluation) With regard to administration feasibility, a flat tax is easier in terms of administration, while the distance-based method is more complicated; in addition, in the latter case there are two different sources of revenue, thereby increasing the administration complexity.

Subcriterion C4.2: Political feasibility (criterion type: qualitative evaluation) In terms of political feasibility, two points about Plan B are noteworthy. Firstly, the introduction of a flat tax with respect to location is the only permitted method within the current UK legal frame. The implementation of the proposed scheme would thus require to be accompanied by necessary changes to the relevant national legal framework. Secondly, imposing a tax on existing residential property might be associated with a high political cost.

Table 5.2 presents the origin of the essential data used in order to form the sub criteria and calculate their values where necessary; moreover, the table includes the sources of the background information used for the creation of the thematic maps.

Both the existed value capture strategy (Plan A) and the proposed in the present study approach (Plan B) are evaluated forming a 5-level hierarchical structure of objectives and sub-objectives (Figure 5.5). In the first level the analysis goal is settled while the fifth level consists of the examined alternatives. In the intermediate levels are formed by the analysis evaluation criteria.

Table 5.2 Origin of data used

Data type	Source	Year
Value of properties liable to the BRS	Greater London Authority	2010
Average house prices	Land Registry UK (Price Index)	2012
Average number of sales	Land Registry UK (Price Index)	2012
Total estimated annual transport and economic benefits from Crossrail	Regeneris Consulting	2010
Annual median personal income and tax	Greater London Authority	2007/8
Low – paid residents %	London's poverty profile	2012
London Boroughs shapefile	Greater London Authority	2012
Crossrail route and stations shapefile	Crossrail (digitization by authors)	2012

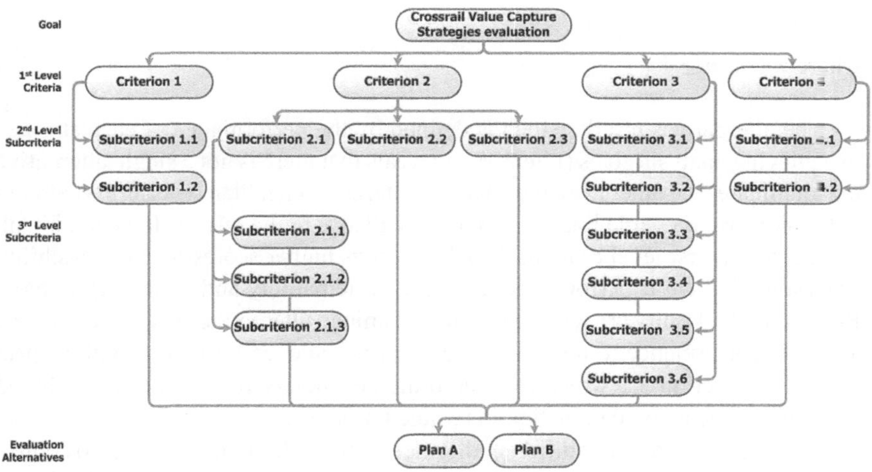

Figure 5.5 Decision analysis hierarchy

The most significant and noteworthy results of the performed analysis are presented in the next section.

Results Analysis and Discussion

As soon as the decision analysis hierarchy is formed, criterion weights elicitation process is taking place. As it is already mentioned, criterion weights are derived by performing pairwise comparisons of all the elements in each cluster with respect to the contribution in achieving the objective settled in the parent node. Using AHP, local weights vector is obtained as the right principal eigenvector of the pairwise comparison matrix. Global weights, that is the weights of each factor with respect to the contribution of achieving the analysis overall goal, are obtained by weighting the criterion scores from the top to the bottom of the hierarchy. Alternatives are evaluated according to their performance to the analysis decision criteria. In order to obtain coherent results standardisation (normalisation) of the attributes row data has been performed using the maximum score approach. As a result the process for synthesizing the performance per criterion of the alternatives is given on Equation 7. The same approach has been adapted for the evaluation of the intangible alternatives performances by providing 100 units of preference to the alternative that it is considered to perform efficiently to the examined criterion and lower grade to the other.

$$L_p^S = \left[\sum_{j=1}^{m} \left(\frac{\left| f_j^+ - f_j(x) \right|}{f_j^+} \right)^p \right]^{1/p} \tag{7}$$

where $x \in A, p \geq 1$

The alternatives profile which is defined by the performance of each strategy in achieving analysis goals (Figure 5.6) reveals that there is not a single alternative that dominates the other to all the analysis criteria. In fact, Plan A performs better in 4 decision criteria while it is outbalanced by Plan B to the rest 11. In general, with respect to the first level criteria, Plan A succeeds higher scores for the Feasibility criteria while Plan B performs better to the Sustainability and Efficiency criteria. Regarding the Equity criteria none of the examined alternatives takes precedence. The best performance to the analysis criteria is achieved by Plan B with respect to the Subcriterion 3.2 which reflects to the beneficiary role of a distance-based approach to the income reduction in remote boroughs.

The implementation of different distance metrics allows the analysis to control the levels of compensation among the analysis criteria. For example, when Manhattan distance metric ($p = 1$) is considered that the final rankings are the same with the ones derived by any other additive utility norm, thus allowing full compensation among the analysis criteria. On the other hand, the implementation of Chebyshev distance metric ($p = \infty$) provides a non-compensatory decision rule where the distance of the alternative scenarios is defined strictly by the worst estimated performance to all the analysis criteria. Consequently, intermediate values of p express gradient levels of compensation. In our study, the worst performance of Plan A and Plan B are indicated for the Sustainability and Feasibility criteria respectively. As a result, the higher the value of the parameter p, the higher the impact of the worst performed criterion gets, to the distance from the ideal solution formation (Figure 5.7).

The implementation of Manhattan distance metric revealed that Plan B is better that Plan A when full compensation among the analysis attributes is allowed. In particular, when this option is considered as decision strategy, the distance of Plan B from the ideal solution is the 40.5 per cent of the total distance.

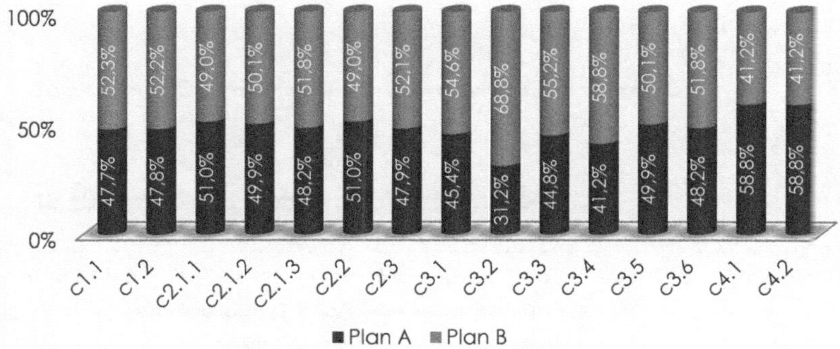

Figure 5.6 Alternative scenarios performance to the analysis criteria

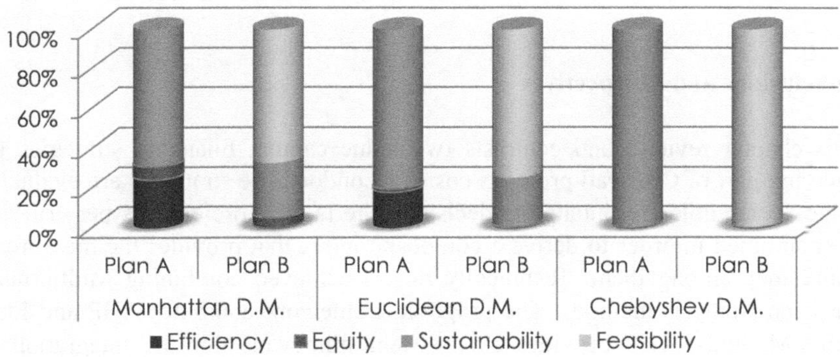

Figure 5.7 Contribution of the 1st level criteria to the final outcome for each one of the distance metrics

On the contrast, Plan A sustains a distance that represents the 59.5 per cent of the alternatives' total distance. To investigate the impact of the low performance that Plan B receives to the Feasibility decision criteria, a variety of distance metrics are examined. Chebychev distance metric derives for $p = 16$, where the alternative strategies score equals with the worst performance that they receive to all the analysis criteria. As shown in Figure 5.8, rank reversal is obtained already by the Euclidean distance option implementation and it is maintained, as expected, until the Chebyshev distance metric implementation. The rank reversal reveals that in our model, Plan B cannot counterbalance its weak performance to the Feasibility criterion category, when low or non-compensatory approaches are considered as decision rules.

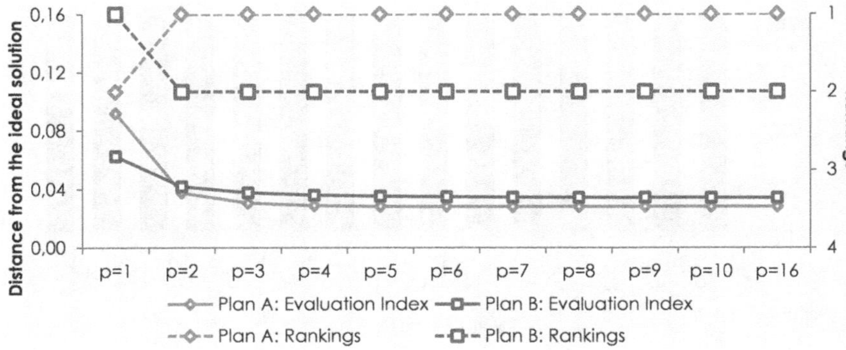

Figure 5.8 Alternative strategies rankings and their corresponding performance for a variety of distance metrics

Conclusions and Perspectives

This chapter reviews and contrasts two value capture financing strategies for covering part of Crossrail project's costs in London. The strategies are evaluated using 15 tangible and intangible decision criteria. The preferences per criterion are combined in order to derive a composite index that provides the measure of dominance among them. Technically this is achieved combining Multicriteria Decision Analysis methods. The proposed framework combines AHP and Ideal Point Methods. AHP provides a consistent framework for the integration of decision-makers' preferences aiming to support the criterion weights elicitation procedure. On the other hand, Ideal Point Methods provide the analysis with a variety of operators that allows policy makers' to handle the level of compensation among the analysis criteria. As a result, significant information for decision making can be derived based on the ability of each alternative to counterbalance their worst performances.

The proposed funding scheme that combines distance based variations of the implemented tax policy (Plan B) presents significant advantages which results to be considered as a more attractive option when weighted summation is performed as decision rule. However as long as the level of compensation among the decision criteria is reduced the implemented strategy (Plan A) seems to be more preferable. This is owned mainly to the fact that the proposed approach is not capable to counterbalance its weaknesses related to the feasibility factors when the compensation among the analysis criteria is reduced.

Overall, the present chapter indicates the importance of considering the impact of transportation infrastructure on property values and using the causality between property and infrastructure in terms of distance in planning evaluation. Therefore, it emphasises and clearly shows the need to improve the efforts of establishing a place – based approach to planning evaluation.

References

Ahlfeldt, G.M. and Feddersen, A. 2010. 'From Periphery to Core: Economic Adjustments to High Speed Rail'. Working Paper, LSE Research Online, London School of Economics and Political Science.

Banai-Kashani, R. 1989. A New Method for Site Suitability Analysis: The Analytic Hierarchy Process. *Environmental Management*, 13(6), pp. 685–693.

Banister, D. and Berechma, Y. 2002. 'The economic development effects of transport investments'. In Pearman A., Mackie P., Nellthorp J. and Giorgi L. (eds) 2002. *Transport Projects, Programmes and Policies: Evaluation, Needs and Capabilities*. Ashgate: Aldershot.

Chesterton Property Market Study. 2002. *Final Report prepared for the Jubilee Line Extension*. Impact Study Unit. London: University of Westminster.

Doherty, M. 2008. *Funding Public Transport Development through Land Value Capture Programs*. Institute for Sustainable Future.

Drennan, M.P. and Brecher, C. 2012. Does public transit use enhance the economic efficiency of urban areas? *The Journal of Transport and Land Use*, 5(3), pp. 53–67.

Geurs, K.T. and van Wee, B. 2004. Accessibility evaluation of land – use patterns and transport strategies: Review and research directions. *Journal of Transport Geography*, 12, pp. 127–140.

Gibbons, S. and Machin, S. 2004. *Valuing Rail Access Using Transport Innovations*. Centre for Economic Performance, London School of Economics and Political Science.

Greater London Authority. 2010. 'Intention to levy a business rate supplement to finance the Greater London Authority's contribution to the Crossrail Project – Final Prospectus' [online]. Available at: http://www.london.gov.uk/sites/default/files/finalprospectus.pdf [accessed 10 February 2013].

Greater London Authority. 2010. 'Personal Incomes of UK Tax – Payers in 2007/08 – Survey of Personal Income', GLA Intelligence Update, DMAG Social Exclusion Team [online]. Available at: http://data.london.gov.uk/documents/Intelligence_Update_052010_Survey_of_Personal_Incomes.pdf [accessed 5 March 2013].

GVA. 2012. 'Crossrail Property Impact Study, London' [online]. Available at: http://www.crossrail.co.uk/assets/library/document/c/original/crossrail_property_impact_study_main-_small.pdf [accessed 17 January 2013].

Huxley, J. 2009. *Value Capture Finance: Making Urban Development Pay its Way*. London: Urban Land Institute.

Hwang, C.L., Lai, Y.J. and Liu, T.Y. 1993. A new approach for multiple objective decision making. *Computers in Operations Research*, 20(8), pp. 889–899.

Iacono, M., Levinson, D., Zhirong, Z. and Lari, A. 2009. *Value Capture for Transportation Finance*. Report to the Minnesota Legislature, the Center for Transportation Studies, University of Minnesota.

Kemp, A., Mollard, V. and Wallis, I. 2012. 'Value capture mechanisms for funding transport infrastructure'. NZ Transport Agency Research Report 511, p. 107 [online]. Available at: http://www.nzta.govt.nz/resources/research/reports/511/docs/511.pdf [accessed 20 January 2014].

Levinson, D. and Istrate, E. 2011. *Access for Value: Financing Transportation through Land Value Capture*. Washington, DC: Brookings Institute.

Mathur, S. and Smith, A. 2012. 'A decision – support framework for using value capture to fund public transit: Lessons from Project – Specific Analyses'. MTI Report 11–14 [online]. Available at: http://transweb.sjsu.edu/PDFs/research/1004-decision-support-framework-value-capture-public-transit-funding.pdf [accessed 12 January 2014].

Medda, F. 2009. 'Land value tax as an investment mechanism for public transport assets'. In *European, IBJTF* (ed.) 2009. European Regional Science Association, Lodz.

Medda, F. 2012. Land value capture finance for transport accessibility: A review. *Journal of Transport Geography*, 25, pp. 154–161.

Medda, F. and Modelewska, M. 2010. 'Land Value Capture as a funding source for urban investment: The Warsaw Metro System'. UCL QASER Lab, ERNST & YOUNG: Better Government Program 2009–10 [online]. Available at: http://www.ucl.ac.uk/qaser/LAND_VALUE_CAPTURE.pdf [accessed 7 March 2013].

Mejia, L. and Muñoz, R. 2009. 'The Colombian approach to urban mass transit infrastructure financing through value capture: progress to date'. TRANSyT Working Paper 2009-01-EN. Universidad Politécnica de Madrid, Spain.

OECD. 2002. 'Impact of Transport Infrastructure Investment on Regional Development'. Programme of Research on Road Transport and Intermodal Linkages (RTR) [online]. Available at: http://www.internationaltransportforum.org/pub/pdf/02RTRinvestE.pdf [accessed 7 March 2013].

Pereira, J.M.C. and Duckstein, L. 1993. A multiple criteria decision-making approach to GIS-based land suitability evaluation. *International Journal of Geographical Information Systems*, 77(5), pp. 407–424.

Reconnecting America's Center for Transit – Oriented Development. 2008. 'Capturing the Value of Transit'. Report Prepared for: United States Department of Transportation, Federal Transit Administration [online]. Available at: http://www.reconnectingamerica.org/assets/Uploads/ctodvalcapture110508v2.pdf [accessed 15 March 2013]

Riley, D. 2001. *Taken for a Ride: Taxpayers, Strains and HM Treasury*. Centre for Land Policy Studies.

Roukouni, A. and Medda, F. 2012. Evaluation of Value Capture mechanisms as a funding source for urban transport: The case of London's Crossrail. *Procedia – Social and Behavioral Sciences*, 48, pp. 2393–2404.

Roukouni, A., Basbas, S. and Giannopoulou, M. 2012. Land use around metro stations: A case study. *International Journal of Civil, Architectural Science and Engineering*, 6, pp. 1–5.

Saaty, T.L. 1977. A scaling method for priorities in hierarchical structures. *Journal of Mathematical Psychology*, 15, pp. 234–281.

Saaty, T.L. 1996. *Decision Making with Dependence and Feedback: The Analytic Network Process*. RWS Publications, Pittsburgh, Pennsylvania.

Salon, D. and Shewmake, S. 2011. 'Opportunities for value capture to fund public transport: A comprehensive review of the literature with a focus on East Asia'. Institute for Transportation and Development Policy [online]. Available at: http://www.itdp.org/uploads/Salon_and_Shewmake_Opportunities_for_Value_Capture_to_Fund_Public_Transport.pdf [accessed 19 March 2013].

Smith, J., Gihring, T. and Litman, T. 2012. 'Financing Transit Systems through Value Capture – An Annotated Bibliography'. Victoria Transport Policy Institute [online]. Available at: http://www.vtpi.org/smith.pdf [accessed 7 February 2013].

Zeleny, M. 1982. *Multiple Criteria Decision Making*. New York: McGraw-Hill.

Websites

Crossrail Official: http://www.crossrail.co.uk/ [accessed 5 January 2013].

HM Revenue and Customs: http://www.hmrc.gov.uk/ [accessed 5 January 2013].

Greater London Authority: http://www.london.gov.uk/who-runs-london/greater-london-authority [accessed 5 January 2013].

Land Registry UK: http://www.landregistry.gov.uk/ [accessed 20 January 2013].

London's poverty profile: http://www.londonspovertyprofile.org.uk/ [accessed 7 February 2013].

Valuation Office Agency: http://www.voa.gov.uk/ [accessed 5 January 2013].

PART II:
Understanding the Evaluation of Impacts and Space

Chapter 6

Current Trends in Social Impact Assessment: Implications for Infrastructure Developments

Frank Vanclay and Ana Maria Esteves

Introduction

Social impact assessment (SIA) developed alongside environmental impact assessment (EIA) in the early 1970s as an ex-ante mechanism to consider the social impacts of planned interventions (Burdge and Vanclay, 1995; Vanclay, 2014). Early understandings of SIA were narrowly conceptualised, tending to apply SIA only at the project level, only considering a narrow selection of first-order impacts, with the role of SIA being limited to the predictive assessment of the negative consequences of projects in the context of a regulatory framework triggered only by a change in the biophysical environment (Vanclay, 2006) This limited understanding of SIA pervaded and continues to dominate the legislation, policy, procedures and organizational cultures of the environmental management agencies of many countries as well as many environmental consultancies. In contrast, nowadays most SIA professionals consider that SIA is more than a technique or step, it is philosophy about development and democracy. As such, ideally it considers the pathologies of development (i.e. impacts), the goals of development (such as poverty alleviation), and the processes of development (e.g. participation, capacity building) (Vanclay, 2003, 2004). Thus, SIA should also be involved in assisting communities to determine their development priorities, as well as a process for incorporating social dimensions into projects (Esteves and Vanclay, 2009; Esteves et al., 2012).

The contemporary understanding is that SIA is about 'the processes of managing the social issues associated with planned interventions' (Esteves et al., 2012, p. 35), and is equivalent to what is often called 'social performance' in the corporate world. An elaboration of that definition is:

> Social Impact Assessment includes the processes of analysing, monitoring and managing the intended and unintended social consequences, both positive and negative, of planned interventions (policies, programs, plans, projects) and any social change processes invoked by those interventions. Its primary purpose is to bring about a more sustainable and equitable biophysical and human environment. (Vanclay, 2003, p. 6)

Although arguably SIA applies to policies, plans and programs, the practice and thinking of SIA tends to be at the project level because this is where the demand for SIA has been. However, a major change over time is from SIA being seen as a regulatory tool required by regulatory agencies and resented by proponents, to being an internal corporate process actively embraced by leading corporations. This change has occurred for many reasons, including: the neoliberal turn in notions about the role of governments; the growing acceptance by companies of the corporate social responsibility and sustainability agendas, and their desire to be a 'developer of choice'; the increasing expectations, activism and empowerment of communities; an increasing acceptance of the concept of 'social licence to operate'; high profile litigation cases; as well as the fact that the SIA community has actively promoted the business case for doing SIA (Vanclay and Esteves, 2011; Kemp and Vanclay, 2013, van der Ploeg and Vanclay, 2013; Vanclay, 2014).

Unfortunately, such a shift has not guaranteed that SIA will always be done properly, or that the SIA process is able to adequately influence company operations (Kemp, 2011). Many structural limitations affecting SIA remain, including the lack of training or accreditation of SIA practitioners, the lack of routine use of peer review processes, and greenwashing by companies. The funding and timing allocated for social issues also continue to be inadequate (Vanclay, 2004; Kemp, 2011; Baines et al., 2013; Vanclay et al., 2013).

SIA is of benefit not only to regulatory agencies as has been traditionally conceived, but also to communities, government developers and the private sector. Traditionally, SIA has been a proponent-led process to meet regulatory requirements. There will still be an ongoing need for that, but SIA can do much more. SIA can also be a proponent-led process to assist companies to develop impacts and benefits agreements with communities in fairness and good faith, ensure an ongoing social license to operate, represent best practice in corporate social responsibility, as well as influence all business drivers (Esteves, 2008). Furthermore, SIA can also be a community-led process to assist affected communities in developing a thorough understanding of the social and environmental issues associated with a proposed development which is needed for them to decide whether or not to accept a project and to negotiate sustainable outcomes with a project developer (Esteves and Vanclay, 2009).

This chapter discusses the relevance of social impact assessment to infrastructure projects. It argues that SIA is useful and necessary irrespective of whether it is a legal requirement or not. After a general introduction to current good practice in SIA, it then highlights current trends that are impacting not only on SIA practice but on infrastructure projects and other developments generally. The purpose of the chapter is to increase awareness of SIA and of these trends in the hope that there will be more systematic use of SIA in infrastructure projects.

Current Good Practice Social Impact Assessment

The underlying philosophy of the current paradigm of SIA seeks to maximize the positive outcomes while minimizing harm, which it does by being an ongoing process of managing the social issues of development, in other words, an adaptive management approach. Thus, in addition to considering mitigation measures, there is a role for SIA consultants to assist in the development of opportunities for benefit attainment through, for example, local procurement arrangements the provision of employment opportunities for locals, and other ways of enhancing the benefits of projects including infrastructure sharing arrangements and local social investment. Therefore, what is important is not just the ability to predict impacts, but also having a social awareness and the ability to manage social issues effectively. Instead of a statement of social impacts somewhat akin to an Environmental Impact Statement, what regulatory authorities should consider is a Social Impact Management Plan (SIMP) (see Franks and Vanclay, 2013).

The process of doing an SIA is a composite of many steps depending on the particulars of the specific project. A wide range of social research methods are used. It is not feasible to fully describe the practical details of how to do an SIA in a chapter such as this. However, to give a general impression of what is included, it can be stated that SIA addresses the topics listed in Table 6.1. These steps can be arranged conceptually as in the diagram in Figure 6.1. It should be noted that a participatory approach is applied throughout the whole process.

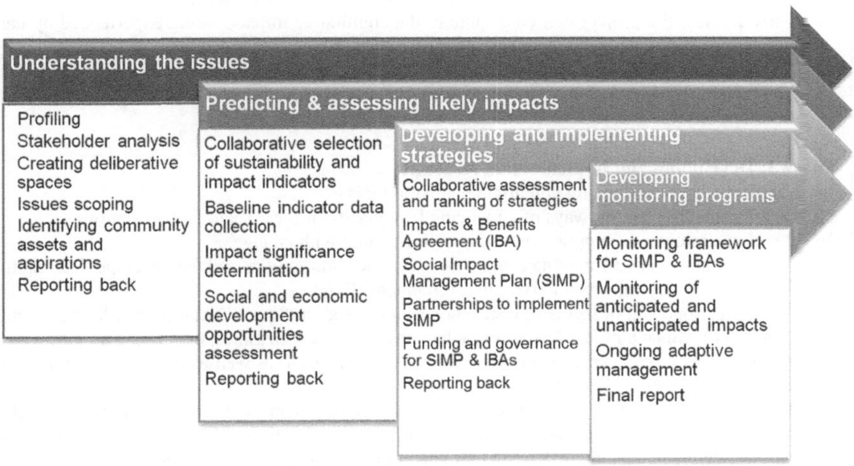

Figure 6.1 Schema representing the activities in social impact assessment
Source: Modified from Esteves and Vanclay (2009).

Table 6.1 Activities undertaken in a typical social impact assessment

Phase 1: Understanding the issues

1. Gain a good understanding of the proposed project, including all subsidiary activities necessary to support the project's development and operation.
2. Clarify the responsibilities and roles of all involved in or associated with the SIA, including relationships to the other specialist studies being undertaken, and establish what national laws and/or international guidelines and standards are to be observed.
3. Identify the preliminary 'social area of influence' of the project, likely impacted communities (nearby and distant), and stakeholders.
4. Gain a good understanding of the communities likely to be affected by the project by preparing a community profile which includes: (a) a thorough stakeholder analysis; (b) an assessment of the differing needs, interests and values of the various subgroups of those communities including a gender assessment; (c) an assessment of their impact history, that is their experience of past projects and other relevant historical events; (d) a discussion of trends happening in those communities; (e) a discussion of the varying needs, aspirations and values in the impacted communities, as well as their strengths and assets. (this step is typically called profiling.)
5. Inform communities about: (a) the project; (b) similar projects elsewhere to give them a sense of how they are likely to be affected; (c) how they can be involved in the SIA; (d) their procedural rights in the regulatory and social performance framework for the project; and (e) their access to grievance mechanisms.
6. Devise inclusive participatory processes and deliberative spaces to help communities: (a) understand how they will be impacted; (b) determine the acceptability of likely impacts and proposed benefits; (c) make informed decisions about the project; (d) facilitate community visioning about desired futures; and (e) prepare for change.
7. Identify the key social and human rights issues (i.e. scoping).
8. Assemble relevant baseline data for key social issues.

Phase 2: Predicting and assessing likely impacts

9. Determine the likely social changes and impacts that may result from the project.
10. Determine how the various affected groups and communities will likely respond.
11. Consider how the project will contribute to the cumulative impacts being experienced by the host communities.
12. Establish the significance of the predicted changes (i.e. prioritize them).
13. Actively contribute to the design and evaluation of project alternatives including no go and other options.

Phase 3: Developing and implementing strategies

14. Identify ways of mitigating potential negative impacts.
15. Develop and implement ways of enhancing benefits and project-related opportunities.
16. Develop strategies to support communities for coping with change.
17. Facilitate an agreement-making process between the communities and the developer leading to the drafting of an Impacts and Benefits Agreement (IBA).
18. Assist the proponent in facilitating stakeholder input and drafting a Social Impact Management Plan (SIMP) which puts into operation the benefits, mitigation measures, monitoring arrangements and governance arrangements that were agreed to in the IBA, as well as plans for dealing with any ongoing unanticipated issues as they may arise.
19. Put processes in place to enable proponents, government authorities and civil society stakeholders to implement arrangements implied in the SIMP and IBA and to develop their own respective management action plans and embed them in their own organizations, establish respective roles and responsibilities throughout the implementation of those action plans, and maintain an ongoing role in monitoring.
20. Assist the proponent in developing and implementing ongoing social performance plans that address contractor obligations, community relations, social monitoring, and the management of grievances.

Phase 4: Design and Implement monitoring programs 21. Develop indicators to monitor change over time. 22. Develop a participatory monitoring plan. 23. Consider how adaptive management will be implemented. 24. Undertake evaluation and periodic review (audit).

New Developments impinging on SIA and Project Implementation

In this section, seven contemporary developments are presented. These trends intersect with SIA theory and practice in various ways, and will likely have a profound influence into the future. These trends however will also have important ramifications for the way infrastructure projects unfold. The seven trends are updated from Vanclay and Esteves (2011), Esteves et al. (2012) and Vanclay (2012, 2014).

1. Moving from NIMBY to Social Licence to Operate

NIMBY is an acronym for 'not in my backyard', a colloquial version of what are sometimes technically described as 'locally unwanted land uses' (LULUs). Originally NIMBY referred to the actions and responses of people generally approving of socially-desirably projects so long as it did not personally affect them (i.e. was not in their neighbourhood). Over time, however, NIMBY has become a way by which project proponents attempt to discredit opponents and neutralize the legitimate opposition of local communities. Rather than make their projects socially acceptable, project proponents have sought to denigrate opponents. Such techniques are likely to increase conflict and protest, increase the risk of litigation, and increase the risk of sabotage actions (Prenzel and Vanclay, 2014). The cost of this conflict can be considerable.

An alternative approach is for projects to be conducted in such a way that ensures they have a 'social licence to operate'. A social license to operate can be defined as an informal mechanism relating to 'the degree to which a corporation and its activities meet the expectations of local communities, the wider society, and various constituent groups' (Gunningham et al., 2004, p. 313; Lynch-Wood and Williamson, 2007). Gaining a social licence requires, amongst other things, that the project proponent: (1) treats local communities with respect; (2) justifies the need for the project to affected peoples; (3) adequately addresses all likely social and environmental issues; and (4) makes a positive contribution to local communities. Although there are costs associated with these actions, they may be far less than those experienced as a result of conflict. More importantly, the company develops a positive reputation and can therefore grow as a business.

2. 'Free, Prior and Informed Consent' as Respect for ALL Communities

An important concept becoming established in the field of international law is that of 'free, prior and informed consent' (FPIC). Although it developed in the context of the self-determination of indigenous peoples and is a key feature of the United Nations Declaration on the Rights of Indigenous Peoples (2007), it is becoming a general principle applicable to all project-affected communities (see IIED, 2002; Hill et al., 2010; Vanclay and Esteves, 2011; Hanna and Vanclay, 2013). When conceived this way, FPIC is similar to the concept of a social license to operate (Miranda et al., 2005) in that it is essentially about respect for local communities and would constitute the sort of actions a project would need to do to gain a social licence.

In its strict formal sense, FPIC 'recognizes indigenous peoples' inherent and prior rights to their lands and resources and respects their legitimate authority to require that third parties enter into an equal and respectful relationship with them based on the principle of informed consent. Procedurally, free, prior and informed consent requires processes that allow and support meaningful choices by indigenous peoples about their development path' (United Nations, 2004, p. 5). In both the formal and more general utilization of FPIC, each word contributes meaning to the concept. *Free*, meaning that there must be no coercion, intimidation or manipulation by companies or governments, and that should a community say 'no' there must be no retaliation. *Prior*, meaning that consent should be sought and received before any activity on community land is commenced and that sufficient time is provided for adequate consideration by any affected communities. *Informed*, meaning that there is full disclosure by project developers of their plans in the language acceptable to the affected communities, and that each community has enough information to have a reasonable understanding of what those plans will likely mean for them, including of the social impacts they will experience. *Consent*, meaning that communities have a real choice, that they can say yes if there is a good flow of benefits and development opportunities to them, or they can say no if they are not satisfied with the deal, and that there is a workable mechanism for determining whether there is widespread consent in the community as a whole and not just a small elite group within the community (Colchester, 2010; Hill et al., 2010; Lehr and Smith, 2010; Vanclay and Esteves, 2011).

For an FPIC process to be undertaken properly and fairly, there are some preconditions, primarily that all parties have sufficient resources and capacity to be able to participate effectively, that there is sufficient time to allow adequate consideration of the issues to take place, that there is a proper process of information collection and consideration about the issues (such as an SIA), and that there to be an effective process of agreement making.

FPIC is thus very much aligned with the current understanding of SIA. In fact, in some handbooks and brochures about FPIC (notably Hill et al., 2010), the series of steps on how to do it looks remarkably similar to the basic steps of SIA. Notwithstanding the fact that FPIC may be difficult to implement and/or that there are still some unresolved issues (Cariño, 2005; Macintyre, 2007), potentially the emerging legal requirement for FPIC could provide the universal imprimatur for SIA, something that has always been wanted.

3. Community Agreements and Impacts and Benefits Agreements

Another emerging trend, especially in relation to large resource developments and indigenous communities, is the use of Impacts and Benefits Agreements, also called community-based agreements, community development agreements, partnering agreements, indigenous land use agreements, empowerment agreements, community contracts, or shared responsibility agreements (Gibson and O'Faircheallaigh, 2010; ICMM, 2010). Ideally, these agreements should emerge as an outcome of an SIA process, or at the very least be informed by an SIA process. They are developed usually in a manner compatible with FPIC, and as a way for the project to achieve a social licence to operate. Essentially they are an agreement between a project and the affected communities about what impacts a project is likely to create, how they will be managed, and what benefits will flow to the communities by way of compensation.

Obviously, it is important that any discussion of how the benefits from projects will be distributed be based on a sound knowledge and understanding of what the likely social impacts will be, and on a basis of an understanding of the issues associated with visioning the future for a community. Thus SIA has a central role.

4. Acknowledging the Existence of Social Risks

Project managers have long comprehended the existence of various types of technical risks, and have learnt to manage them with the use of risk assessment frameworks, risk management plans and risk registers. Risks arising from social issues are now also being regarded as a type of risk (sometimes described as non-technical risk) (Joyce and Thomson, 2000; Vanclay, 2014). There are many ways that social issues can be a risk to a project.

Communities have become empowered and better resourced, including by international environmental NGOs. The internet and associated globalization have expanded the reach and influence of individuals and groups. Communities, or disgruntled components within communities, may protest or blockade a project when they are dissatisfied (i.e. when a project does not have a social licence to operate), thus causing disruption and delay as well as gaining national and international media attention. They might also take direct action in the form of sabotage. They might also engage in peaceful, but politically powerful protest that results in loss of political support. Finally, they might also pursue legal action, and whether or not they are successful, the legal case can be costly, time-consuming, delay inducing, and may lead to reputational risk. All of these issues are real risks a business faces. Thus, in the same way that there are technical risks, engineering risks, financial risks, etc., there are also non-technical risks such as social risks that need to be seriously considered by a business. The fact that many businesses have accepted this point means that they are now very appreciative of the need to be aware of and manage these risks. Rio Tinto, for example, makes its policy statement on social risk publicly available: http://www.riotinto.com/documents/Social_risk_analysis_guidance_2011_2014.pdf.

5. Corporate Social Responsibility and Industry Sustainability Guidelines

Although the concept of corporate social responsibility (CSR) arguably dates from the 1950s (Bowen, 1953; van der Ploeg and Vanclay, 2013), over the last two decades or so there has been a marked rise in the attention given to CSR as a result of the international sustainability agenda (Carroll, 1999; Matten and Moon, 2008). Sustainable development and CSR are now part of the corporate mentality and culture. Many industry sectors have adopted a sustainable development ethos (e.g. IIED, 2002; ICMM, 2003) and many industry bodies have developed codes of practice, sometimes specifically on SIA. Companies are willing to do more than the minimum required by governments, and many of them frame their business strategy to be 'the developer of choice', and are being encouraged to do so by leading management consultants (e.g. Accenture, 2012).

As a result of this increasing commitment to CSR, and after a long process of development, the International Organization for Standardization released its ISO 26000 Guidance on Social Responsibility on 1 November 2010 (ISO, 2010a). Unlike ISO's earlier management systems (ISO 9001 Quality Assurance and ISO 14001 Environmental Management Systems), ISO 26000 is a guidance document and is not intended to be used as a certification system, nor is it intended to be cited in legal action (ISO, 2010a, Clause 1). The intention is to help organizations understand what social responsibility (SR) is and how to implement it. It does this by 'developing an international consensus on what SR means and the SR issues that organizations need to address; providing guidance on translating principles into effective actions; [and] refining best practices that have already evolved and disseminating the information worldwide for the good of the international community' (ISO, 2010b, p. 3). The Standard is called 'social responsibility' rather than 'corporate social responsibility' in order to provide advice not only to the private sector but also to all types of public sector organizations.

In ISO 26000, social responsibility is founded on seven principles: accountability, transparency, ethical behaviour, respect for stakeholder interests, respect for the rule of law, respect for international norms of behaviour, and respect for human rights. These principles overlap considerably with the core values given in the *International Principles for Social Impact Assessment*. ISO 26000 addresses seven core subjects including organizational governance, human rights, labour practices, the environment, fair operating practices, consumer issues, and community involvement and development.

Organizations intending to adopt the Standard are advised to become familiar with the concept of social responsibility and to understand and respect the principles.

To integrate social responsibility within an organization 'involves practices such as: making social responsibility integral to its policies, organizational culture, strategies and operations; building internal competency for social responsibility; undertaking internal and external communication on social responsibility; and regularly reviewing these actions and practices related to social responsibility' (ISO, 2010a, p. ix).

6. The Expansion of the Human Rights Domain

A very significant development is the rise of the business and human rights agenda. In 2005, the United Nations appointed a Special Representative of the Secretary-General on Business and Human Rights, Prof John Ruggie, to assist in clarifying the roles and responsibilities of states, companies and other social actors in relation to human rights. In 2008, Ruggie presented a 'Protect, Respect and Remedy' Framework, which was endorsed by the United Nations. It is based on three core principles: 'the state duty to protect against human rights abuses by third parties, including business; the corporate responsibility to respect human rights; and greater access by victims to effective remedy, both judicial and non-judicial' (Ruggie, 2008, p. 1). He then went on to develop a set of Guiding Principles on Business and Human Rights to assist in implementing the Protect, Respect and Remedy Framework. These were adopted by the United Nations in late 2011.

Although the minimum standards for human rights observance were made explicit by Ruggie, the UNGP also requires businesses to consider additional standards as may be appropriate depending on host and home state circumstances. This has led to much discussion. Also relevant is the notion of 'emerging rights'. Two previously-considered emerging rights, the right to water and the right to sanitation, have now been confirmed as basic rights by the United Nations (UNGA, 2010). Other emerging rights relate to the right to food, the right to a livelihood, the right to housing, the right to live in a healthy environment, the right to health and healthcare, the right to an adequate standard of living, the right to cultural integrity, the right to education, the right to adequate information, and the right to participation (Kemp and Vanclay, 2013; Sanz-Caballero, 2013). Whether confirmed as fundamental human rights or not, project developers will need to consider all these issues. An implication of the growth of the human rights agenda is that social impacts on people are now potentially infringements of their human rights, and impacted stakeholders are now rights holders who may have legal redress. SIA will of necessity have to incorporate human rights, and will gain an increased mandate because of the human rights agenda. The legal standing of human rights and the increasing awareness that business has an obligation to respect human rights increases corporate interest in ensuring that they are meeting their responsibilities and their interest in SIA as a mechanism to do this.

7. Expectations around Local Procurement and Local Content

A final development is the rise of local content and local procurement arrangements that can be triggered by: commitments in Impacts and Benefits Agreements with local communities; donor requirements; in some cases by host country legislation; and by corporate commitment to benefit enhancement. However, the extent to which local communities actually benefit from local content requirements is dependent on their capacity to take up the opportunities, the extent to which these opportunities align with community values and aspirations, and their ability to adapt to the business cycle of the project and changing circumstances. Planning for local content and ensuring that communities gain maximum value from the opportunities can be enhanced considerably by SIA.

Conclusion: The Business Case for Social Impact Assessment

While the goal of SIA will always be to ensure better development outcomes for people and communities, it is increasingly obvious that this can only be achieved when companies and all stakeholders see the benefits of embracing SIA. Thus, to achieve its objectives, both of minimizing harm and maximizing benefits, SIA must be aligned with corporate processes and time scheduling, and companies need to see the value of SIA so that they are fully committed to doing SIA properly and to implementing any recommended mitigation and enhancement strategies in good faith.

Rather than seeing SIA as a cost, it should be seen as an investment in risk management. There are many ways in which SIA can assist, including by:

- enabling more effective risk identification and risk management processes;
- gaining enhanced access to land and resources;
- gaining increased access to local workforce and suppliers;
- increasing the ability to identify and act on issues much earlier;
- achieving cost reductions (e.g. through earlier remediation);
- achieving full cost consideration of all likely costs;
- creating greater certainty for investors, government, and society;
- enhancing the company's legal and social license to operate and to grow;
- improving stakeholder relations; and
- maintaining and enhancing company and product reputation.

All of these have the consequential effects of: improving efficiencies; avoiding delays, shutdowns, and potentially the untimely closure of projects; and increasing the short-term and long-term success of the project. In fact, it is likely that through a comprehensive analysis of risks, SIA can contribute to all business drivers, including to: cost savings and productivity; enhanced access to capital; improved risk management; increased access to human capital; increased reputation; all of which will lead to revenue growth and enhanced access to markets.

Acknowledgement

Some sections in this chapter draw on previous published material, including from Vanclay and Esteves 2011; Esteves et al., 2012; Vanclay 2014.

References

Accenture. 2012. 'Achieving High Performance for Mining in 2020: Six Strategies for Safe, Sustainable and Superior Operations' [online]. Available at: http://www.accenture.com/SiteCollectionDocuments/PDF/Accenture-Achieving-High-Performance-Mining-2020.pdf [accessed 28 August 2013].

Baines, J., Taylor, C.N. and Vanclay, F. 2013. Social impact assessment and ethical social research principles: Ethical professional practice in impact assessment Part II. *Impact Assessment and Project Appraisal*, 31(4), pp. 254–260.

Burdge, R. and Vanclay, F. 1995. 'Social impact assessment'. In F. Vanclay and D. Bronstein (eds) 1995. *Environmental and Social Impact Assessment*. Chichester: Wiley, pp. 31–65.

Cariño, J. 2005. Indigenous peoples' right to Free, Prior, Informed Consent: Reflections on concepts and practice. *Arizona Journal of International and Comparative Law*, 22(1), pp. 19–39.

Carroll, A.B. 1999. Corporate social responsibility: Evolution of a definitional construct. *Business and Society*, 38, pp. 268–295.

Colchester, M. 2010. *Free, Prior and Informed Consent: Making FPIC work for Forests and Peoples*. New Haven: The Forests Dialogue, Yale University.

Esteves, A.M. 2008. Evaluating community investments in the mining sector using multi-criteria decision analysis to integrate SIA with business planning. *Environmental Impact Assessment Review*, 28(4–5), pp. 338–348.

Esteves, A.M. and Vanclay, F. 2009. Social Development Needs Analysis: Adapting SIA methods to guide corporate-community investment in the minerals industry. *Environmental Impact Assessment Review*, 29(2), pp. 137–145.

Esteves, A.M., Franks, D. and Vanclay, F. 2012. Social impact assessment: The state of the art. *Impact Assessment and Project Appraisal*, 30(1), pp. 35–44.

Franks, D. and Vanclay, F. 2013. Social Impact Management Plans: Innovation in corporate and public policy. *Environmental Impact Assessment Review*, 43, pp. 40–48.

Gibson, G. and O'Faircheallaigh, C. 2010. *IBA Community Toolkit: Negotiation and Implementation of Impact and Benefit Agreements*. Ottawa: Walter & Duncan Gordon Foundation.

Gunningham, N., Kagan, R. and Thornton, D. 2004. Social license and environmental protection: Why businesses go beyond compliance. *Law and Social Inquiry*, 29(2), pp. 307–341.

Hanna, P. and Vanclay, F. 2013. Human rights, Indigenous peoples and the concept of Free, Prior and Informed Consent. *Impact Assessment and Project Appraisal*, 31(2), pp. 146–157.

Hill, C., Lillywhite, S. and Simon, M. 2010. *Guide to Free Prior and Informed Consent*. Carlton Vic: Oxfam Australia.

ICMM. 2003. 'ICMM Sustainable Development Framework: ICMM Principles'. London: International Council on Mining and Metals [online]. Available at: http://liveassets.iucn.getunik.net/downloads/minicmmstat.pdf [accessed 28 August 2013].

ICMM. 2010. *Indigenous Peoples and Mining: Good Practice Guide*. London: International Council on Mining and Metals.

IIED (International Institute for Environment and Development). 2002. *Breaking New Ground: The Report of the Mining, Minerals and Sustainable Development Project*. London: Earthscan [online]. Available at: http://www.iied.org/mmsd-final-report [accessed 28 August 2013].

ISO. 2010a. *ISO 26000 Guidance on Social Responsibility.* Geneva: International Organization for Standardization.

ISO. 2010b. *ISO 26000 Project Overview.* Geneva: International Organization for Standardization.

Joyce, S. and Thomson, I. 2000. Earning a social licence to operate: Social acceptability and resource development in Latin America. *Canadian Institute of Mining Bulletin*, 93(1037), pp. 49–53.

Kemp, D. 2011. 'Understanding the organizational context'. In F. Vanclay and A.M. Esteves (eds) 2011. *New Directions in Social Impact Assessment: Conceptual and Methodological Advances*. Cheltenham: Edward Elgar, pp. 20–37.

Kemp, D. and Vanclay, F. 2013. Human rights and impact assessment: Clarifying the connections in practice. *Impact Assessment and Project Appraisal*, 31(2), pp. 86–96.

Lehr, A. and Smith, G. 2010. *Implementing a Corporate Free, Prior, and Informed Consent Policy: Benefits and Challenges*. Boston: Foley Hoag.

Lynch-Wood, G. and Williamson, D. 2007. The social licence as a form of regulation for small and medium enterprises. *Journal of Law and Society*, 34(3), pp. 321–341.

Macintyre, M. 2007. Informed consent and mining projects: A view from Papua New Guinea. *Pacific Affairs*, 80(1), pp. 49–65.

Matten, D. and Moon, J. 2008. Implicit and explicit CSR: A conceptual framework for a comparative understanding of corporate social responsibility. *Academy of Management Review*, 33, pp. 404–424.

Miranda, M., Chambers, D. and Coumans, C. 2005. 'Framework for Responsible Mining: A Guide to Evolving Standards' [online]. Available at: www.frameworkforresponsiblemining.org.

Prenzel, P. and Vanclay, F. 2014. How social impact assessment can contribute to conflict management. *Environmental Impact Assessment Review*, 45, pp. 30–37.

Ruggie, J. 2008. 'Promotion and Protection of all Human Rights, Civil, Political, Economic, Social and Cultural Rights, including the Right to Development – Protect, Respect and Remedy: A Framework for Business and Human Rights Report of the Special Representative of the Secretary-General on the Issue of Human Rights and Transnational Corporations and Other Business Enterprises, John Ruggie'. UN Doc. A/HRC/8/5.

Sanz-Caballero, S. 2013. Children's rights in a changing climate: A perspective from the United Nations Convention on the Rights of the Child. *Ethics in Science and Environmental Politics*, 13, 11–14.

United Nations Declaration on the Rights of Indigenous Peoples. 2007. United Nations General Assembly Resolution 61/295 adopted 13 September 2007.

United Nations Economic and Social Council. 2004. 'Commission on Human Rights, Sub-Commission on the Promotion and Protection of Human Rights. Working Group on Indigenous Populations'. UN Doc. E/CN.4/Sub.2/AC.4/2004/4.

van der Ploeg, L. and Vanclay, F. 2013. Credible claim or corporate spin: A checklist to evaluate corporate sustainability reports. *Journal of Environmental Assessment Policy and Management*, 15(3), 1350012.

Vanclay, F. 2002. Conceptualising social impacts. *Environmental Impact Assessment Review*, 22(3), pp. 183–211.

Vanclay, F. 2003. International principles for social impact assessment. *Impact Assessment and Project Appraisal*, 21(1), pp. 5–11.

Vanclay, F. 2004. The Triple Bottom Line and Impact Assessment: How do TBL, EIA, SIA, SEA and EMS relate to each other?. *Journal of Environmental Assessment Policy and Management*, 6(3), pp. 265–288.

Vanclay, F. 2006. Principles for Social Impact Assessment: A critical comparison between the International and US documents. *Environmental Impact Assessment Review,* 26(1), pp. 3–14.

Vanclay, F. 2012. The potential application of Social Impact Assessment in integrated coastal zone management. *Ocean & Coastal Management,* 68, pp. 149–156.

Vanclay, F. 2014. Developments in Social Impact Assessment: An introduction to a collection of seminal research papers. In: F. Vanclay ed. *Developments in Social Impact Assessment*, Cheltenham: Edward Elgar, pp. xv–xxxix.

Vanclay, F. and Esteves, A.M. 2011. 'Current issues and trends in social impact assessment'. In F. Vanclay and A.M. Esteves (eds) 2011. *New Directions in Social Impact Assessment: Conceptual and Methodological Advances*, Cheltenham: Edward Elgar, pp. 3–19.

Vanclay, F., Baines, J. and Taylor, C.N. 2013. Principles for ethical research involving humans: Ethical professional practice in impact assessment Part I. *Impact Assessment and Project Appraisal*, 31(4), pp. 243–253.

Chapter 7

Identifying Community Engagement Research Techniques and Sampling Strategies via Health Impact Assessments (HIAs): Applied Planning Lessons on Generating Emic and Etic Community Observations

John Gaber and Tammy Overacker

Introduction

The planning profession has a long tradition of engaging the public in the plan-making process (Brody, Godschalk and Burby, 2003, p. 245). Despite decades of experience with participatory decision making, planning scholars and practicing planners are still lacking a clear understanding on the definitions and methods used when they engage the community in the planning process (Laurian and Shaw, 2009, p. 294). One reason for this problem is that 'empirical planning research' has not been able to provide enough detail to help planners address technical applied situations like how should planners contact the community during the participatory process (Lauria and Wagner, 2006, p. 365; Forsyth, 2012, p. 164; Laurian and Shaw, 2009, p. 295). To be specific, the planning literature lacks clear direction informing practicing planners on their community engagement technique options (e.g. focus group meetings or surveys) and which type of sampling strategy (e.g. city wide, specific neighbourhoods, or a sub-group of local residents) best fits each community engagement technique. For instance, a planner will use a very different community engagement technique and sampling strategy if they want to generate qualitative data from a very small population of 25+ local residents who are about to get a new neighbourhood park in comparison to a planner who wants to generate quantitative data from a 2,500+ community within a larger metropolitan area who are about to get a new large regional park.

The two primary reasons why there is very little consensus on how planners can obtain local community observations is because the majority of empirical based planning research is either derived from single-case investigations which makes generalizability of observations to practical planning applications extremely difficult, and/or is heavily theoretical providing little technical insight to practicing planners who are looking for answers to 'how-to' questions (Deyle and Slotterback, 2009, p. 26; Lauria and Wagner, 2006, p. 376; Laurian and Shaw, 2009, p. 294). This lack of connection from planning research to planning practice is less a failure on the part of planning researchers and is more of a methodological conundrum. As easy as it sounds for Lauria and Wagner (2006) to suggest that planning researchers need to 'consider conducting studies with more than one case' (p. 376), it is methodologically very hard for planning researchers to isolate and have access to a larger number of planners all engaging the community for the same reason and are adhering to a similar set of participatory decision guidelines or research strategies.

In the last few years two empirical planning research projects come close to analysing a larger number of practicing planners all engaging the community within a very similar participatory planning process. They are Laurian's (2007) five case studies of environmental clean-ups and Laurian and Shaw's (2009) study of 761 AICP planners. Laurian's (2007) analysis looked at five case studies of military and civilian environmental clean-ups of Superfunds sites sponsored by the US Environmental Protection Agency (EPA), Department of Defence (DOD), and the Department of Energy (DOE). This study does a great job isolating planners conducting comparable participatory planning process (environmental clean-ups of Superfund sites). However, Laurian only accessed five different case studies and she only looked at Citizen Advisory Boards (CAB) community engagement technique, and provides no discussion of planners accessing different community engagement techniques. The Laurian and Shaw (2009) study has a large sampled population of 761 surveyed AICP planners and does a great job identifying how planners access multiple community engagement techniques (e.g. workshop, public meeting, task force, public hearing, survey, focus groups), type of sampling strategy (e.g. neighbourhood, citywide, countywide, regional, state-wide, multistate, national), and, focus of the project (land use, transportation, housing, economic development, environmental). In this study, the authors focus on how AICP planners evaluate their public participation processes. Laurian and Shaw point out that one limitation in their investigation was they were unable to observe how planners connected their community engagement techniques with specific citizen participation goals or other parts of participatory planning process like sampling strategy (Laurian and Shaw, 2009, p. 306).

In this chapter we generated an international data base of health impact assessments (HIAs) conducted from 1999 to 2011 to get a better understanding how community health planners go about engaging the community for the goal of empirical observations. We sampled publicly published HIAs conducted around the world and were able to analyse a total of 95 assessments. HIAs are a standardized policy assessment strategy that combines community input with a series of data sets to evaluate the health impacts of proposed and existing projects with a particular interest in reducing health inequalities (Abrahams, 2004, p. 6; Birely, 2003, p. 313; Tamburrini et al., 2011, p. 203; Northridge, Sclar and Biswas, 2003, p. 120; Gaber and Overacker, 2010, p. 70). They are extensively used around the world and are recommended by the World health Organization (WHO) as the preferred methodology to assess health impacts.

HIAs take an 'outcome-based approach' in participatory decision by using community observations as a source of empirical data to generate deeper understandings of problems and identifying unforeseen solutions as a way to improving planning outcomes (Maginn, 2006, p. 21; Laurian and Shaw, 2009, pp. 296–297). There are two types of empirical community vistas: emic and etic observations (Pike, 1967, 1990). Emic community observations are based on the community's insider 'emic' experiences and planners use this type of community data to better understand a particular situation. Etic community observations are based on the community's outsider objective perspective and planner use this type of community data to better understand a particular situation within a larger context. Community-based researchers have historically been heavily 'biased' in favour of planners obtaining emic community observations against obtaining etic community observations resulting in very little awareness that etic observations even exists let alone can be obtained (Harris, 2001, p. 35). In our analysis of HIAs, we pay particular attention to the emic and etic choices planners made in relation to the type of research technique they used to engage the community in the participatory process and their sampling strategy in terms of whose input was needed to improve planning outcomes.

In our analysis of the HIAs, we focused on two technical aspects of empirical citizen engagement: 1) research technique (how did the planers go about gathering community input); and 2) sampling strategy (who in the community did they consider as being important sources of information). In addition we were curious to see if any characteristics particular to the HIA process influenced how planners conducted their community engagement investigations. Here, we looked at three specific sub-questions: 1) Does the type of organization (e.g. non-government organizations (NGOs) that is conducting the HIA predisposes the planner to be more inclined to use a specific type of community input technique (e.g focus group meetings) in their investigations? 2) Does the topic of the HIA study (e.g. human services vs. land use) cause the planner to be more likely to use a specific type of community engagement technique (e.g. survey)? and 3) Are some community input techniques (e.g. stakeholder meetings) more likely to be used by a planner in a short turnaround HIA project (Rapid HIA) than in a long-term study (Comprehensive HIA)? This chapter is divided into four sections: 1) overview of the HIA process; 2) overview of emic and etic approaches to community input; 3) discussion of our HIA citizen engagement research findings; and 4) summary of research findings and new understandings to guide future participatory planning processes.

The HIA Process

Internationally, planners have been using the HIA assessment strategy for well over two decades. In 1999, the World Health Organization (Regional Office for Europe) and European Centre for Health Policy developed the 'Gothenburg consensus paper' that established an international framework so that policy makers can operate from a common understanding of HIAs (Phillips et al., 2010, p. 69). The World Health Organization (WHO) divides the HIA process into five sequential stages (Gaber and Overacker, 2012, pp. 275–276). They are:

- Screening. The selection that determines whether a particular policy, plan, or project has a potential health impact. Is an HIA needed?
- Scoping. If the screening has determined that an HIA is to be carried out, this process identifies the steering group that establishes a work plan for the HIA.
- Appraisal. The research 'engine' of the HIA where impacts are considered, evidence is gathered and recommendations are framed.
- Report. Formal identification of recommendations and production of the final report.
- Monitoring. Assessment of the HIA process: (a) impact of the process; (b) relevance of research; and (c) monitoring of the implemented projects for anticipated and unanticipated health outcomes.

After it is determined that an HIA investigations is needed at the end of the Screening stage, the planner needs to decide what kind of HIA they want to execute. An HIA investigation is organized according to two criteria: level and type. An HIA can be conducted at three different 'Levels' of magnitude depending on how much data and non-empirical information is needed to be processed by the community health planner. The three HIA levels are: *rapid* (also called 'tabletop') – the shortest study (one to two days); *intermediate* – a more detailed investigation (several weeks); and *comprehensive* – an intensive investigation (several months up to a year) (Gaber and Overacker, 2012, p. 276). So, for example, if a planner has very little time and limited resources to conduct her health impact assessment, she will conduct a rapid HIA. At the opposite end of the spectrum, a planner with more resources and a larger time frame will conduct a comprehensive HIA. There is a built-in magnitude of problem assumption in the Levels of HIAs where small and simpler problems will get rapid HIA attention while larger and more complicated problems will warrant comprehensive HIA attention. In reality, close to 40 per cent of all of the HIAs we analysed in this study tended to take on very complicated problems (e.g. health impacts of groundwater contamination), but were executed at the rapid HIA level (Gaber and Overacker, 2012, p. 279).

An HIA can be integrated at three different points in the decision-making process, called 'Types' of HIA investigation. The Type of HIA is a function of when the planner is presented with the opportunity to conduct the HIA. There are three different types of a HIA's: *prospective* – analysing the impact of a proposed policy or plan before it is actually implemented; *concurrent* – investigating current and future impacts of an existing policy or program; and *retrospective* – assessing the post-facto health impacts of a policy or plan (Gaber and Overacker, 2012, p. 276). For example, if a planner is confronted with the opportunity to research the health impacts of a proposed installation of a wind turbine in a neighbourhood park, she will be conducting a prospective HIA investigation. If the wind turbine has already been installed in the park and the planner is asked to post-facto investigate the health impact of the turbine, she will be conducting a retrospective HIA.

The combination of HIA Level and Type criteria tells us as planning researchers the magnitude of the HIA investigation and where in the policy process it took place. For example, a quick HIA study to analyse the possible health impact of a new bicycle path that is to take place before the city council votes on the proposed bicycle path is called a Rapid, Prospective HIA investigation. At the other end of the spectrum would be a large-scale HIA study to analyse the health impact of an existing wind farm that was built to generate an alternate energy source for a local town is called a Comprehensive, Retrospective HIA investigation.

Research and community input data is generated and gathered during the Appraisal stage in the HIA investigation. Methodologically, an HIA is founded on a mixed method research design where 'quantitative and qualitative data sets are combined into one research project' (Gaber and Gaber, 2010, p. 1039). For example, a planner analysing the health impact of a proposed park may combine a series of qualitative focus group meetings with neighbouring residents to learn more about what type of recreation they like with a quantitative census data analysis of the immediate area to estimate how many people may access the park. The mixed method research approach is particularly useful in HIA investigations because it provides the planner with a more comprehensive and holistic assessment of the possible health impacts (Gothenburg consensus paper – World Health Office (Regional Office for Europe) European Centre for Health Policy, 1999, p. 4).

Emic and Etic Approaches to Community Engagement

Embedded within the Appraisal stage of the HIA study is the community engagement process. Here, citizen participation is seen as important 'data slice' within the overall research project (Glaser and Strauss. 1970, p. 108). The ontological foundation to community empirical observations is drawn from Dewey's 'postulate of immediate empiricism' where the community is understood as being able to provide empirical 'non-technical observations' on 'everyday things' that they experienced in their day-to-day lives (Dewey, 1969, p. 151; Dewey, 1905, p. 393; Shusterman, 1999, p. 195). Obtaining the local view of citizen participation is not a new concept in planning. Lisa Peattie (1967) has been talking about the value of gaining the 'anthropologist's' view from the community for over 45 years. The planning idea behind citizen participation is that 'individual participants bring valuable knowledge and innovative ideas about their community that can increase the quality of adopted plans' (Brody, Godschalk and Burby, 2003, p. 246, Dandekar, 1986, p. 45).

There are two types of empirical vistas within the community. Pike (1967, 1990) calls these vistas 'emic' and 'etic' viewpoints. An '*emic viewpoint* is the view point of the group member, the insider … (and) the *etic viewpoint* is that of the 'objective' outsider looking at and studying the group' (Plano Clark and Creswell, 2008, p. 290, original italics). In the HIA literature and the community engagement literature, the planner is striving to get 'emic' observations during their community phase in the HIA investigation (Harris, 2001, p. 35, 1999, p. 41). By listening to what the impacted community residents knows about a topic from their first-hand experiences, the planner is able to gain and understanding from the community experiences and to 'reconstruct events in which (they) did not participate' (Rubin and Rubin, 1995, p. 1; Pike, 1967, p. 41). On the other side of the observational spectrum are 'etic' observations which provide the planner with an 'outsider' perspective on what is going on in the community. Here, the planner will get input from community representatives or leaders that provide more of a bigger picture understanding of the HIA related topic and place it in a larger context. Surprisingly, neither the HIA literature nor the planning engagement literature provide any insight instructing planners on how they can obtain etic observations. 'Neither emic or etic data are better or worse than the other; they simply differ' (Fetterman, 1998, p. 508). The most significant point of distinction between the two observations is that 'bigger the social problem, the less likely that it can be accounted for in emic intentions and are more likely' to be better accounted for by etic accounts (Harris, 1999, p. 47).

The most useful way to think about community input research strategies is along 'a continuum of research techniques, with some methods lying closer to the emic side of the continuum and some techniques lying closer to the etic side' (Fetterman, 1998, p. 507). Community input research techniques that allow for individuals to respond in their own words using their own categorizations and that the planner is able to record each and everything they say are better fit to emic observations. Examples of emic community input research techniques that best fit emic observations would interviews and focus groups. 'Phenomena that are not well understood are often first studied with tools that yield more emic data' (Fetterman, 1998, p. 508). Community input research techniques oriented toward etic observations require individuals to respond in terms that are similar to what are used by planners and that can be conducted in a conversational fashion that builds on a shared basic understanding of the HIA situation. Survey research and meeting-based techniques (e.g. town hall meeting) are better suited to etic community observations. As the HIA research topic becomes better understood by everyone in the community as the result of previous empirical investigations, planners will be more likely to use 'tools that yield more etic types of data' (Fetterman, 1998, p. 508).

HIA Citizen Engagement Research Findings

Research Methodology

We conducted an internet investigation of published HIAs (in English) from 1999 to 2011. A total of 147 HIAs were obtained and analysed. The countries represented in the HIA sample include: Australia, Canada, Germany, India, Italy, Netherlands, New Zealand, Russia, Slovenia, Sweden, UK (England, Ireland, Scotland, Wales), and the United States. We generated our HIA database through a two-step purposeful sample strategy. The HIAs were first found through international collecting HIA websites (e.g. 'The HIA Gateway', maintained by the 'West Midlands Public Health Observatory'). After exhausting these collecting HIA websites, we then looked at specific agencies / organizations websites that have published HIAs found in the collecting HIA websites to see if they have any more HIAs in their archives. Two examples of this secondary search would be our looking at the US HIA organization 'Health Impact Partners (HIP)' and the UK HIA Organization 'Impact' at the University of Liverpool to see if they had any additional HIAs.

The sampling strategy used in this study contains two significant limitations that narrow the explanatory scope of our findings. First, all of the 'collecting HIA websites' that we accessed only collected public HIA documents. This makes sense because private HIA consulting firms do not publish their privately funded HIA products for free public consumption. As a result, none of the HIAs reviewed in this study were published by the private sector which makes our observations relevant to only public sector published HIA reports. Second, since our sample was limited to only those HIAs that were published in English, the countries represented in our sample were more likely to be higher income countries (e.g. Australia, Germany, and United Kingdom). With the exception of India, our sample does not include HIAs conducted in middle income or lower income countries. Our observations regarding HIA methodologies are therefore more relevant to higher income countries.

The HIAs were catalogued and analysed first by a latent content analysis and afterwards by a manifest analysis. A latent content analysis is a qualitative approach to reading the text where the researcher is analysing what was said and by whom while a manifest content analysis is quantitative approach that analyses elements that are countable (Gaber and Gaber, 2007, p. 104, Gray and Densten, 1998, p. 420) The HIAs were read three times through a latent content analysis to obtain comparable bits of information. We first began the study with a simple assessment to determine if the HIA used any research strategy. Surprisingly, a good number (n = 47) of the HIAs either did not have a research methodology, which dropped our sampled HIA population down to 100. Out of the 100 HIAs with research methodologies, five of them did not access community observations or comments. Interestingly, four out of the five HIAs that did not look for citizen input were authored by University research groups (the 5th was an NGO). In the end, a total of 95 HIAs were reviewed in this study and five different citizen input techniques were identified (meeting, workshop, interview, focus group, and other (field observations and photo voice). Since most of the reviewed HIAs used a mixed method research strategy, more than one community input technique were used per HIA resulting in a total of 209 separate research projects.

In the second reading of the HIAs, we collected information on the following variables: date, author, topic (e.g. land use or social services), type of HIA conducted (Rapid, Intermediate, or Comprehensive), and, location in the planning process (Prospective, Retrospective, or Concurrent). For this study we defined 'human services' as HIA studies that focused on the health impacts of a plan, policy, or project in relation to a specific population and/or provision of a health or social service to a particular community. 'Land use' was defined as those HIA investigations that looked at the health impacts of an existing or proposed land use development. Lastly, 'environmental' HIAs are those which analysed the health impacts of specific environmental factors and their immediate health impacts with local residents (Gaber and Overacker, 2012, p. 279).

In the third reading, we focused on the particular citizen participation technique used (e.g. focus group) in each HIA in relation to a particular HIA characteristic (e.g. what type of citizen input technique was used in a Rapid HIA?). In this part of the analysis we were looking at what type of emic or etic research techniques the planners used in their HIA investigations. In addition to the research techniques the planners used, we were curious to look at which citizens were targeted in the civic engagement process and looked at the 'sampling' strategies used in the HIAs. Here, we looked to see if the entire community was sampled or if a particular sub-community (e.g. stakeholders) were sought for their observations. This is pushing the etic/emic continuum research agenda to include sampling strategies: small localized sampled populations would be more inclined toward emic observations while large-scale community-wide sampled populations, or sampling with only community leaders would be more focused on etic observations. Once all of the comparable data was collected, they were dropped-in into a spreadsheet and were analysed via a manifest content analysis in search of emic and etic community input techniques and sampling strategy trends.

Research Findings

In our exploration of the HIA data base, our point of departure was an interconnected two-part question: which community input research techniques did the planners use in their participatory process, and who in the community (sampling strategy) did they consider as being important sources of information? In looking at the reviewed HIAs, some very clear community engagement techniques themes became apparent (see Table 7.1). The community engagement technique of choice tended to be either meetings (29 per cent), interviews (25 per cent), and, workshops (20 per cent) making up 74 per cent of all HIA community input techniques. This loosely conforms to Laurian and Shaw (2009) observations where public meetings were the widely used community engagement technique with surveys and focus groups coming in last (p. 299). When looking at the community input techniques via emic vs. etic lens, it becomes clear that planners are more interested in generating etic observations (55 per cent) than emic observations (36 per cent). Both meetings and workshops focus more on a collective conversation and mutual learning making it hard for planners to clearly identify who specifically said what in each meeting (Laurian, 2007, p. 415). As a result, the community data produced in these sessions tend to be impressionistic and generalized into larger-scale themes. What interviews and focus groups provide planners are the ability to isolate who said what during the meetings. Interviews and focus group techniques are great for generating emic data where planners can literally identify individual observation on a particular phenomenon and cross-reference these observations to other individuals.

Table 7.1 HIA data community input data collection technique

Community input technique	Total	Emic vs. Etic
Meeting	29% (n = 61)	Etic
Interview	25% (n = 53)	Emic
Workshop	20% (n = 41)	Etic
Focus Group	11% (n = 23)	Emic
Survey	6% (n = 13)	Etic
Other (field observations, photo voice)	9% (n = 18)	NA
Total	100% (n = 209)	55% Etic / 36% Emic

The sampling strategies planners used in their HIA community engagement investigations were overwhelmingly (76 per cent) oriented toward obtaining etic observations from the community. Four different sampling strategies were more widely used among the analysed HIAs: stakeholder, steering committee, key informant, and community (see Table 7.2). Over half (51 per cent) of the sampling strategies used a stakeholder approach. A stakeholder is 'any person, group, or organization that can place a claim on an organization's attention, resources, or output or is affected by that output' (Bryson, 1995, p. 27). A stakeholder sampling strategy is a focused purposeful sampling process that yields etic community observations from persons who can influence and be influenced by the context surrounding the HIA study. Coming in at a distant second sampling strategy were steering committees (18 per cent). Steering committee takes the stakeholder strategy one level closer to a smaller sup-population with planners purposefully picking participants who both have a stake in the project and have the capacity to provide some sort of advice to HIA researchers. Similar to Citizen Advisory Boards (CABs), they meet regularly over extended periods of time throughout the planning process (Laurian, 2007, p. 416).These people tend to be community representatives, and are more likely to have an etic 'arms-distance' perspective to the community they are representing.

Table 7.2 HIA sample strategy

Community sample strategy	Total	Emic vs. Etic
Stakeholder	51% (n = 106)	Etic
Steering Committee	18% (n = 39)	Etic
Key Informant	17% (n = 36)	Emic
Community	7% (n = 14)	Etic
Other (youth, native/indigenous)	7% (n = 14)	NA
Total	100% (n = 209)	76% Etic / 17% Emic

Coming in at a close third were key informant sampling strategies (17 per cent). Here, planners focus on an even smaller population than the steering committee sample strategy. However, unlike the steering committee strategy that looks at a more representative community of participants, the key informant sample strategy purposefully samples specific persons in the community who have first-hand experiences that are more important to the study than their ability to provide a big-picture observation about how the HIA fits within a larger context. Key informant sampling is the quintessential emic sampling strategy because it solely focuses on the individual insider community perspective. Surprising to us, the community sample strategy was the largest scale etic sample strategy identified among the reviewed HIAs and was the least commonly used at 7 per cent (n = 14). This sampling strategy looks at the entire group of persons directly or indirectly related to the HIA topic at hand and is purposefully identified by location and/or by shared community characteristics.

Community Input Techniques by HIA Author

Five types of organizations authored the HIAs in our study. They are: government, university, NGO, government and university as co-authors, and other (co-authors between NGO and university, government and NGO, and government, NGO, and university). The majority (95 per cent) of the authored HIAs in our study were government (35 per cent), university (32 per cent), NGO (20 per cent), and government and university (8 per cent) (see Table 7.3). Sixty-three per cent of government authored HIAs and 56 per cent of NGO authored studies used etic community input techniques (meeting, workshop, survey). Meetings were the most popular community engagement technique used by planners in government (39 per cent (n = 28)) and NGO's (29 per cent (n = 12)) authored HIAs. This was in contrast to university authored HIAs where planners used interviews as their emic method of choice at 30 per cent (n = 20). This is not to say that university authored HIAs ignored using etic techniques for citizen input, meetings came in at second at 21 per cent (n = 14). Forty-three per cent of all university authored HIA studies used emic community input techniques (meetings and workshops) in comparison to NGOs at 37 per cent and government at 30 per cent. In fact, focus groups were much more widely used among university authored HIA projects with 39 per cent (n = 9) of all focus groups studies in comparison to government studies making up 26 per cent (n = 6) and NGOs investigation at 17 per cent (n = 4) of the reviewed HIAs.

There are two simple reasons for this emic/etic divergence in community input strategies per author. First, universities because of their research orientation may be more inclined to take on emic HIA projects that are under-researched or relatively new on the HIA radar. As a result, university planners take to more emic community input techniques in order to learn more about what is going on directly from the involved community. Second, universities are more likely than not to have more advanced technical training and extensive research facilities than do government and NGO authors, which allows them to more easily accommodate and manage the sampling and analysis requirements in interviews and focus groups projects.

Table 7.3 Data collection strategy by HIA author

	Government	University	NGO	Government and university	Other*	Total
Meeting	39% (28)	21% (14)	29% (12)	38% (6)	8% (1)	29% (61)
Interview	22% (16)	30% (20)	27% (11)	25% (4)	17% (2)	25% (53)
Workshop	20% (14)	16% (11)	22% (9)	25% (4)	25% (3)	20% (41)
Focus group	8% (6)	13% (9)	10% (4)	0	34% (4)	11% (23)
Survey	4% (3)	7% (5)	5% (2)	12% (2)	8% (1)	6% (13)
Other**	7% (5)	13% (9)	7% (3)	0	8% (1)	9% (18)
Total	100% (72)	100% (68)	100% (41)	100% (16)	100% (12)	100% (209)

Note: Other* = NGO and University, Government and NGO, Government and University and NGO; Other** = field observations, photo voice.

Breaking the authored community engagement techniques HIA data down by sampling strategies yielded a few obvious observations that were anticipated and a couple of surprises (Table 7.4). The more obvious observations were in meetings, workshops, and surveys and the type of sample strategies planners used per technique. When government planners used meetings as a community input technique, they usually applied either a steering committee (57 per cent) or stakeholder (26 per cent) sample strategy. Both steering committee and stakeholder groups are better suited sampled populations to provide planners with

Table 7.4 Data collection by technique and sample strategy by HIA author

	Government	University	NGO	Government and university	Other	Total
Meeting						
Stakeholder	32% (9)	29% (4)	25% (3)	0	0	26% (16)
Steering Committee	50% (14)	50% (7)	67% (8)	100% (6)	0	57% (35)
Key Informant	7% (2)	14% (2)	8% (1)	0	100% (1)	10% (6)
Community	0	0	0	0	0	0
Other*	11% (3)	7% (1)	0	0	0	7% (4)
Total	100% (28)	100% (14)	100% (12)	100% (6)	100% (1)	100% (61)
Interview						
Stakeholder	44% (7)	55% (11)	64% (7)	25% (1)	100% (2)	53% (28)
Steering Committee	0	0	0	0	0	0
Key Informant	56% (9)	45% (9)	36% (4)	75% (3)	0	47% (25)
Community	0	0	0	0	0	0
Total	100% (16)	100% (20)	100% (11)	100% (4)	100% (2)	100% (53)
Workshop						
Stakeholder	79% (11)	91% (10)	100% (9)	75% (3)	100% (3)	88%(36)
Steering Committee	0	0	0	0	0	0
Key Informant	14% (2)	9% (1)	0	25% (1)	0	10% (4)
Community	0	0	0	0	0	0
Other*	7% (1)	0	0	0	0	2% (1)
Total	100% (14)	100% (11)	100% (9)	100% (4)	100% (3)	100% (41)

Focus group						
Stakeholder	100% (6)	89% (8)	75% (3)	0	100% (4)	92% (21)
Steering Committee						
Key Informant	0	0	25% (1)	0	0	4% (1)
Community	0	0	0	0	0	0
Other*	0	11% (1)	0	0	0	4% (1)
Total	100% (6)	100% (9)	100% (4)	0	100% (4)	100% (23)
Survey						
Stakeholder	0	20% (1)	23% (1)	0	0	15% (2)
Steering Committee	0	0	0	0	0	0
Key Informant	0	0	0	0	0	0
Community	100% (3)	80% (4)	67% (2)	100% (2)	0	85% (11)
Total	100% (3)	100% (5)	100% (3)	100% (2)	0	100% (13)

Note: Other* = field observations, photo voice.

a larger context observation concerning the HIA topic than what an individual key informant can provide based on their lived experiences. These give-and-take series of informational meetings, especially with stakeholders, provide an ideal conversational format where observations from a particular group tend to be more in a for-your-information (FYI) format on what is new with them than what is going on in relation to the HIA topic.

The vast majority of workshops that planners used in their HIAs tended to access stakeholders (88 per cent). In 100 per cent of NGO‘s and 91 per cent of university authored workshops community engagement projects accessed stakeholder segments of the community. This makes sense since workshops tend to be one day informational engagement with the community where the planner is providing more information than receiving from the participant (Laurian, 2007, p. 415). The only slight variation in the sampling of workshop participants were in government authored (14 per cent) and government/university authored (25 per cent) HIAs where planners sometimes sampled key informants in their investigation. Lastly, the application of survey techniques in HIA investigations were mostly used with a community sample strategy (85 per cent). One hundred per cent of government and 80 per cent of government/university HIA community engagement projects sampled the community in their survey investigations. The survey community input technique is tailored-made for sampled populations that are large in number and/or geographically disperse where it is more cost-effective to have individuals complete a survey than it is for the planner to go out and talk to each person in an interview or have all the sample participants come to a meeting or workshop.

The two surprises were the etic sampling strategies used with emic interview and focus group community input techniques in the HIA projects. Both of these community input techniques are particular good at providing planners with the opportunity to record detailed local community observations and connect them to an individual person. Intuitively, emic interviews and focus groups approach to community engagement would be well suited to work with smaller and very local key informant sampling strategy. The opposite was true; planners using emic community input techniques were more likely to go with larger context etic sampling strategies. Overall, stakeholders are more commonly sampled by planners in HIAs interviews (53 per cent) than key informants (47 per cent). University (at 55 per cent) and NGO (64 per cent) planners are more likely to access stakeholders in their interview investigations than key informants. A slight variation to this was in government/university authored (75 per cent) and government authored (56 per cent) HIAs were more likely to interview key informants than stakeholders. The biggest sampling surprise was in the application of focus group meetings in the HIA studies. Here, 92 per cent of all HIA focus group studies planners sampled stakeholders and only one study (by an NGO) accessed key informants. Again, it is apparent that the type of emic observations desired in HIA investigations tended to be geared at a larger context perspective than at the individualized person-on-the street perspective.

Community Input Technique by HIA Topic

The reviewed HIAs were fairly evenly split between Human Service assessments at 51 per cent (n = 107) and Land Use assessments at 45 per cent (n = 94) with a small smattering of Environmental oriented investigations at 4 per cent (n = 8) (see Table 7.5). Etic community input strategies were much more widely used in Land Use and Environmental HIA studies than in Health Service projects. Over half (54 per cent) of all Land Use projects and 76 per cent of all Environmental studies were based on meetings and workshops community input techniques. Less than half (42 per cent) of planners working on Human Services HIAs used etic meeting and workshop community input techniques. Instead, planners working in Human Service HIAs were slightly (47 per cent) more likely to either use emic interviews at 33 per cent (n = 35) or focus group investigations at 14 per cent (n = 15). This is particularly clear in focus group investigations. Although focus group investigations were only used 15 times in Human Services studies they represent over 65 per cent (n = 23) of all focus groups conducted among the surveyed HIAs. In comparison, planners in Land Use HIA investigations only made up 35 per cent of the focus group projects (n = 8). It is very clear that Human Service HIAs were more interested in getting emic fine-grain detail from individual members of the community than in Land Use and Environmental studies that were more interested in getting etic general input from the impacted population.

Table 7.5 Data collection strategy by topic

	Land use	Human services	Environmental	Total
Meeting	36% (34)	22% (24)	39% (3)	29% (61)
Workshop	18% (17)	20% (21)	38% (3)	20% (41)
Interview	18% (17)	33% (35)	12% (1)	25% (53)
Focus group	9% (8)	14% (15)	0	11% (23)
Other*	10% (10)	6% (7)	12% (1)	9% (1[illegible])
Survey	9% (8)	5% (5)	0	6% (13)
Total	100% (94)	100% (107)	100% (8)	100% (209)

Note: Other* = field observations, photo voice.

As shown in Table 7.6, the topic of the HIA plays a little difference in which type of sampling strategy planners applied in their investigations. For both Land Use and Human Service HIAs, planners working in meeting community techniques tended to apply in a similar fashion etic stakeholder (26 per cent) and steering committee (57 per cent) sampling strategies. In their application of emic Interview community input techniques, planners were almost identical in their etic sampling strategies going with stakeholders (53 per cent) and key informants (47 per cent) sampled populations in Land Use and Human Service investigations.

Table 7.6 Data collection by technique and sample strategy by HIA topic

	Land use	Human services	Environmental	Total
Meeting				
Stakeholder	29% (10)	25% (6)	0	26% (16)
Steering Committee	50% (17)	67% (16)	67% (2)	57% (35)
Key Informant	12% (4)	4% (1)	23% (1)	10% (6)
Community	0	0	0	0
Other*	9% (3)	4% (1)	0	7% (4)
Total	100% (34)	100% (24)	100% (3)	100% (61)
Interview				
Stakeholder	53% (9)	54% (19)	0	53% (28)
Steering Committee	0	0	0	0
Key Informant	47% (8)	46% (16)	100% (1)	47% (25)
Community	0	0	0	0
Total	100% (17)	100% (35)	100% (1)	100% (53)
Workshop				
Stakeholder	88% (15)	90% (19)	67% (2)	88% (36)
Steering Committee	0	0	0	0
Key Informant	6% (1)	10% (2)	23% (1)	10% (4)
Community	0	0	0	0
Other*	6% (1)	0	0	2% (1)
Total	100% (17)	100% (21)	100% (3)	100% (41)

Focus group				
Stakeholder	76% (6)	100% (15)	0	92% (21)
Steering Committee	0	0	0	0
Key Informant	12% (1)	0	0	4% (1)
Community	0	0	0	0
Other*	12% (1)	0	0	4% (1)
Total	100% (8)	100% (15)	0	100% (23)
Survey				
Stakeholder	0	40% (2)	0	15% (2)
Steering Committee	0	0	0	0
Key Informant	0	0	0	0
Community	100% (8)	60% (3)	0	85% (11)
Total	100% (8)	100% (5)	0	100% (13)

Note: Other* = field observations, photo voice.

Lastly, planners heavily leaned on stakeholder sampled populations (88 per cent) in their Land Use and Human Service application of Workshop community input techniques. Where the sampling strategies diverge among Land Use and Health Service HIAs is in the application of focus group and survey community input techniques. In both circumstances, planners were more likely go with a stakeholder sample strategy in Human Service investigations than in Land Use studies. In fact, 100 per cent of all sampled focus group HIA projects only go with stakeholder sampling strategies. This goes against what we anticipated. We anticipated that planners conducting Human Service HIAs would be more interested in obtaining more detailed 'local' emic views from key informants instead of more larger-scale context understandings that are provided by stakeholders.

Community Input Technique by HIA Level

Some community input techniques were more commonly used by planners for shorter investigations (Rapid HIA) while other techniques were more likely used for longer projects (Comprehensive HIA) (Table 7.7). Workshops were by far the most widely used community input techniques in Rapid HIA projects constituting 33 per cent (n = 20) of all citizen input techniques at this level of HIA analysis. This input technique can require very little front-end sampling strategy and can be executed quickly in a single meeting. At the opposite end of the spectrum were survey community input strategies which were largely conducted in long-term Comprehensive HIA studies. Sixty-nine per cent of all survey projects (n = 9) were conducted in Comprehensive HIA investigations and are almost non-existent at a little less than 2 per cent (1 out of 60) of all Rapid HIA projects. Surprisingly, the logic of community input techniques that require little front-end preparation or back-end analysis that were executed in Rapid HIAs did not apply to the focus group technique. Focus group research strategies were consistently used in roughly 11 per cent among all three of the analysed levels of HIAs.

Table 7.7 Data collection strategy by level

	Rapid	Intermediate	Comprehensive	Total
Meeting	30% (18)	30% (22)	28% (21)	29% (61)
Interview	18% (11)	33% (24)	24% (18)	25% (53)
Workshop	33% (20)	15% (11)	13% (10)	20% (41)
Focus group	12% (7)	11% (8)	11% (8)	11% (23)
Other*	5% (3)	7% (5)	13% (10)	9% (18)
Survey	2% (1)	4% (3)	11% (9)	6% (13)
Total	100% (60)	100% (73)	100% (76)	100% (209)

Note: Other* = field observations, photo voice.

The HIA level in which planners were conducting their investigations played a more significant role in the type of community sampling strategy they applied (see Table 7.8). This was seen in HIA Meetings where it initially looked like meetings were fairly evenly spread over the Rapid, Intermediate, and Comprehensive Levels of HIA studies (Table 7.7). As seen in Table 7.8, among the top two meeting sampling strategies (steering committee at 57 per cent and stakeholders at 26 per cent), stakeholder sampling strategies were more widely used in Comprehensive HIA investigations at 43 per cent in comparison to Rapid investigations at 6 per cent. Conversely, steering committee sampling strategies were more widely applied in Rapid meetings at 72 per cent than in Comprehensive HIA studies that used meetings at 43 per cent. In the application of interview community input strategies, planners were more likely to access stakeholder community members in Intermediate HIAs (61 per cent) and Comprehensive HIAs (56 per cent) than in Rapid HIA investigations at 33 per cent. Interestingly, in Rapid HIA investigations where planners used an interview community input technique, they were more likely to access key informants (67 per cent) than stakeholders (33 per cent). This caught us a little off-guard because we anticipated that key informants sampling strategy would better fit with the deep emic observations made possible with the interview technique in longer-term HIA studies like Intermediate and Comprehensive investigations so the planners could have more time to cultivate key informants data. What this tells us is that planners working in Rapid HIA situations are more likely to get quick emic observations through an interview technique and key informant sample strategy. Lastly, planners were more likely to apply stakeholder sampling strategies in Rapid (90 per cent) and Intermediate (93 per cent) HIA workshop techniques. Where the Level of the HIA investigation had no impact was in focus group community input techniques where planners evenly went with a stakeholder sampling strategy across the board among the three HIA levels.

Summary of Research Findings and Future HIA Investigations

In our analysis of community engagement techniques and sampling strategies in the reviewed HIAs, we found that the majority (55 per cent) of planners were more likely to use etic community engagement techniques (e.g. meetings and workshops) and were primarily (76 per cent) sampling segments of the community (e.g. stakeholders and steering committees) who can only provide larger-context etic observations. Our conclusion is consistent with Laurian and Shaw's (2009, p. 299) observation that three-quarters of the participatory methods used by planners in their study used etic research techniques (meetings and workshops) and sampling strategies (task force an advisory group). This tells us two things for practicing planners: planners conducing community engagement research within the HIA process are mostly looking at the possible impacts to large-scale problems, and, most of the anticipated health impacts are largely known by the

Table 7.8 Data collection by technique and sample strategy by HIA level

	Rapid	Intermediate	Comprehensive	Total
Meeting				
Stakeholder	6% (1)	27% (6)	43% (9)	26% (16)
Steering Committee	72% (13)	59% (13)	43% (9)	57% (35)
Key Informant	11% (2)	10% (2)	10% (2)	10% (6)
Community	0	0	0	0
Other*	11% (2)	4% (1)	7% (1)	7% (1)
Total	100% (18)	100% (22)	100% (21)	100% (61)
Interview				
Stakeholder	33% (4)	61% (14)	56% (10)	53% (28)
Steering Committee	0	0	0	0
Key Informant	67% (8)	39% (9)	44% (8)	47% (25)
Community	0	0	0	0
Total	100% (12)	100% (23)	100% (18)	100% (53)
Workshop				
Stakeholder	90% (17)	93% (13)	75% (6)	88% (36)
Steering Committee	0	0	0	0
Key Informant	5% (1)	7% (1)	25% (2)	10% (4)
Community	0	0	0	0
Other*	5% (1)	0	0	2% (1)
Total	100% (19)	100% (14)	100% (8)	100% (41)

Focus group				
Stakeholder	100% (6)	88% (7)	100% (8)	95% (21)
Steering Committee	0	0	0	0
Key Informant	0	0	0	0
Community	0	0	0	0
Other*	0	0	0	0
Total	100% (6)	100% (7)	100% (8)	100% (22)
Survey				
Stakeholder	0	0	22% (2)	15% (2)
Steering Committee	0	0	0	0
Key Informant	0	0	0	0
Community	100% (1)	100% (3)	78% (7)	85% (11)
Total	100% (1)	100% (3)	100% (9)	100% (13)

Note: Other* = field observations, photo voice.

community. Here, community engagement is seen as an opportunity for planners to learn more about the larger-context impact of a possible planning decision than on learning new information about unforeseen health impacts from the community that would be impacted by proposed planning decision.

A good HIA example of an etic oriented community engagement project is the health impact assessment of Central Plains Water Scheme in Canterbury, New Zealand. This was a Rapid, Prospective environmental HIA conducted by a local government agency (Community and Public, Canterbury District Health Board) that looked at the possible health impacts of proposed water project (Central Plains Water Scheme (CPWS)) designed to provide irrigation for 60 hectares of land in Canterbury (Humphrey, 2008, p. 4). The anticipated effects of CPWS on water quality were widely known by the planners and the impacted community. Three workshops were conducted with stakeholders who were 'experts and people who were knowledgeable about the local community, and/or the population groups of interests' (Humphrey, 2008, p. 21). During the workshop, stakeholders were asked if they knew of any 'gaps in the analysis' in the government's research findings and for information about key factors that may encourage, prevent, or mitigate' the health impact of the CPWS project.

We realized the following research conclusions in regards to the three sub-questions introduced at the start of this chapter. In terms of the type of public agency conducting the HIA, we found that the author of the study tended to be more pre-disposed to certain type of community input techniques versus other techniques. This was dramatically seen in university authored HIA studies (tables 7.3 and 7.4). University authored studies tended to use a wider range of emic and etic techniques (interviews, workshops, focus groups, and surveys) than did government or NGO projects. This is not to say that government or NGO's did not access these other community engagement techniques, they did, but at a much smaller number. Government and NGO HIA projects tended to stick to more etic community techniques (meetings and workshops). Overall, Government and NGO planners were more interested in stakeholder and steering committee sampled projects than key informant meetings.

In terms of the topical focus of the HIA, we found the research topic tended to dictate the type of community engagement techniques (but not sample strategy) that planners picked when they are seeking community input. Human Service HIA studies went with emic interviews as their go-to technique (Table 7.5). This makes sense since Human Service studies are more likely to be in-tune to more detailed information about people's lives. What was surprisingly with the Health Service HIA community engagement research was that more than half of them were conducted with the etic stakeholder sampling strategy (54 per cent) than with the emic key informant sampling strategy (46 per cent) (Table 7.6). Land Use studies on the other hand largely utilized meetings and workshop community engagement techniques (Table 7.5). Land Use meetings tended to use steering committees (50 per cent) while workshops largely went with stakeholder (88 per cent) sampling formats (Table 7.6).

Lastly, the level of the HIA played a very little role in dictating what type of community engagement technique planners implemented, but was much more influential in the sampling strategy of who planners talked to in their meetings with the community. This was seen in mostly short-term Rapid HIA studies. For example, in Rapid HIA studies, planners executing interview community engagement technique are more likely to sample key informants (67 per cent) than stakeholders (33 per cent) (Table 7.8).

The planning literature (nor the larger social science literature) provides little guidance on how planners can go about executing etic community engagement research projects. Based on our HIA research, it appears that planners have a good handle on picking relevant etic community engagement techniques (e.g. meetings) with logical etic sampling strategies (e.g. stakeholders). One area that needs more research is analysing why planners are using emic interviews with the etic sampled stakeholders. From what we can see from our data, stakeholder interviews are more commonly completed in Intermediate and Comprehensive HIAs, so the interviews may be long-term and may be running concurrently during the HIA investigations. One possible explanation can be that these stakeholder interviews are being used by planners to provide concurrent assessments of the overall HIA process. If this is the case, the stakeholder interviews may explain why planners are not actively trying to assess the success of their participatory process at the end of their planning projects (Laurian and Shaw, 2009, p. 301) because they are concurrently assessing the participatory process in real time.

References

Abrahams, D. 2004. *European Policy Health Impact Assessment: A Guide*. Geneva: World Health Organization.

Birley, M. 2003. A fault analysis for health impact assessment: Procurement, competence, expectations, and jurisdictions. *Impact Assessment Project Appraisal*, 25(4), pp. 281–289.

Brody, S., D. Godschalk and R. Burby. 2003. Mandating Citizen Participation in Plan Making: Six strategic planning choices. *Journal of the American Planning Association*, 69(3), pp. 245–264.

Bryson, J. 1995. *Strategic Planning for Public and Nonprofit Organizations*. New Jersey: Jossey-Bass.

Dandekar, H. 1986. *The Planner's Use of Information*. Chicago: APA Planners Press.

Dewey, J. 1905. The Postulate of Immediate Empiricism. *The Journal of Philosophy, Psychology, and Scientific Method*, July, II(15), pp. 393–399.

Dewey, J. 1969. *Early Works, 1882–1898: 'The Postulate of Immediate Empiricism'*. Vol. 3, Carbondale: Southern Illinois Press, pp. 158–167.

Deyle, R. and Slotterback, C. 2009. Group Learning in Participatory Planning Process: An explanatory quasiexperimental analysis of local mitigation planning. *Journal of Planning Education and Research*, 29(1), pp. 23–38.

Fetterman, D. 1997. 'Ethnography'. In L. Bickman and D. Rog (eds) 1997. *Handbook of Applied Social Research Methods*. Los Angeles: Sage Publications.

Forsyth, A. 2012. Commentary: Alternative Cultures in Planning Research – From Extending Scientific Frontiers to Exploring Enduring Questions. *Journal of Planning Education and Research*, 32(2), pp. 160–168.

Gaber, J. 2010. 'Applied research'. In N. Salkind (ed.) 2010. *Encyclopedia of Research Design*. London: Sage, pp. 35–37.

Gaber, J. and Gaber, S. 2010. 'Planning Research'. In N. Salkind (ed.) 2010. *Encyclopedia of Research Design*. London: Sage, pp. 1039–1042.

Gaber, J. and Gaber., S. 2007. *Qualitative Analysis for Planning & Policy*. Chicago: APA Planners Press.

Gaber, J. and Overacker, T. 2012. Establishing mixed method research design guidelines in health impact assessment investigations. *Journal of Impact Assessment and Project Appraisal*, 30(4), pp. 275–283.

Glaser, B. and Straussss, A. 1970. 'Theoretical sampling'. In N. Denzin (ed.) 1970. *Sociological Principles*. New York: McGraw-Hill.

Gray, J. and Densten, I.L. 1998. Integrating quantitative and qualitative analysis using latent and manifest variables. *Quality & Quantity*, 32, pp. 419–431.

Harris, M. 1999. *Theories of Culture in Postmodern Times*. Walnut Creek: AltaMira Press.

Harris, M. 2001. *Cultural Materialism: The Struggle for a Science of Culture*. Walnut Creek: AltaMira Press.

Humphrey, A., Porteous, W., Quigley, P. and Thronley, L. 2008. *Health Impact Assessment of Central Plains Water Schedule*. Canterbury: Canterbury District Health Board.

Kaufman, R., Oakley-Browne, H., Watkins, R. and Leigh, D. 2003. *Strategic Planning for Success*. San Francisco: Jossey-Bass.

Lauria, M. and Wagner, J. 2006. What Can We Learn from Empirical Studies of Planning Theory?: A comparative case analysis of extant literature. *Journal of Planning Education and Research*, 25(4), pp. 364–381.

Laurian, L. 2007. Deliberative Planning through Citizen Advisory Boards: Five case studies from military and civilian environmental clean-ups. *Journal of Planning Education and Research*, 26(4), pp. 415–434.

Laurian, L. and Shaw, M. 2009. Evaluation of Public Participation: The practice of certified planners. *Journal of Planning Education and Research*, 28(3), pp. 293–309.

Northridge M., Sclar, E. and Biswas P. 2003. Sorting out the connections between the built environment and health: A conceptual framework for navigating pathways and planning healthy cities. *Journal of Urban Health*, 80, pp. 556–568.

Patton, M. 2002. *Qualitative Research & Evaluation Methods* (3rd edition). Los Angeles: Sage Publications.

Peattie, L. 1967. The social anthropologists in planning. *Journal of the American Institute of Planners*, XXXIII(4).
Phillips, C., McCarthy, M. and Barrowcliffe, R. 2010. Methods for quantitative health impact assessment for an airport and waste incinerator: Two case studies. *Impact Assessment Project Appraisal*, 26(1), pp. 69–75.
Pike, K. 1967. *Language in Relation to a Unified Theory of the Structure of Human Behavior*. The Hague: Mouton & Co.
Pike, K. 1990. *Emics and Etics: The insider/outsider debate*. Los Angeles: Sage Publications.
Plano Clark, V. and Creswell, J. 2008. *The Mixed Method Reader*. Los Angeles: Sage Publications.
Rubin, H. and Rubin, I. 1995. *Qualitative Interviewing: The Art of Hearing Data*. Los Angeles: Sage Publications.
Shusterman, R. 1999. 'Dewey on Experience: Foundation or Reconstruction?' In Casey, H. and Seiple, D. (eds) 1999. *Dewey Reconfigured: Essays on Deweyian Pragmatism*. New York: State University of New York Press.
Tamburrini, A., Gilhuly, K. and Harris-Roxas, B. 2011. Enhancing benefits in health impact assessments through stakeholder consultation. *Impact Assessment Project Appraisal*, 29(3), pp. 195–204.
World Health Office. 1999. *Health Impact Assessment: Main Concepts and Suggested Approach*. Brussels: European Centre for Health Policy.

Chapter 8
The Plan-Process-Results (PPR): Methodology for Evaluation and Planning

Vítor Oliveira

Introduction

This chapter presents the Plan-Process-Results (PPR) methodology. The first part of the chapter proposes PPR as a methodology to evaluate planning and plan implementation. It starts by describing its theoretical background and its main methodological aspects. It then describes the application of PPR to the *Plano Director Municipal* (PDM) for Porto. This evaluation exercise was developed in an academic context in 2006 in the beginning of plan implementation. The second part of the chapter explores the use of PPR as a planning framework and methodology. This chapter draws on an exercise developed together with planning practitioners in the Porto local authority, in 2012, in the middle of the PDM implementation. It starts with a discussion on the use of each evaluation criterion as a design criterion, moving then to an illustration of two of these criteria.

The Plan-Process-Results (PPR) Methodology for Evaluation

The Design of PPR

The Plan-Process-Results (PPR) was originally designed as a methodology for evaluating planning and plan implementation. The design of PPR was framed by seven principles to structure and guide the evaluation process in urban planning (Table 8.1) established after a comprehensive literature review covering a wide range of topics, namely the evolution of evaluation theory and methods, the contemporary debate on planning evaluation, and the practice of planning evaluation. Part of this literature review is presented in the next section. For more detail on the review and on the design of these principles see Oliveira and Pinho (2010).

These principles are as follows: i) planning practice should be evaluated; ii) the design of a methodology to assess this practice must be clearly linked with theory on evaluation in planning; iii) the evaluation methodology should suit the object under appraisal; iv) the planning practice must be evaluated as a whole; v) evaluation and planning processes should be developed together, right from the beginning; vi) the evaluation methodology must have a balanced development over time; and finally, vii) the presentation of evaluation results and the analysis of their use should be included.

Table 8.1 Principles for evaluation in planning and the PPR methodology

Principles for evaluation in planning	PPR
Planning practice should be evaluated	Is a methodology for evaluating planning and plan implementation
The design of an assessment methodology must be clearly linked with planning evaluation theory	Reflects a view of evaluation in planning, integrating different approaches
The evaluation methodology should suit the object under appraisal	Is adapted to the Portuguese planning context in order to assess its main tool, the *Plano Director Municipal*
The planning practice should be evaluated as a whole	Focuses on plans, processes and results
Evaluation and planning processes should be developed together, right from the beginning	Aims to be integrated in the planning practice since early stages
The evaluation process must have a balanced development in time (can include different methods)	Comprises the *ex-ante* (with some limitations), *on-going* and *ex-post* dimensions
The presentation of evaluation results and the analysis of their use by planning, should be included	Intends to find a balance between communication and technical knowledge, and establish the terms by which a successful utilization may be judged

Different Types of Evaluation

The sixth principle presented in Table 8.1 establishes that the process of evaluation in planning must be concurrent with implementation, which is contrary to much of the current practice of the ex-ante assessments. This continuous evaluation process may include different methods to assess different objects, with different purposes, in different moments of the planning process.

The next paragraphs will briefly describe two different approaches, conformance-based evaluation and performance-based evaluation. Conformance-based evaluation means judging the success or failure of planning, using one or two criteria – the conformance degree between the actual outcomes on the ground and the initial plan proposals, and the promotion of planning goals and objectives through the implementation of the instruments available (Alexander, 2006).

After the development of some fundamental research works in the end of the 1970s (Alterman and Hill, 1978; Calkins, 1979) and in the second half of the 1990s (Baer, 1997; Talen, 1996, 1997), a number of interesting studies on a conformance-based approach have been produced throughout the last 15 years, mainly in the United States.

The Plan Implementation Evaluation (PIE), presented by Laurian et al. (2004) is a conformance-based methodology that relies on the analysis of plans and planning permits and offers a rigorous, quantitative, and systematic way of assessing the degree to which land use plans are implemented. Plan implementation is defined as the degree to which plan policies are implemented through the application of specified development techniques in planning practice. Two aspects of plan implementation are measured, 'breadth' and 'depth'. PIE has been applied to six New Zealand plans and to almost 400 land development permits. So far, it has largely focused on storm water and urban amenity management.

Brody, Highfield and Thornton (Brody et al., 2006a) examine the spatial pattern of wetland development permits in Florida, verifying its conformance with the proposals of local plans. These authors analyse how and where have wetlands been developed over a 10-year period, if wetland permits clustered in areas designated for high-density development or if they deviate from the plan's original spatial designation, and if the quality and content of the original plan relate to its degree of implementation. In another paper, Brody, Carrasco and Highfield (Brody et al., 2006b) use the same methodology and the same case study to analyse the effective influence on the territory of five sprawl-reduction planning policies included in local plans.

In 2008, Chapin, Deyle and Baker published two papers on the evaluation of planning policies to reduce exposure to hurricane flooding (Chapin et al., 2008; Deyle et al., 2008). In the former article, a parcel-based GIS method for measuring land use changes, as the basis to assess the implementation of local land use policies, is presented. In the later paper, Deyle et al. (2008) explore the relationships between the process of implementation and the quality of the maps and policies of local plans.

Performance-based evaluation follows from defining a plan as a decision framework (Alexander, 2006). The plan performance expresses its usefulness in filling this role. It is important to understand if, and how, the plan was consulted for subsequent decisions. What happens to the plan is crucial to its assessment (Faludi, 2000).

Based on the work of Fudge and Barrett (1981), who have highlighted the differences between a performance and a conformance view, the Dutch school of planning evaluation has been developing a considerable body of research on performance. Faludi (2000) and Mastop and Faludi (1997) sustain that strategic plans – as opposed to project plans – should provide a frame of reference for operational decisions, and do not necessarily have to produce direct impacts on the physical development process. As such, the evaluation of strategic plans should correspond to a detailed analysis of the decisions and actions of a number of actors that are supposed to receive the plan messages.

Faludi (2006) extends further the performance-based approach to the evaluation of the European Spatial Development Perspective (ESDP). Drawing on the distinction between planning as a technical exercise and as a learning process, the author contrasts the concept of 'application' of plan messages with the traditional concept of plan implementation, and presents a method to evaluate the success of the former. In the context of policy implementation analysis, Stame (2008) proposes the concept of 'promotion' somehow similar to the Faludi's concept of 'application'. The purpose of applying ideas in a document such as the ESDP is to provide the professionals involved in European spatial planning processes with better knowledge of their working contexts and of the directions to follow.

Rivolin (2008) extends the idea of 'performance of plans' to 'performance of planning systems'. Rivolin sustains that the main question at stake is not whether performing strategic plans are preferable to conforming regulative plans, but how the strategic and regulative functions of the planning activity should be differently correlated in a planning system addressed to performing rather than to conforming aims.

Some studies on plan implementation explore the potentialities of an integrated use of both approaches (Alexander and Faludi, 1989), or of their simultaneous application to obtain comparable results (Altes, 2006; Berke et al., 2006).

The Main Influences of PPR

The design of PPR involved a focus on 10 existing methodologies,[1] most of them with an on-going dimension and clearly positioned in the debate between conformance and performance. The most important influence in the design of PPR was the Policy-Plan/Programme-Implementation-Process, PPIP proposed by Alexander and Faludi (1989). PPR shares with PPIP a common vision of both evaluation and planning, and common definitions of the most general elements (rationality ex-ante, performance of the planning process and conformance of results). PPR also shares with a wider set of methodologies – that includes the MEANS (European Commission, 1999), the PIE, and the model developed by Norton (2005) – the definition of most of the criteria. The PPR has also a number of similarities with this set of evaluation methodologies in the use of particular techniques such as the SWOT analysis or impact matrices (tables 8.3 and 8.4). Also in common, and as far as the main data sources are concerned, we can find the same use of the plan and of content analysis of interviews to the main stakeholders involved in local development processes. Three innovative aspects distinguish PPR from PPIP and from the other methodologies: i) comparatively to the PPIP model, PPR places a greater emphasis on the physical aspects of cities (urban form and structure, land uses, housing, environment); ii) PPIP has not been submitted to an effective validation through an application to real case studies,

1 These methodologies are: the Policy-Plan/Programme-Implementation-Process, PPIP (Alexander and Faludi, 1989), the Means for Evaluating Actions of a Structural Nature, MEANS (European Commission, 1999), the Plan Implementation Methodology, PIE (Laurian et al., 2004), and other seven methodologies without a specific designation, presented by Norton (2005), Talen (1996), Baer (1997), Lange et al. (1997), Khakee (2000), Burby (2003), and finally, Brody and Highfield (2005).

while PPR, so far, has already been applied to plans in Lisbon and Porto; and iii) comparatively to other evaluation methodologies, PPR proposes a higher number of criteria and of data sources.

The PPR Methodology

The PPR methodology highlights the role of the plan, the process and the results, but also of a set of key elements in the urban development process – the city users (residents and workers), the local politicians, the planning framework and other plans prepared for the city. The general elements and criteria of evaluation emerge from the main relationships among these elements (Figure 8.1). PPR is composed of three general elements – rationality ex-ante, performance of the planning process, and conformance of results – corresponding to nine criteria: internal coherence of the plan, plan relevance, plan interpretation of the planning system, external coherence between the plan and other key documents, public participation, plan utilization in decision making, commitment of resources, effectiveness (plan-results), and finally, direction of the urban development process.

PPR uses 10 main data sources – Table 8.2 identifies which data sources are used in the evaluation of each specific criteria. The data sources are as follows: i) the local plan; ii) other plans covering the whole or parts of the city (e.g. regional plans, strategic plans); iii) newspapers with a local focus (to reconstruct large periods of the urban development process); iv) interviews of key actors involved in the urban development process – local politicians, professionals from the local planning department and from the central government planning agency; v) statistical information on society, economics, and environment; vi) the national legal planning framework (the extension of laws and decree laws under analysis depends on each particular case); vii) the physical form of the city; viii) a fair amount of planning applications, covering different parts of the municipal territory; ix) cartographic material (preferably from different time periods to reconstruct the evolution of the city form and structure); and finally, x) other official documents prepared by the local authority, such as municipal budgets and activity plans.

The Evaluation of the Plano Director Municipal for Porto

This part of the chapter elaborates on each of the nine criteria through an application to a case study – the *Plano Director Municipal* (PDM) for Porto. PPR has already been applied to the plans of the two most important cities in Portugal, Lisbon and Porto. In the case of Porto (and in contrast to Lisbon, due to the ratification of a new plan for the capital city in 2012), PPR has been applied to the same plan both as an evaluation methodology and as a planning methodology.

The PDM is the main instrument of the Portuguese planning system. It establishes the model for the spatial structure of the municipal area and defines the strategy for local development. The plan is composed of a regulatory code, maps

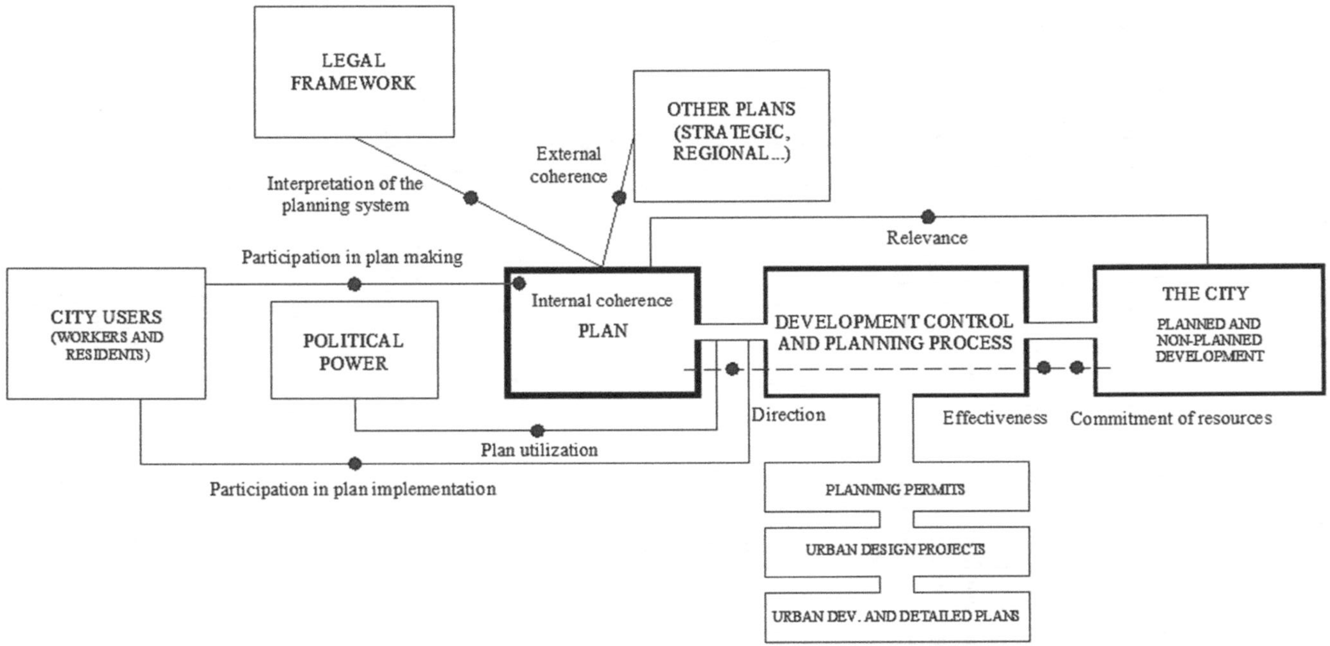

Figure 8.1 **The urban development process and the PPR evaluation criteria**

Table 8.2 Evaluation criteria and data sources

	The plan	Other plans	Newspapers	Interviews	Statistical information	Planning system	The physical form of the city	Planning applications	Cartography	Other documents (local)
Internal coherence	x									
Relevance	x		x	x	x					
Interpretation	x					x				
External coherence	x	x								
Participation	x	x	x	x						
Plan utilization	x		x	x						
Commitment of resources	x			x						x
Effectiveness	x	x					x	x	x	
Direction	x		x	x	x		x			

identifying the different land uses, urban systems, and operative units for planning and management, and a further map depicting local rights of way and planning restrictions. These are complemented by a report and a number of planning studies which are not legally binding. The preparation of the plan is undertaken by the local authority. After its approval by the municipal assembly it has to be ratified by Central Government. The Porto plan was approved by the City Council in 2005 and ratified by the Central Government in January 2006.

Internal coherence The assessment of the first PPR criterion, its internal coherence, focuses on the robustness of the linkages between the main parts of the plan, which normally are the objectives, the territorial model (including land uses and the different urban systems), and the mechanisms for plan implementation. The reading of the plan allows for the identification of these parts and for an initial qualitative appraisal. This appraisal can be represented in an impact matrix. In this matrix, a white dot means that X has no impact on Y, a black dot means that X has an indirect impact on Y, and two black dots mean that X has a direct impact on Y (Table 8.3). The next paragraphs present the case of Porto and of its PDM.

The strategic goals of the PDM are the enhancement of the urban identity of Porto (O1), the rehabilitation of public space and the built environment (O2), the rationalization of transport systems (O3), the reduction of existing social and territorial imbalances between the eastern and western parts of the city (O4), and, finally, the rehabilitation of historical quarters and the central area (O5).

The PDM identified 10 particular form-based zones: Historical Areas; Areas of Continuous Building Frontages and Largely Replete Plots; Areas of Continuous Building Frontages and Plots in the Process of Repletion; Single-Family Housing Areas; Areas of Isolated Buildings; Areas of Public Services; Special Urban Development Areas; Business Park; Green Areas; and, finally, Transport Areas.

The plan defined three urban systems: Environmental System (US1), Built Heritage System (US2), and Mobility System (US3). From these systems, two particular features should be highlighted, the definition of a municipal ecological structure and, for the first time in the plans for Porto, a clear preference of public transport over private cars.

Finally, the mechanisms for plan implementation included a number of Urban Actions and Investments (IM1) and the identification and definition of a number of smaller Operational Units for Planning and Management (UOPG) for further analysis to later support detailed planning proposals (IM2).

The linkages between the objectives and the territorial model are quite strong. This means that the different parts of the territorial model were designed in a way that contributes for the achievement of the plan objectives. This is the case with the proposed definition of land uses based on a form-based zoning. According to the zoning proposal, the existing urban forms in Porto suggest the definition of different zones and provide the criteria for the design of new urban forms, contributing to the achievement of three objectives: 1, 2 and 5.

The linkage between the objectives and the mechanisms of plan implementation is less robust. This means that the two mechanisms of plan implementation the Urban Actions and Investments and the Operational Units for Planning and Management, are not offering a sound contribution to the achievement of the objectives of the plan. One of the reasons for this might be that the former (the objectives) corresponds to the most general level, and the latter (the mechanisms) to the most specific level of plan proposals. An analysis of the impact on each objective shows that the plan attains the highest scores in objectives 1 and 2, the enhancement of the urban identity of Porto and the rehabilitation of public space and the built environment. This is mainly due to their relationships with all the classes of space and with two urban systems, environment and urban heritage (Table 8.3). On the contrary, objectives 3 and 4 – the rationalization of transport systems and the reduction of the urban imbalances – obtain the lowest scores.

Table 8.3 The relations between the PDM objectives and urban systems

	O1	O2	O3	O4	O5
Environmental System (US1)	••	••	••	•	••
Built Heritage System (US2)	••	••	•	•	••
Mobility System (US3)	•	••	•	•	•

Relevance of the plan After assessing the internal coherence of the plan, PPR evaluates its relevance to the main needs and aspirations of the city. This involves the reconstruction of the baseline situation for plan making purposes, using newspapers, statistical information, experts and politicians interviews, and planning reports on different aspects of city life. All this information is gathered and synthesized in a SWOT table (Table 8.4) that should be as representative as possible (eventually, due to the complexity of the evaluation object, not all needs and aspirations of all stakeholders are included). Then, PPR verifies if the plan (each of the main parts of the plan) provides solid answers to the main weaknesses of and threats to the city and if it considers and develops the main strengths of and opportunities for the city.

Table 8.4 presents a SWOT of the existing situation in Porto in the beginning of the last decade. The relationships between each of the main components of the PDM and the fundamental problems of the city proved to be quite positive. Two of these problems deserved particular attention from the plan – the piecemeal destruction of the built forms and built heritage, and the steady degradation of environmental quality.

Table 8.4 SWOT analysis of Porto in the beginning of the last decade

Strengths	Weaknesses
The most important city in the north of Portugal. Part of the city is a UNESCO World Heritage Site. It comprises the largest Portuguese university and one of the main museums of modern art in the country. It has an excellent road network, a modern airport and a dynamic seaport.	The decline in resident population. The growing social and territorial unbalance between the eastern and the western areas. The piecemeal destruction of the built forms and of the built heritage. The steady degradation of the environmental quality. The traffic congestion problems.
Opportunities	**Threats**
The implementation of a light rail system (the largest investment programme in transport systems in Europe during the first half of this decade). The development of the high speed train (TGV) linking Lisbon, Porto and the north of Spain.	The current national adverse economic context. A highly centralized system. The lack of a strong political leadership at city-region level.

Plan interpretation of the planning system The evaluation of plan interpretation involves two different types of analysis. Close to a checklist, the assessment of interpretation in terms of form verifies the presence, in the plan, of all elements required by the planning framework. In terms of substance, all these elements are assessed against the general guidance provided by that framework. The reading of the plan and of the main planning legislation should allow for the identification of the main links between these documents. This criterion helps to explain the differences between plans prepared in different time periods and under different legal frameworks, and to identify elements of innovation and creativity – exploring new directions in planning – in the comparison between plans prepared under the same legal framework.

Against its legal framework, the PDM reveals a very positive performance. One of the most interesting aspects is the innovative interpretation of the article on the strategy for land use qualification through the establishment of typological zoning (Figure 8.2). According to this comprehensive approach, the existing urban forms in Porto (and not the existing and proposed land uses) suggest the definition of different zones and provide the criteria for the design of new urban forms, adapted to each particular urban tissue, contributing as such for the achievement of the three objectives of the plan. Oliveira (2006) and Oliveira et al. (2014) have argued for the innovative character of this typological approach, in the international context, not only for the effective preservation of built heritage but also for the sensitive design of new built forms.

External coherence Some of the city needs and ambitions the plan is expected to address may also be targeted by other plans, policies or programmes implemented in the same territory. It is therefore necessary to ensure that plan objectives, territorial model (the form-based zones and the urban systems), and implementation mechanisms, do not contradict the proposals contained in these other planning instruments.

In Porto the focus was on the coherence between the PDM and an urban regeneration programme. The evaluation of this criterion, structured around three different aspects, expresses different degrees of articulation between the plan and the programme. In terms of territorial model, the relation is very weak, reflecting some incoherencies of the programme (definition of the intervention area, methodology for buildings rehabilitation). The articulation of objectives is positive, despite the reduced expression of an economic dimension in the plan and the lack of a strategy on mobility and transports in the programme. The third aspect of this relation is a multicriteria information system linking the proactive regeneration of old tissues with the design of new areas, through the transference of building rights.

Public participation The evaluation of public participation involves three different levels: the promotion of public participation by the planning authority, the effective participation, and the influence of public participation in the plan. The main data sources to assess this criterion are the plan report on public participation (including all written comments), the newspapers (containing information on the most publicized cases, usually undertaken by development interests and organized groups) and the interviews to experts and politicians.

The analysis of the first level focuses on some choices of the planning authority, such as the type of information made available, the ways to provide and receive information (effective opportunities for dialogue, in which planners inform citizens about planning issues and listen to citizen concerns, should be provided), the selection of stakeholder groups (a variety of groups should be invited to take part in the planning process), and the timing for public involvement (participation should be promoted right from the beginning of plan making and not when the main options are already taken).

In the second level, participation is considered in quantitative terms and is measured by the number of written comments produced during the public participation period. The length of the time period and the number of inhabitants of the city (and their participation rates) should be taken into consideration in the comparison between different case studies.

The third level focuses on the substance of the comments. It analyses the type of issues that are raised (and the extent to which some comments are repeated) and tries to understand which comments lead to changes in the plan content. This exercise involves the reconstruction of the planning process, based on the written comments and on the different versions of the plan prepared over the period of plan making.

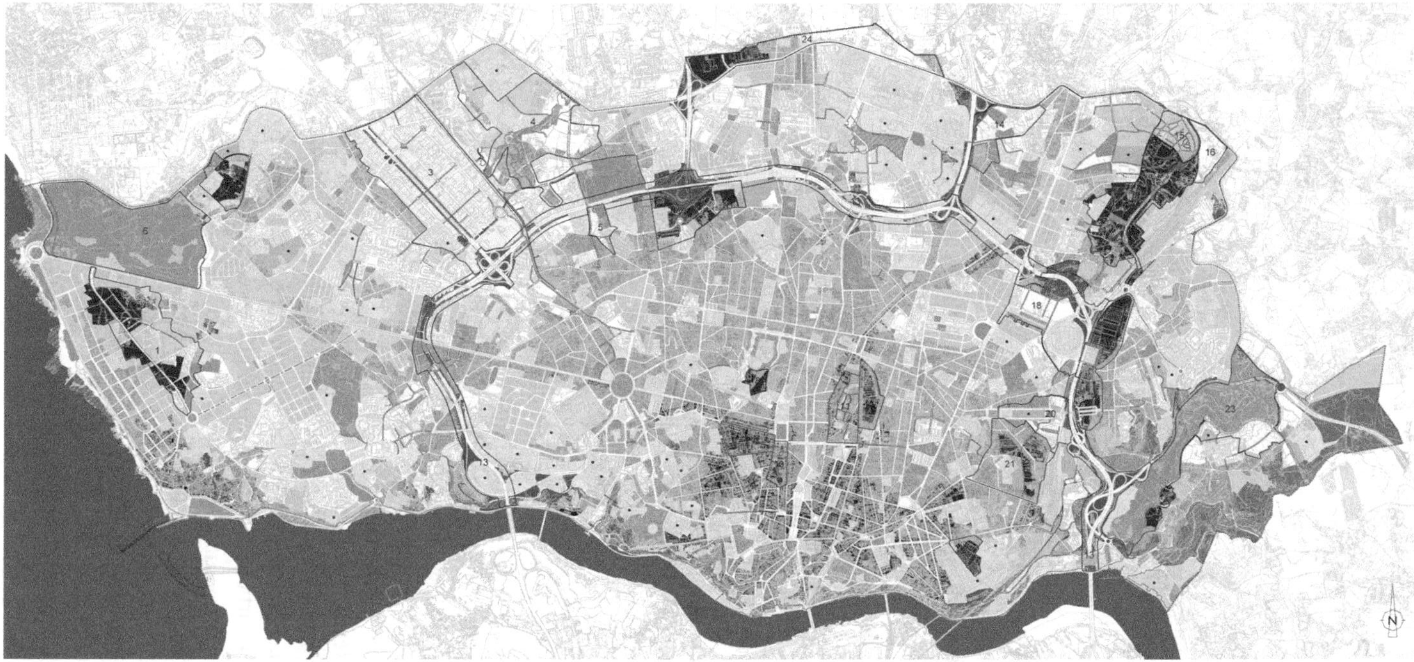

Figure 8.2 The PDM typological zoning

Public participation in the PDM had occurred during three periods between October of 2003 and January of 2005. The consultation of the plan took place in the City Hall, in the Municipal Parishes, and in the local libraries. A forum in the internet and a number of debates, promoted by the City Council and by other institutions, complemented the dissemination of the plan. Throughout the process of public participation 632 written comments were recorded (348, 131 and 153 in the different periods). The comments, mainly by individual citizens, focused on the plan proposals on land use (half of these comments focused on building rights), built heritage (proposing the classification, or the exclusion, of particular buildings), street system, environment (most of these comments were presented by environmental groups), the business park, the central area of the city, and on two particular cases – the plots of a local football team and of a working-class neighbourhood. The analysis of this process reveals three main findings: i) the number of interventions is relatively low; ii) the quality of these interventions is quite diverse, although it should be highlighted that 14 per cent of these have been accepted by planners and were incorporated into the plan; and iii) there was a reasonable stimulus to public participation by the local authority.

Plan utilization Plan utilization is a key criterion in a methodology that intends to integrate elements from a decision-centred view of planning (Faludi, 1987) and from a performance-based approach to evaluation (Mastop and Faludi, 1997). From both perspectives, the plan is important because it provides, in anticipation, a useful frame of reference for political decision makers. The assessment of this criterion has some difficulties, mainly due to the need to reconstruct decision-making processes and find answers to the following questions: What did the decision makers know about the situation? What were their motives? What was the context for their actions?

In the case of Porto, the evaluation of plan utilization reveals three main findings: i) the presidency change of the Porto Council had interfered in the plan preparation schedule and had led to changes in the plan contents; ii) this interference is most evident when the final version of the plan is compared with the 2003 version, the last to be prepared by the original planning team; iii) there was an obvious lack of identification of the local politicians and the PDM.

Commitment of Resources

The evaluation of commitment of resources involves three different levels, as follows: the evolution of the availability of resources over a particular period in time, the type and the diversity of available resources, and the articulation between planning performance and the allocation of resources. The first level corresponds to an appraisal of the evolution of the total number of the technical staff in the local authority and of the evolution of the municipal budgets. The second level, corresponds to an appraisal of the relative importance of the planning department's technical staff (human resources), of the relationships between capital and running costs, and of the relative weight of the resources allocated to planning issues (financial resources). The third level corresponds to an appraisal of the relationships between the specific type of plan under analysis and the human and financial resources needed for its effective implementation.

In Porto, the administration has not been able to promote a continuous positive growth of municipal budgets, with a balanced relation between capital costs and running costs, although it has allocated a reasonable part of financial resources to the PDM. In terms of planning staff, the interesting work of the Office for Project Appraisal, in a process of development control framed by an innovative plan, should be highlighted.

Effectiveness The comparison between the plan proposals and the results on the ground comprises three different stages. In the first stage, the analysis focuses on the plan development through lower level plans: i) comparing these lower level plans with priority areas defined in the main plan, in terms of location and area; ii) verifying if these plans follow or contradict the general guidance of the main plan. In the second stage, the analysis focuses on the implementation of plan proposals on the public space. In the third stage, the process of development control is assessed, based on a detailed analysis of a representative sample of planning permissions.

Besides the PDM, only a detailed plan was approved and implemented, reflecting an obvious inattention on the production of lower level plans. In the end of the first year of the PDM implementation, a positive appraisal of the proposed street network could be made: 11.1 per cent of the streets had been built and 16.7 per cent had been partially built. The last part of the assessment of effectiveness is an analysis of 50 applications developed under the umbrella of the PDM (Figure 8.3) and it showed that the new process of development control is contributing to a higher quality building design.

Direction of the urban development process The last criterion is the definition of a direction of the urban development process. Its assessment, mainly based on the real impacts on the city of the plan proposals, involves a comparison between how the city was before plan making, how the city would be without the plan, and how the city actually is, due to the process of urban development that occurred under the umbrella of the plan. This assessment focuses on three dimensions of urban development, physical, social and economic (although with a stronger emphasis on the physical dimension).

Evaluation Results

In the end of the evaluation exercise, results from the assessment of the nine criteria have to be brought together. These are presented on a table format (Table 8.5) and followed by recommendations for decision-making and planning practice. Both table and recommendations should balance the tensions between communication

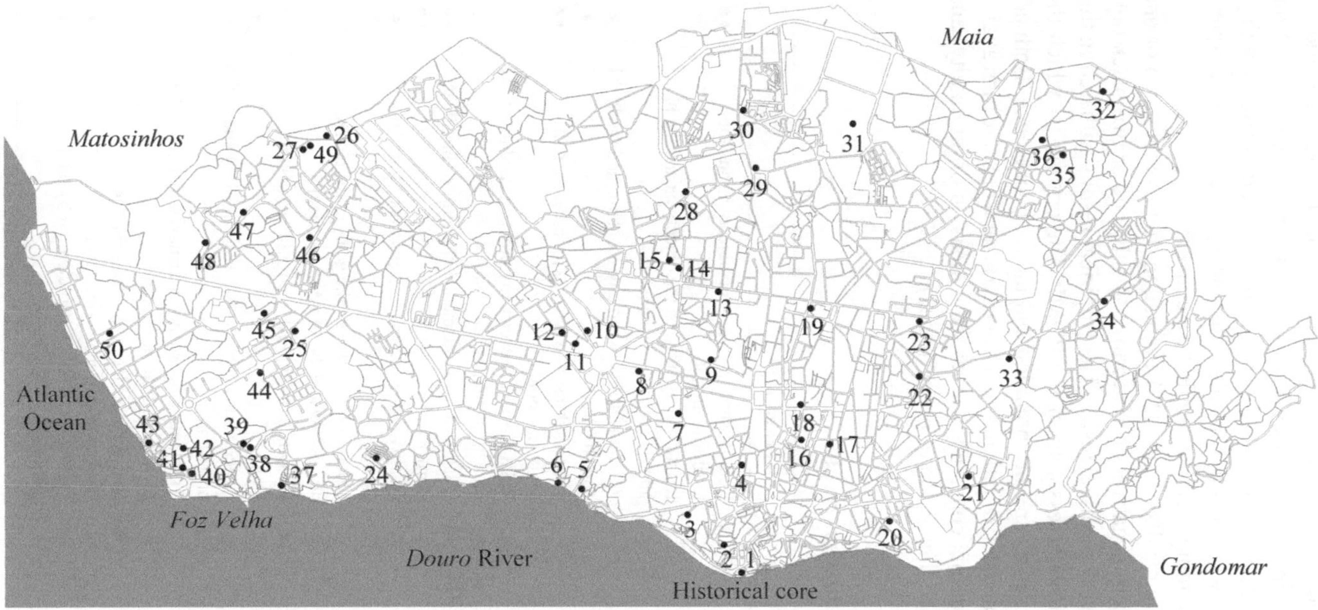

Figure 8.3 Sample of building applications

and technical knowledge and be written in a community-based 'evaluation vocabulary'; they should offer information that is useful to the stakeholders; and address an effective integration of evaluation results in the dynamics of the planning process. PPR rests on a cyclical view of both evaluation and planning. As such, the way these evaluation results are incorporated in subsequent decision-making and planning practice should be a concern.

Table 8.5 presents the results of the PDM evaluation. In this table, two or one white dot represent a 'very negative' or a 'negative' result; one or two black dots mean a 'positive' or a 'very positive' result. The empirical results gathered in this exercise point to a positive assessment of the PDMP. The criteria on which the plan attains the highest scores are interpretation and internal coherence. Both are closely related to the dimension of rationality. The criteria on which the plan attains the lowest scores are external coherence (under the dimension of rationality) and plan utilization (under the dimension of performance).

Table 8.5 PDM evaluation results

Criteria		Sub-criteria	
Internal coherence	**A**	Relationships between the objectives and the territorial model	●●
		Relationships between the objectives and the plan implementation mechanisms	●
Relevance	**B**	Relationships between city problems and the PDM objectives	●
		Relationships between city problems and the territorial model	●
		Relationships between city problems and the PDM implementation mechanisms	●●
Interpretation of planning system	**A**	Quality of the fundamental elements of the PDM (form)	●●
		Quality of the secondary elements of the PDM (form)	●●
		Relationships between the goals of the decree no. 380/99 and the PDM (substance)	●
External Coherence	**C**	Existence of regional, metropolitan or strategic plans	○
		Articulation between objectives of PDM and of the programme	●
		Articulation between the territorial models of the PDM and of the programme	○
Participation in plan making	**B**	Number of citizens' written comments	○
		Quality of citizens' written comments	●
		Promotion of public participation by the local authority	●

Criteria		Sub-criteria	
Plan utilization	C	Identification between the Mayor and the PDM	○
		Introduction of changes that weaken the PDM objectives	○
		Maintenance of the fundamental elements of the PDM	●
Commitment of resources	B	Evolution of the municipal budgets	○
		Relationships between capital costs and running costs	○
		Relative weight of the resources allocated to planning matters	●
		Relationship between number of graduates and workers in the planning department	●
		Diversity of planning officials' backgrounds	●
		Utilization of the PDM by planning officials	●
Effectiveness	B	Development of PDM through urban development plans and detailed plans	○
		Urban design projects	●
		Process of development control	●
Direction	–	–	–

The Plan-Process-Results Methodology for Planning

The second part of this chapter explores the use of PPR as a framework, and as a methodology, for planning. It draws on an exercise developed together with planning practitioners in Porto local authority, in 2012, in the middle of the PDM implementation. This part of the chapter starts with a discussion on the use of each evaluation criterion of PPR as a design criterion, moving then to an illustration of two of these criteria – internal coherence and relevance – focusing on one particular planning issue, urban form.

Discussion

Three years after the design of PPR, a three-year project was launched involving the local planning department of Porto, to explore the use of PPR, not as an evaluation methodology, as in the former academic exercise, but as a planning methodology. The main goal is to offer a methodology that can be able to shape, in a more direct way, the design of plans and of planning practices developed in Portugal. This project has included a set of monthly working meetings with planning practitioners from the local authority. The next paragraphs offer a discussion around the PPR criteria, framed by this collaborative work with the local authority of Porto and by a reflection on the planning practice in Portugal.

Internal coherence The PDMs that are being prepared in Portugal are, by nature, difficult and complex documents. In many cases, this complexity leads to situations where different parts of the plan are not coherent with and on occasion even contradict each other. Inconsistencies among the parts must be detected (through the use of simple techniques, such as the impact matrixes presented above) and reduced in the process of plan making. In the end, the different parts of the PDM – the objectives, the land uses, the different urban systems (built heritage, environment, mobility and transports), the mechanisms for plan implementation, and the planning restrictions – should form a coherent whole, and contribute to *one* sound strategy.

The reflection developed together with the planning practitioners on the Portuguese planning practice leads us to recommend a simplification of these documents, avoiding the temptation to include every issue under the planning scope. As Wildavsky (1973) reminded us, 40 years ago, 'if planning is everything, maybe it's nothing'. The challenge is to prepare coherent and simple documents.

An aspect that can be related to the coherence of the parts is the flexibility of *some* parts. A higher flexibility of the planning process, advocated by different authors, should be introduced based on the assumption that the PDM is more selective in what it defines in an inflexible way. In other words, the PDM should be able to identify what are the most essential and structural aspects, linking them to a set of inflexible principles, and allow a higher flexibility in a number of elements where uncertainty is dominant.

Relevance of the plan The importance of 'place' is crucial in planning. The design of place-based planning proposals depends on the ability of the planner to 'look' at the city and to learn to interpret and understand more accurately the local spatial characteristics (e.g. institutional capacities, local economic potential, social benefits). The planner proposals should be relevant to the city (to a variety of public and private stakeholders), and we should be able to understand if these are relevant – and correct them, if necessary – even in the process of plan making.

The preparation of the PDM should use, from the beginning, updated information and the most appropriate techniques for the description, explanation and prescription of urban phenomena. Moreover, these techniques can be used both in analysis and in subsequent synthesis (this issue is developed by Lichfield et al. 1975). The permanent availability of updated information raises the issue of monitoring. Some key indicators (environmental, social and economic) should be selected by the local authority and continuously measured to allow this monitoring. A well-designed monitoring procedure warrants a baseline for future analysis and a continuous movement between the 'city' and the 'plan'.

Plan interpretation of the planning system An 'evaluation culture' is also a 'learning culture'. When preparing a new plan, a planning team should develop a creative approach as opposed to an uncritical answer to an established legal framework. On the other hand, the preparation of a local plan can provide inputs for the improvement of the weakest aspects of the national planning system. It is argued that the central government can also learn from the experience of local authorities, combining 'top-down' inputs with 'bottom-up' inputs.

The process of plan review should also be improved. The ability for measuring and evaluating plans must go along with the ability to correct and change plans in the right timing. Indeed, planning practice should be dynamic.

External coherence The existence of other documents or plans at other territorial scales (usually municipal or regional) or approaches (such as those with a strategic nature) must be identified at early stages of PDM preparation. Although it has a central role in the Portuguese planning system, PDM is only *one* part of the planning process. As such, it is necessary to ensure that the design of new planning proposals do not contradict or duplicate other proposals for the same territory.

Public participation As mentioned above, the PDM is, by nature, a complex and difficult document to be read (see Khakee, 2003, on the subject of reading plans). Accordingly, it is difficult for citizens to participate in the construction of the plan. However, this difficulty cannot lead to the withdrawal of citizen involvement. Simplicity must be one of the main characteristics of the PDM. The involvement of citizens should be faced as a key challenge, crucial for legitimizing the plan and the planning activity built around the plan. The process of building a meaningful public participation should consider different levels: the effective promotion of public participation (ensuring that there is a broad perspective represented among the involved public and that no one group dominates the process); the monitoring of the development of the public participation process (as opposed to waiting for the inevitability of 'another process without citizen involvement'); and the incorporation of the inputs of public participation into the plan.

Plan utilization Throughout the plan implementation process, it should be possible to understand if, and how, the plan is being used by politicians and professionals in decision making. The success of plan implementation depends on a clear identification by politicians and professionals and the document that frames the process of city building (an issue that has been extensively debated in literature since the seminal text by Forrester, 1989).

This perception on the use of the plan involves three different levels: i) the influence of politicians in plan making (in terms of processes and in terms of structures); ii) the effective use of the plan by politicians, and finally; iii) the use of the plan by the professionals over plan implementation and development control.

The last aspect raises the issue of the composition of planning teams It would be desirable that the local authorities promote the establishment of qualified staff that could effectively internalize a significant part of the planning process (without denying the possibility of working together with academics and the private sector, and to explore innovative planning approaches). As such, the local authority's professionals could participate in a more continuous way in this process. This would certainly contribute to coordination among professionals, the plan and the planning process.

Commitment of resources It should be possible for the local authority, in each particular moment of plan implementation, to understand if the human and financial resources that are needed to achieve the expected results are actually being involved, and eventually, to have the means to correct a negative situation. The emphasis placed on preparing a good plan should be followed by an equivalent effort for ensuring that the team and the financial means that are needed to support plan implementation are in fact mobilized. The evolution of the availability of resources over time, the type and diversity of resources that are available, and the relationship between planning practice and allocation of resources are crucial aspects. And so it is the way 'time' (as a valuable resource) is managed over the whole process of preparing and implementing the plan. The introduction of a timetable, with some flexibility, but also with a set of milestones will, undoubtedly, bring advantages to this process. An effort to identify those elements that might introduce disturbances (in terms of time) into the planning process should also be developed.

Effectiveness Overall, plan proposals are implemented through direct action of the local authority (through the development of the PDM proposals in lower order plans and in projects on the street network and on public spaces) and through indirect action involving the guidance of private activity throughout the process of development control. In both cases, it is desirable to monitor the implementation of plan proposals.

Direction of the urban development process The definition of a vision for the process of urban development is one of the fundamental purposes of the PDM. Although this is the most difficult (evaluation and design) criterion, in each moment, the planning department should be able to monitor plan implementation and to have a perception of the direction that the process of urban development is taking.

Application This section illustrates how PPR can be used as a planning methodology. It focuses on two specific criteria of PPR – internal coherence and relevance of the plan – and on one particular planning issue, urban form. To design a plan with high internal coherence, PPR proposes the definition of sound and place-based objectives; the acknowledgment of the objectives' central role within the plan and the planning process; and finally, the definition of how to measure the objectives achievement throughout the planning process. To prepare a relevant plan, PPR proposes a new perspective on the relationships between evidence and decision making, and sustains the need to develop specific planning support techniques to deal with particular aspects of planning.

The *Morpho* technique has been designed to deal with urban form (see Oliveira, 2013, for the formulation of *Morpho*, and Oliveira and Silva, 2013, for a description of its application to Porto). It focuses on the fundamental elements of urban form: streets, plots and buildings. It includes seven parameters: one for each of the elements individually; one for each relation between pairs of elements; and finally, one linking form and function. The application of *Morpho* provides a quantitative assessment of the 'morphological basis' of an urban area, identifying its strengths and weaknesses.

This exercise draws on the assumption that objectives must assume an increasingly central role in the plan, and that the different parts composing the plan should contribute more clearly to achieve these objectives. How could *Morpho* contribute to the achievement of one (or more) objective of PDM?

The most obvious contribution of Morpho is on the achievement of Objective 1, the enhancement of urban identity of Porto. According to the PDM this enhancement should be done through the dynamic conservation of existing tissues and the design of new and coherent tissues, the control of urban densities, and the safeguarding and promotion of the built heritage and of the image of the city.

It is argued that it should be possible to measure, throughout the implementation of the plan, the degree of achievement of objectives (or a deviation from these, in order to design alternative solutions). Accordingly, it was assumed that the enhancement of the urban identity of Porto should be based on the enhancement of the quality of the morphological basis of the city. It was also assumed that the progressive realization of this objective can be measured by comparing the results of the morphological analysis of: i) the city, and ii) each planning proposal.

Figure 8.4 shows a comparison, for a specific part of the city, between the results of the analysis of the integration of existing streets and the results that would be obtained after the construction of the streets proposed by the plan for this part of city. The analysis reveals that the realization of the plan proposals would result in increased average integration (impacts are calculated considering the city is taken as a whole).

Conclusions

Drawing on a description of PPR as an evaluation methodology, this chapter explored the possibility of using the Plan-Process-Results as a framework and as a methodology for planning, using the evaluation criteria as design criteria, and investigating the integration of new methods and techniques. The main goal of this

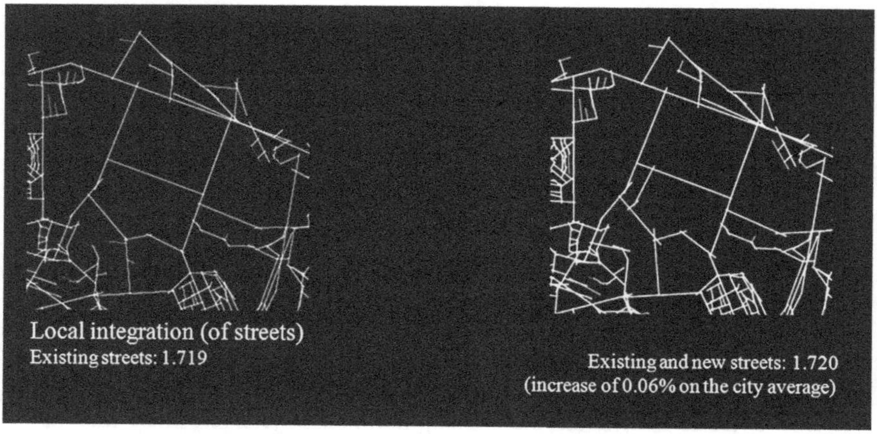

Figure 8.4 Application of *Morpho*: Integration of streets

recent development of PPR – designed in close collaboration with practitioners from the planning department of Porto local authority – is to offer a methodology that is able to directly shape the design of plans and of planning practices that are being prepared in Portugal.

Although it is deeply grounded in the Portuguese planning context, where the *Plano Director* Municipal is the main planning instrument, the framework offered by PPR has valuable insights for plans prepared in other geographical contexts. Planning practice reviews developed by Alexander (1998), Carmona and Sieh (2005), Khakee (2003) and Seasons (2003) make us believe that the planning activity developed in Portugal shares a number of weaknesses and threats with professional practice carried out in many different countries. It should also share a line of research and practice aiming at producing better plans, planning processes and results. The design of the Portuguese planning system and of the PDM is strongly influenced by the French system and by the *Plan d'Occupation des Sols* (POS). It is our strong believe that, in the short term, PPR has the conditions to be tested in the context of the French planning system as well as in other countries that are part of the so-called 'Napoleonic legal family' (Newman and Thornley, 1996).

The development of PPR, and the monthly meetings with the planning practitioners, highlighted the need to produce plans: with a higher simplicity (plans that invite citizen participation and that use a community-based vocabulary more accessible for the public) and coherence between their different parts; where goodness in relation to the cities and societies they address can be easily found; that take the planning system as the basis for the design of creative and innovative processes, and have the ambition to improve the system itself; that reflect the aspirations of politicians, professionals from the local authority, and of a wide variety of public and private stakeholders; that are able to build a consistent path linking the document proposals to their effective implementation 'on the ground';

and finally, that can make the best use (in difficult financial times as the ones we are living) of human resources, of financial resources, and of time.

The improvement of PPR has also revealed the need to develop specific methods and techniques to address particular planning issues, as opposed to those techniques that can be applied to deal with a variety of issues (e.g. SWOT analysis, impact matrices). That is the case of *Morpho* (the previous section exemplified its use in the evaluation of internal coherence and of relevance), applied in Porto under the umbrella of the PPR, dealing with the physical form and structure of the city. Finally, the application of PPR to the PDM revealed that some criteria are easier to evaluate than others. One of the main challenges for future applications of PPR should be the improvement of four criteria: relevance, plan utilization, commitment of resources and direction of the urban development process.

References

Alexander, E. 1998. 'Evaluation in Israeli spatial planning'. In N. Lichfield, A. Barbanente and D. Borri, et al. (eds) 1998. *Evaluation in Planning: Facing the Challenge of Complexity*. Dordrecht: Kluwer, pp. 299–310.

Alexander, E. 2006. Dilemmas in evaluating planning, or back to basics: What is planning for? *Planning Theory & Practice*, 10, pp. 233–44.

Alexander, E. and Faludi, A. 1989. Planning and plan implementation: Notes on evaluation criteria. *Environment and Planning B: Planning & Design* 16, pp. 127–40.

Alterman, R. and Hill, M. 1978. Implementation of urban land use plans. *Journal of the American Institute of Planners*, 33, pp. 274–85.

Altes, W. 2006. Stagnation in housing production: Another success in the Dutch planner's paradise? *Environment and Planning B: Planning & Design*, 33, pp. 97–114.

Baer, W. 1997. General plan evaluation criteria. *Journal of the American Planning Association*, 63, pp. 329–44.

Berke, P., Backhurst, M. and Day, M., et al. 2006. What makes plan implementation successful? *Environment and Planning B: Planning & Design*, 33, pp. 581–600.

Brody, S. and Highfield, W. 2005. Does planning work? *Journal of the American Planning Association*, 71, pp. 159–75.

Brody, S., Highfield, W. and Thornton, S. 2006. Planning at the urban fringe: An examination of the factors influencing nonconforming development patterns in Southern Florida. *Environment and Planning B: Planning & Design*, 33, pp. 75–96.

Brody, S., Carrasco, V. and Highfield, W. 2006. Measuring the adoption of local sprawl reduction planning policies in Florida. *Journal of Planning Education and Research*, 25, pp. 294–310.

Burby, R. 2003. Making plans that matter: Citizen involvement and government action. *Journal of the American Planning Association*, 69, pp. 33–49.

Calkins, H. 1979. The planning monitor: An accountability theory of plan evaluation. *Environment and Planning A*, 11, pp. 745–58.

Carmona, M. and Sieh, L. 2005. Performance measurement innovation in English planning authorities. *Planning Theory & Practice*, 6, pp. 303–33.

Chapin, T., Doyle, R. and Baker, E. 2008. A parcel-based GIS method for evaluating conformance of local land-use planning with a state mandate to reduce exposure to hurricane flooding. *Environment and Planning B: Planning and Design*, 35, pp. 261–79.

Deyle, R., Chapin, T. and Baker, E. 2008. The proof of the planning is in the platting: An evaluation of Florida's hurricane exposure mitigation planning mandate. *Journal of the American Planning Association*, 74, pp. 349–70.

European Commission. 1999. *MEANS Collection – Evaluating Socio-Economic Programmes*. Luxemburg: Official Publications Office of the European Communities.

Faludi, A. 2000. The performance of spatial planning. *Planning Practice & Research*, 15, pp. 299–318.

Faludi, A. 1987. *A Decision-Centred View of Environmental Planning*. Oxford: Pergamon.

Faludi, A. 2006. 'Evaluating plans: The application of the European Spatial Development Perspective'. In E. Alexander (ed.) 2006. *Evaluation and Planning, Evolution and Prospects*. Aldershot: Ashgate, pp. 119–43.

Faludi, A. and Altes, W. 1994. Evaluating communicative planning: A revised design for performance research. *European Planning Studies*, 2, pp. 403–18.

Forrester, J. 1989. *Planning in the Face of Power*. Berkley: University of California Press.

Fudge, C. and Barrett, S. 1981. 'Reconstruction of the field of analysis'. In S. Barrett and C. Fudge (eds) 1981. *Policy and Action: Essays on the Implementation of Public Policy*. London: Methuen, pp. 249–78.

Khakee, A. 2000. Reading plans as an exercise in evaluation. *Evaluation*, 6, pp. 119–36.

Khakee, A. 2003. The emerging gap between evaluation research and practice. *Evaluation*, 9, pp. 340–52.

Lange, M., Mastop, H. and Spit, T. 1997. Performance of national policies. *Environment and Planning B: Planning & Design*, 24, pp. 845–58.

Laurian, L., Day, M., Berke, P., et al. 2004. Evaluating plan implementation. *Journal of the American Planning Association*, 70, pp. 471–80.

Lichfield, N., Kettle, P. and Whitbread, M. 1975. *Evaluation in the Planning Process*. Oxford: Pergamon Press.

Mastop, H. 1997. Performance in Dutch spatial planning. *Environment and Planning B: Planning & Design*, 24, pp. 807–13.

Mastop, H. and Faludi, A. 1997. Evaluation of strategic plans: The performance principle. *Environment and Planning B: Planning & Design*, 24, pp. 815–32.

Mastop, H. and Needham, B. 1997. Performance studies in spatial planning: The state of the art. *Environment and Planning B: Planning & Design*, 24, pp. 881–88.

Newman, P. and Thornley, A. 1996. *Urban Planning in Europe*. London: Routledge.

Norton, R. 2005. More and better local planning. A state-mandated local planning in Coastal North Carolina. *Journal of the American Planning Association*, 71, pp. 55–71.

Oliveira,V. 2006. The morphological dimension of municipal plans. *Urban Morphology*, 10(2), pp. 101–113.

Oliveira, V. 2013. Morpho, a methodology for assessing urban form. *Urban Morphology*, 17, pp. 149–61.

Oliveira, V. and Pinho, P. 2010. Evaluation in urban planning: Advances and prospects. *Journal of Planning Literature*, 24, pp. 343–61.

Oliveira, V. and Silva, M. 2013. Morpho: Urban morphological research and planning practice. *Revista de Morfologia Urbana*, 1, pp. 31–44.

Oliveira, V., Silva, M. and Samuels, I. 2014. Urban morphological research and planning practice: A Portuguese assessment. *Urban Morphology*, 18(1), pp. 23–39.

Rivolin, U. 2008. Conforming and performing planning systems in Europe: An unbearable cohabitation. *Planning Practice and Research*, 23, pp. 167–86.

Seasons, M. 2003. Monitoring and evaluation in municipal planning. *Journal of the American Planning Association*, 69, pp. 430–40.

Stame, N. 2008. *Evaluation and Policy Implementation in Multi-Level Governance*. European Evaluation Society Biennial Conference, Lisbon, 1–3 October 2008.

Talen, E. 1996. After the plans: Methods to evaluate the implementation success of plans. *Journal of Planning Education and Research*, 16, pp. 79–91.

Talen, E. 1997. Success, failure and conformance: An alternative approach to planning evaluation. *Environment and Planning B: Planning & Design*, 24, pp. 573–87.

Wildavsky, A. 1973. If Planning is everything, maybe it's nothing. *Policy Sciences*, 4, pp. 127–53.

Chapter 9
Assessing Present Space with the Help of Future Scenarios

Abdul Khakee and Laura Grassini

Introduction

Space is a rich concept. Lefebvre's (1974) pioneering work distinguishing spatial practices (perceived space), representation of space (conceived space) and spaces of representation (lived space) inspired many post-modern geographers (Soja, 1989; Gregory, 1994) to show the limited perspective of space in professional practices including urban and regional planning.

In conventional spatial planning, besides the morphological aspects, the value of land is related to the development perspective more generally and production perspective in particular. Legislative devices to regulate the use of land are normally based on economic as well as social consideration. Planning evaluation has in this context followed conventional planning tradition. To paraphrase post-modern geographers, the concept of space in planning and planning evaluation is limited to some aspects of 'representation of space' (Soja, 2003).

Visual images play an important role in both traditional and new discourses on space. Photographs, text analyses, imaginary geography, semiotics and future scenarios are important tools in analysing representations of space. These methods have been used in spatial planning. However, they have not been used to combine all the three elements of space.

This chapter is an attempt to evaluate the richness of the space perception with the help of future scenarios. The use of scenarios in strategic planning in order to identify future possibilities and to make stakeholders aware of uncertainties has been increasingly recognized in planning research and practice. However the use of scenarios to assess current planning preferences has not received similar attention in planning evaluation. The participatory scenario-building workshop was a way to foster the perception of space that is neither intrinsically factual nor exclusively mental. It is instead – as we try to demonstrate – a complex perception of space, where past and present experiences merge with desires and concerns and unveil possible futures, which would be blind to pure scientific investigation. It is at the same time the outcome of mind and heart explorations, of individual and collective reasoning.

Besides the introduction, the chapter is divided into four sections. In the first one we briefly discuss the major theories conceptualizing the multiple perspectives of space. In the following section we make an attempt to appreciate the multiple aspects of space as a product of social, political and economic relations with the help of one case study, Izmir (Turkey). We show how future scenarios can provide deeper and richer appreciation of present space and thereby improve planning practice. The third section presents the empirical findings of the scenario exercise, and in the concluding section we discuss the relevance of the scenario method for evaluating the multiple space perspectives.

Theorizing Space Perception

How to theorize space and how to represent it has been a highly contested issue during the twentieth century and has aroused a strong debate among geographers and social scientists, with considerable echoes in planning theory and practice. Geographers have traditionally treated space as a 'static object' to classify and measure. What ancient cartographers and colonial explorers traditionally did, i.e. mapping and measuring space with the most accurate means they had at their disposal, a majority of quantitative geographers more recently did with the help of new technological advancements in remote sensing and geographical information systems. The pervasive identification of the world with a map and the underlying simplistic Euclidean conception of space strongly influenced mainstream planning (Friedmann, 1992), which often restricted its scope to the definition of rational means to pursue given objects (Graham and Healey, 1999). Models of urban agglomerations and regional growth, notions of central place hierarchies, studies on urban morphology were some new ideas that were adopted in planning in its attempt to develop the society towards a progressive and rational development (Cheshire and Evans, 1991). Cities and regions were planned in such a way as to create the appropriate physical infrastructures to support efficiency and growth, social and moral order (Harvey, 1996).

The increasing attention to global processes (Sassen, 1991, 1994) and the appreciation of the impact of intertwining between global and local dynamics of change on territorial reconfiguration (Brenner, 1999; Swyngedouw, 1992) added further complexity to the analysis of space and led to the redefinition of the urban question as a scale question. While the discovery of the existence of a 'space of flow' shaping the global information network (Castells, 1996) challenged the prevailing Euclidean space concept and led to the definition of the new geography of space-time compression, actual tools for planning evaluations did not change much. Concepts like proximity and distance, core and periphery, central locations, were called in doubt by the emergence of 'network', space-time and other concepts, so planners were increasingly compelled to take into consideration in their regional plans global forces shaping broader social and economic transformations (Amin and Thrift, 1994). Despite these challenges, the concept of space within mainstream geography and in planning remained bound to the objectivist perception of space, divorced from cultural or social order and whose attributes needed to be analysed and then shaped by rational means (Gregory, 1994; Graham and Healey, 1999).

Theories elaborated within Marxist geography could only partly challenge this static conception of space as their main driving historical forces, i.e. class struggles, were only accidentally linked to space issues. In the words of Soja both Marxist and liberal traditions fuelled de-spatializing historicism and thereby 'successfully occluded, devaluated, and depoliticized space as an object of critical social discourse (so) that even the possibility of an emancipatory spatial praxis disappeared from view for almost a century' (Soja, 1989, p. 4).

Interesting challenges to this limited conception of space came from the Marxist geographer, David Harvey, who underlined the importance of focusing on the relationship between space, time and society. For him, space was not only an outcome of social actions but it was a dialectic force in the social realm and not autonomous from main historical and social forces shaping society (Harvey, 1973, 1985). It somehow internalized and translated experiences, structures and power relations, but at the same time it was part of a process of social reproduction. While the history of cartography and spatial measurement from feudalism to capitalism has been part of a power system aimed to clarify property and territorial rights in the interests of political and economic power, Harvey prompted the development of new 'cartographies of resistance' rooted in the political translation of a new materialist relational theory of space and time (Harvey, 1996).

Foucault further criticized historicism as 'the great obsession of the nineteenth century', with its issues about crisis and cycle, ever-accumulating past, and its great preponderance of dead men (Foucault, 1986). Instead he shared the structuralists' claim of the 'spatialization of history' as the making of history entwined with the social production of space, the structuring of a historical geography. In his book *Power/Knowledge* he stated that 'A whole history remains to be written of spaces – which would at the same time be the history of powers (*both of these terms in the plural* – our remark) from the great strategies of geopolitics to the little tactics of the habitat' (Foucault, 1980 cited in Soja, 2003, p. 21). In so doing, he tried to reassert a deeper concept of space within critical social theory, overcoming the traditional perception of space and uneven geographical development as a product of modernity and not as a structural force in itself.

One of the major impulses on the new conceptualization of space came from the works of the French social philosopher, Henri Lefebvre, who emphasized the decisive and pre-eminent role of spatial structural forces in modern capitalist society (Lefebvre, 1970). In his book, *The Production of Space* (1974) Lefebvre presented a critical theorization of the social production of space. He distinguished three types of space: a 'physical' space, as the locus of 'spatial practice' objectively decipherable from its physical structures; a 'mental' space, as the embodiment of all subjective 'representations of space', including logical and formal abstractions of scientists, planners, technocrats; a 'social' space, as the 'space of representations', as the space of a 'complex symbolism, sometimes coded, sometimes not', linked also to 'clandestine or underground side of social life', including art. This third concept of space is thus important for Lefebvre as it breaks the 'double illusion' of both objectivist and subjectivist conceptions of space in favour of a 'lived' space, and underlines a more complex social approach to space perception, linked to abstract and non-verbal communication more than to science.

Lefebvre's ideas inspired a whole range of research on the intertwining of space and social practices and prompted a deep restructuring of the concept of space far beyond the field of social sciences. His theorization of space paved way for much of the critical postmodern thinking in general and in post-modern geography in particular. Following Lefebvre, Soja elaborated further what he calls 'thirding' process, as the identification of a 'Third space: the space where all places are, capable of being seen from every angle, each standing clear; but also a secret and conjured object, filled with illusions and allusions, a space that is common to all of us yet never able to be completely seen and understood' (Soja, 1996, p. 56). According to him, this is 'the terrain for the generation of "counterspaces", spaces of resistance to the dominant order arising precisely from their subordinate, peripheral or marginalized positioning [...] as a *strategic location* from which to encompass, understand, and potentially transform all spaces simultaneously' (Soja, 1996, p. 68).

Interesting contributions to enrich the concept of space by linking it to social and cultural practices also came from research made by sociologists and anthropologists in the second half of the twentieth century. Among them, de Certeau elaborated on the distinction between 'geometrical space' and 'anthropological space' (de Certeau, 1984) and stressed the importance of making sense of space through everyday life actions instead through maps. Just as a speaker appropriates and takes on language, it is through walking tours across the urban space that pedestrians make sense of the city. 'It is a spatial acting out of a place (just as the speech-act is an acoustic acting out of the language)' (de Certeau, 1984, p. 98). Space is thus for de Certeau a social concept that is understandable only by reading everyday practices and through spatial tactics people use to appropriate space and give it various meanings.

Bourdieu introduced the concept 'habitus' as the mediating concept between space and social practices. It is a system of dispositions that relates to our perception of the world and its structures and influences the way we act and interact with people (Bordieu, 1972). It thus implies that a web of complex processes links the physical, the social and the mental (Hillier and Rooksby, 2002). These issues were further developed by Giddens, whose major contributions on structure and agency led to the development of the structuration theory (Giddens, 1979, 1984). Its impact on the concept of space was significant because space could be defined as both a medium and an outcome of social action.

These discourses imply that space has been increasingly conceived of as contextual and culturally embedded in shared experiences and interpretations; at the same time the creation of specific kinds of spaces a result of complex spatial practices. The inseparable nature of social and spatial structures has been increasingly accepted by scholars in geography and sociology since the 1980s

(Massey, 1984; Gregory and Urry, 1985; Soja, 1989). For planners and planning evaluators the challenge is how to interpret the multiple aspects of space. How can these new physical, mental and symbolic perceptions of space be used in planning changes in cities and regions? The following table summarizes the various perceptions of space, and some of these are used in our case study.

Table 9.1 Space conceptualization

Numerical representation	Approach to space examination	Lefebrev's denotations	Substantive aspects	Planning relatedness
First space	Perceived space	Spatial practices	Spatial patterns of everyday life Analogical	Measured and described within limits. Traditional focus of attention in spatial disciplines
Second space	Conceived space	Representation of space	Tied to relations of production but also utopian and visionary thinking. Cosmological	Space conceptualized by planners, urbanists, technocrats
Third space	Lived space	Spaces of representation	Complex symbolic imaginary ideological Counter-hegemonic	Not used in planning. Increasing significance in political activism

Scenario for Space Evaluation

Methodological Reflections

Scenario is often regarded as the most essential aspect of futures studies. In fact there is often a tendency to equate futures studies with scenarios. This is not entirely correct. However, one cannot but agree with Wendell Bell when he writes that a scenario is the 'end product of all methods of futures research', that it gives 'methodological unity to futures studies' and that 'scenarios can be produced by any and all the specific methods used by futurists' (Bell, 1997, pp. 316–17). The latter implies that scenarios can be based on highly quantitative methods with a high level of expert or professional input. They can just as well be based on qualitative methods requiring intuitive reflections from the participants. In the case of participatory scenarios a broad range of stakeholders are involved in producing images of the future (Khakee, 1999). They produce these images from their own experience, with the help of fantasy and imagination and by deliberately combining, rearranging and adapting ideas. This requires divergent as well as convergent thinking (Parnes, 1992). Suitable techniques for creating images and enhancing them are brainstorming and brainwriting (May, 1996). While brainstorming is a process of oral sharing of ideas (Osborn, 1957), brainwriting involves a group of people silently writing and sharing their written ideas (VanGundy, 1983). Nevertheless, actual techniques used for generating creative ideas within organizations often entail a combination of the oral brainstorming and silent brainwriting as they may involve group discussion of written ideas during the idea-generation process (VanGundy, 1983). In participatory scenario writing, it is often the case of combining these two methods – a kind of 'brainstorm-writing'. In this chapter we shall use the term 'brainstorming' to imply both activities.

Two essential principles of these methods are to encourage creative thinking and postpone judgmental thinking until a later stage in brainstorming process, and to foster as many ideas as possible. The first principle requires that 'criticism and evaluation during the generation of ideas is ruled out' and expression of 'wild and wacky ideas' is encouraged. The second principle requires rapid flow of ideas from participants and building on and improving ideas within the group are encouraged (May, 1996, pp. 191–2). In brainstorming, contributions of all participants are recorded and displayed in order to meet these requirements.

Thus brainstorming is a creative process. It stimulates towards interactive learning and mutual appreciation of ideas. At the same time the method allows participants to work in a group as well individually. Traditionally, phases are carried out through oral brainstorming sessions with the help of handwritten notes taken throughout the interaction. The use of groupware products allows electronic brainstorming, by setting a platform for idea organization and even enabling multiple criteria analysis.

The Izmir Study

The Izmir study was part of the Concerted Action's R&D project entitled 'Agriculture and Urbanisation in the Mediterranean Region' financed by the European Union. The general objective was to co-ordinate research efforts on sustainable use of natural resources, in particular land and water in the Mediterranean region. In the case of Izmir the more specific objective was to analyse the impact of tourism on coastal zones and provide strategies for coastal zone management. The major focus of the Concerted Action's work was on participatory scenario building in local/global policy making involving private and public stakeholders. The basic idea was to be proactive in order to encourage stakeholders to create their own images of the future and design strategies to realize these images. The specific approach in the case of Izmir was Future Workshop (Jungk and Mullert, 1996; Khakee et al., 2002) which was further adapted to the case study as described in the following paragraph.

Future Workshop aims at building scenarios, with a strong commitment to action. The method allows the involvement of non-experts in designing future because it does not require any specific professional knowledge. It provides outlets for participants' creativity and has at the same time a clear orientation towards action and the building of shared visions. It is a collaborative approach encouraging interactive learning in the methodological set-up and in the derivation of the substantive results. Future Workshop is flexible and can be easily adapted to different situations with regard to time, resources and other contextual requirements. (See next section for the various steps of Future Workshop.)

Future Workshop in Izmir

In its classical form, the Future Workshop starts with a preparatory phase followed by three operative phases: critique, fantasy and implementation. The preparatory phase is devoted to the definition of the issue and the practical organization of the brainstorming sessions. An underlying assumption with future envisioning is that there is a general dissatisfaction with something in the present and a strong desire to change the current premises of development (Ziegler, 1991). In the critique phase, participants investigate the issue selected by focusing on problems and negative experiences as well as positive premises of the present situation. In the fantasy phase they are then asked to disregard temporarily current limitations, restrictions and premises, and to generate free ideas in the form of desires, dreams, fantasies and opinions concerning the future. In the implementation phase, the participants are asked to go back to the current reality, with its power structures and constraints and analyse the actual feasibility of the ideas put forward in the previous phase and possible ways that seem to be available to overcome them (Jungk and Mullert, 1996).

The procedure designed for the Izmir case involved some changes in the Future Workshop approach in order to structure the issues relating to the environmental and socioeconomic impact of tourism, holiday homes and clandestine land-use in the coastal areas of Izmir Province. In particular, the first two phases were prolonged over a period of four months. Many of the issues can be described as 'wicked problems' that were affected by many interrelated factors. Problem-structuring takes time if they are to be amenable to scenario analysis (see e.g. Schnaars, 1987).

The critique phase started with a questionnaire survey. Stakeholders were asked to describe three major structural changes that have occurred in the Province of Izmir in the last 10 years and three major problems that characterize the prevailing situation. A modified version of Delphi method was applied in order to acquire a common knowledge base on structural changes and current problems. The results of the survey were then compiled in a memo that was distributed to the stakeholders for collective discussions.

The fantasy and implementation phases were carried out in the form of brainstorming sessions over two consecutive days. The time horizon of the scenarios was 2030. The participants were divided into two groups; one worked entirely in the 'traditional' way, through oral brainstorming, whereas the other group was to make use of a computer programme for electronic brainstorming (Khakee, et al. 2002). The group, that used the traditional way worked with removable self-stick notes and flip-charts. Participants wrote their ideas and proposals on self-stick notes (individual brainstorming) and pasted them on the flip-charts. Then they gathered around these flip-charts for discussions and make additional comments and organized the ideas and proposals in various policy areas (collective brainstorming).

The software platform, *MeetingWorks©* for Windows, for electronic brainstorming includes not only tools for brainstorming and idea organization but also for ranking, voting, cross impact analysis, and multiple criteria analysis.

The implementation phase was devoted to transform visions into operational strategies. Thus the participants were solicited to reflect on the obstacles that would impede getting their group's visions being implemented, to single out measures to remove these obstacles and mobilize necessary resources.

Empirical Findings

The stakeholders were looking for a scenario for the sustainable management of coastal resources. The expectation was the generation of a scenario depicting an *eco-city* or *green city*. But in fact the stakeholders developed the image of Izmir as a *democratic city*, insofar they did not relate the achievement of sustainable management of the city to changes in industrial and services production or to the implementation of technical changes decided by planners and technocrats; instead, they rooted changes in deep modifications in grassroots politics and everyday life.

'Democratic city' is a contested concept and there are few agreed upon criteria for defining a democratic city (Buser, 2013). The disagreement has become even more manifest with the recent shift from government to governance. The notion of democratic city has been given a large number of characteristics including accountable institutions, informal decision making, multisector-collaboration, horizontal networks, civic involvement and other collaborative measures (Gross and Hambleton, 2007). Under such circumstances interactive knowledge-producing activities like Future Workshop stick to characteristics that participants themselves propose as they have done below.

Reduction of air and water pollution as well in household refuse and industrial waste, monitoring of environmental policies, improvement of urban environments are not part of formal policy measures but they are a result of neighbourhood cooperation, voluntary efforts, people's sensitiveness to environmental issues and their active engagement in decision making and democratic changes. This example shows how conventional perceptions of space are subsumed political processes and struggles to define and enforce social realities (see e.g. Philip Fisher's discussion of 'democratic social space'). Also Foucault's (1980) concern in interrogating space as both materiality and ideology is as a means of understanding how power is constituted. These ideas transcend geographers' notions of space and place as we have discussed in an earlier section in this chapter.

Though the two groups worked separately with two different techniques they arrived at a *desired* scenario that was similar in its emphasis on democratic space. In our analysis below we present the major findings of the group that followed the traditional brainstorming approach.

Table 9.2 Vision of a democratic city and spatial interpretations

The current issues	The scenario	Spatial assessment
Izmir lacked an effective and sustainable management of the coastal zone in face of uncontrolled migration from rural areas, unconstrained exploitation of land for holiday housing for Turkish households living abroad and for hotels and other tourism establishments.	In 2030 Izmir is the democratic city in Turkey. Grassroots politics dominate the political system. Politicians are elected on the basis of their attachment to neighbourhoods and districts. Politicians and bureaucrats are accountable to people in the districts. Public policy is territorially integrated reflecting local conditions and requirements. 'Rolling planning' characterize land-use development and environmental plans and they are changed as local conditions change. Family planning and immigration controls are based on districts' carrying capacity. Voluntary efforts, neighbourhood cooperation and increased awareness among people have resulted in a desired reduction of air and water pollution as well in household refuse and industrial waste. Monitoring of environmental policies is carried out by district councils and local NGOs. Izmir is envisioned by outsiders as a friendly city characterized by inter-personal trust, helpfulness and sincerity.	Instead of an obvious vision in the form of 'green city' or 'eco-city' the Stakeholders selected 'democratic city' routed in grass-roots politics. Stakeholders emphasize ideological and counter hegemonic aspects of space as the ideal way to achieve sustainable management of the city. Sustainability issues are not tied to relations of production as would have been conceptualized by planners and technocrats but to day-to-day discourse. Sustainability is a question of everyday life. The stakeholders imagine sustainable management in territorial terms rather than in terms of policy areas. Grassroots, neighbourhoods, NGOs, are the major elements in such management. The latter is a spatially dynamic process where dynamism is related to the changing local conditions, preferences and concerns. [continued overleaf]

The current issues	The scenario	Spatial assessment
	Urban environment is distinguished by low-rise buildings, open public spaces, plenty of meeting places – markets, gardens, playgrounds and recreational facilities.	The stakeholders see an obvious interplay between system and life world – between conceived and lived space. For them the spatial patterns of everyday life are tied to more formal relations of political economy and social development. People's sensitiveness to environmental issues, control of population growth, monitoring of environmental policies are tied to local places' carrying capacity. These issues do not easily amend to measuring, describing and forming policy measures but are part of both spatial practices and spaces of representation.

If we examine some specific statements from the scenario we find similar results (see Table 9.3 below).

Table 9.3 Specific aspects of the scenario and their spatial interpretation

Issue	Scenario statement	Spatial interpretation
Haphazard and uncontrolled exploitation of land for tourism, holiday homes and industries	Local conditions and grassroots' politics determine appropriate use of land	Spatial patterns of daily life, utopian and visionary thinking together with ideology and counter-hegemonism combine various aspects of perceived, conceived and lived space in this as well as the other two scenario statements described below.
Local government is arbitrary, un-modern and maintains status quo in policy making	Local government is pro-active and attentive of local conditions. Policy making is based on local knowledge.	The representations of space are part of a political process subsumed to prevailing power relations. By changing these power relations, prevailing space representations can be altered.

Issue	Scenario statement	Spatial interpretation
School education, social and cultural services and public amenities are in poor shape.	As a result of decentralization of economic and training resources to districts and neighbourhoods education, social and cultural services are vibrant and dynamic.	Restructuring educational, social and cultural practices is not primarily a public management issue. It requires restructuring of spatial political relations through decentralization.

Concluding Reflections

The use of scenarios in planning evaluation enables us to perceive the social processes that constitute the space. The case study shows how rich the conception of space is and that the representations of space need to transcend institutionalized notions of space. It calls for more imaginative production of space.

Soja (1989) has often called for the substitution of space-based analysis for time-based analysis. Ligget and Perry (1995) put the question: How the two – space-based and time-based analyses – can be joined. The case study shows how these two forms of analysis can be successfully combined. It also provides a valid support of urban imagining, making use of novel approaches like scenarios in planning and planning evaluation.

In conventional spatial planning the emphasis is on empirical studies examining various problems currently existing in the use or misuse of space. The emphasis in such a scrutiny is on past and current trends in development. The emphasis on current problem conception hinders more innovative and completely different uses of space. Our case study shows that the importance of the three space concepts is not only in their richness but also in the way they are produced, through collaborative workshops and discursive processes. The participatory scenario is a means to employ and operationalize a more relational concept of space in real planning strategies and governance processes. This is important as far as current practices of strategic planning fall short in the actual restructuring of traditional planning approaches and a relational understanding is underdeveloped and often displaced by more traditional ways of seeing place/space and governance process (Healey, 2004, 2006).

Visualising spatial transformation with the help of scenarios or other forms of futures studies redresses some of the built-in restrictions in the perception and subsequently in the transformation of space. Visionary knowledge provides a challenge and paves way for innovative spatial planning. Given the richness and originality of the scenario produced in the Izmir case, the scenario-building approach seems to be more useful and relevant for the real cognitive and behavioural attitudes of participating agents. It seems to have stimulated a higher adherence to real life complexity and social expectations. For researchers in planning evaluation the use of scenario poses a challenge for assessing the multiple aspects of space.

References

Amin, A. and Thrift, N. (eds) 1994. *Globalization, Institutions, and Regional Development in Europe*. New York: Oxford University Press.

Bell, W. 1997. *Foundations of Futures Studies. Human Science for a New Era.* Vol. 1. New Brunswick: Transaction Publishers.

Bourdieu, P. 1972. *Esquisse d'une théorie de la pratique. Précédé de trois études d'ethnologie kabyle*. Genève: Droz (English version: *Outline of a Theory of Practice.* Cambridge: CUP, 1977).

Brenner, N. 1999. Globalisation as reterritorialisation: The re-scaling of urban governance in the European Union. *Urban Studies*, 36(3), pp. 431–51.

Buser, M. 2013. Democratic accountability and metropolitan governance: The case of South Hamshire. *Urban Studies*, doi: 10.1177/004209801348531.

Castells, M. 1996. *The Information Age: Economy, Society and Culture. Volume 1: The Rise of the Network Society*. Oxford: Blackwell.

Cheshire, P. and Evans A.W. (eds) 1991. *Urban and Regional Economics.* Cheltenham: Edward Elgar.

de Certeau, M. 1984. *The Practice of Everyday Life*. Berkeley: University of California.

Foucault, M. 1980. *Power/Knowledge: Selected Interviews and Other Writings, 1972–1977*. Brighton: Harvester.

Foucault, M. 1986. Of other spaces. *Diacritics*, 6(1), pp. 22–7.

Friedmann, J. 1992. *Empowerment: The Politics of Alternative Development.* Cambridge: Basil Blackwell.

Giddens, A. 1979. *Central Problems in Social Theory: Action, Structure and Contradiction in Social Analysis*. Los Angeles: University of California Press.

Giddens, A. 1984. *The Constitution of Society.* Cambridge: Polity Press.

Graham, S. and Healey, P. 1999. Relational concepts in time and space: Issues for planning theory and practice. *European Planning Studies*, 7(5), pp. 623–46.

Gregory, D. 1994. *Geographical Imaginations*. Oxford: Blackwell.

Gregory, D. and Urry, J. (eds) 1985. *Social Relations and Spatial Structures.* Basingstoke: Macmillan.

Gross, J.E. and Hambleton, R. 2007. *Governing Cities in a Global Era. Urban Innovation, Competition and Democratic Reform.* New York: Palgrave.

Harvey, D. 1973. *Social Justice and the City*. London: Edward Arnold.

Harvey, D. 1985. *Consciousness and the Urban Experience.* Baltimore: Johns Hopkins University Press.

Harvey, D. 1996. *Justice, Nature & the Geography of Difference.* Maiden, MA and Oxford: Blackwell.

Hayden, D. 1997. *The Power of Place: Urban Landscapes as Public History.* Cambridge, MA: MIT Press.

Healey, P. 2004. The treatment of space and place in the new strategic spatial planning in Europe, *International Journal of Urban and Regional Research,* 28(1), pp. 45–67.

Healey, P. 2006. Relational complexity and the imaginative power of strategic spatial planning. *European Planning Studies*, 14(4), pp. 525–46.

Hillier, J. and Rooksby, E. (eds) 2002. *Habitus: A Sense of Place*. Aldershot: Ashgate.

Jungk, R. and Mullert, N. 1996. *Future Workshops. How to Create Desirable Futures*. London: Institute for Social Inventions.

Khakee, A. 1999. Participatory scenarios for sustainable development. *Foresight*, 1(3), pp. 229–40.

Khakee, A., Barbanente, A., Camarda, D. and Puglisi, M. 2002. With or without: Comparative study of preparing participatory scenarios for Izmir with computer-based and traditional brainstorming. *Journal of Futures Studies*, 6(4), pp. 45–64.

Lefebvre, H. 1970. *La Révolution urbaine*. Gallimard: Paris (English version: *The Urban Revolution*. Minneapolis: University of Minnesota Press, 2003).

Lefebvre, H. 1974. *La Production de l'espace*. Anthropos (English version: *The Production of Space*. Oxford: Blackwell, 1991).

Liggett, H. and Perry, D. 1995. *Spatial Practices*. Thousand Oaks, CA: Sage

Massey, D.B. 1984. *Spatial Divisions of Labour: Social Structures and the Geography of Production*. Basingstoke: Macmillan.

May, G.H. 1996. *The Future is Ours. Foreseeing, Managing and Creating the Future*. Westport, CT: Praeger.

Osborn, A.F. 1957. *Applied Imagination: Principles and Procedures of Creative Problemsolving*. New York: Charles Scribner's Sons.

Parnes, S.J. 1992. *Visioning: State-of-the-Art Process for Encouraging Innovative Excellence*. Buffalo, NY: Creative Education Foundation Press.

Sassen, S. 1991. *The Global City: New York, London, Tokyo*. Princeton: Princeton University Press.

Sassen, S. 1994. *Cities in a World Economy*. Thousand Oaks, CA: Pine Forge Press.

Schnaars, S.P. 1987. How to develop and use scenarios. *Long Range Planning*, 20(1), pp. 105–14.

Soja, E.W. 1989. *Postmodern Geographies: The Reassertion of Space in Critical Social Theory*. London and New York: Verso.

Soja, E.W. 1996. *Thirdspace: Journeys to Los Angeles and Other Real-and-Imagined Places*. Oxford: Blackwell Publishers.

Soja, E.W. 2003. Writing the city spatially. *City*, 7(3), pp. 269–81.

Swyngedouw, E. 1992. 'The mammon quest: "Glocalisation", interspatial competition and the monetary order: The construction of new scales'. In M. Dunford and G. Kafkalas (eds) 1992. *Cities and Regions in the New Europe*. New York: Belhaven Press.

VanGundy, A.B. 1983. Brainwriting for new product ideas: An alternative to brainstorming. *Journal of Consumer Marketing*, 1, pp. 67–74.

Ziegler, W. 1991. Envisioning the future. *Futures*, 23(6), pp. 516–27.

Chapter 10
Spatial Synergies and Conflicts: Monitoring Government Policies and Programmes in England

Cecilia Wong, Brian Webb, Andreas Schulze-Baing,
Mark Baker and Stephen Hincks

Introduction

The spatial complexity of area-based urban initiatives in Britain was famously captured by the Audit Commission (1989, p. 1) as '*a patchwork quilt of complexity and idiosyncrasy*'. The 1990s had witnessed a trend towards more strategic thinking in co-ordinating regeneration activities in Britain; for instance, the Single Regeneration Budget was set up in 1994 to simplify the funding regime by bringing together a variety of programmes and initiatives from several government departments. Unfortunately, a decade later, the Labour Government regeneration minister, Lord Rooker, still described the chaotic policy maze as '*a bowl of spaghetti*' (Lords Hansard, 2003). In spite of the promise made by the Labour Government, the haphazard approach to policy planning and coordination continued throughout the 2000s.

The 2004 Planning and Compulsory Purchase Act (HM Government, 2004) introduced a series of reforms to the statutory planning system in England that were heralded as transforming the focus from a narrow, more regulatory, concern with 'land use' to a more holistic and integrated 'spatial' approach that emphasized critical thinking about space and places as the basis for land use intervention to encapsulate a broader meaning of planning concerned with 'place-making'. Better co-ordination and integration of policy, horizontally across different policy sectors and vertically across different spatial scales from the national to the local, was at the heart of this spatial planning vision. However, despite the promise of more joined-up policy, there remained a continuation of the historic disjunction between economic development and strategic planning across different spatial scales (Baker and Wong, 2012). The irony is that a single integrated regional strategy was eventually proposed to co-ordinate economic and land use planning policies (DCLG, 2007) towards the end of the Labour administration, but this was stillborn. The Coalition Government that came to power in 2010 very quickly revoked all regional planning strategies via the 2011 Localism Act which fundamentally shifts spatial policies from the previous top-down, target-driven approach towards an 'open source', local oriented style. While this gravity shift to a more localized approach provides opportunities for developing more contextualized planning, this can arguably only be achieved if there is also a clear and well-articulated national planning framework that, in turn, can provide the parameters within which local planning authorities can develop and deliver their planning policies.

The goals of spatial planning are by their nature broad, varied and complex and their implementation and delivery are heavily reliant upon the action of a plurality of actors and agencies across operationally independent policy sectors (Albrechts, 2001; Wong and Watkins, 2009). In a nutshell, spatial planning deals with what Rittel and Webber (1973) call wicked problems, embedded in a dynamic social context, which makes each problem unique but also difficult or impossible to solve. To Rittel and Webber, only by laying out alternative understandings of problems, competing interests, priorities and constraints, is it then possible to apply more formal analytical tools. The endeavour of monitoring and evaluating spatial planning has been largely focused on the challenge of identifying complex outcomes, but not the process of policy development and coordination in terms of competing interests and practices across different policy sectors. This is closely associated with the departmental silos across different government departments and agencies and the long-standing 'spaghetti bowl' syndrome. However, the emphasis on vertical and horizontal policy coordination with a spatial perspective has underpinned the rationale of spatial planning and should therefore be included in any formal policy monitoring process.

Since the start of the millennium, the Royal Town Planning Institute (RTPI) has commissioned three studies, conducted by the authors, to ascertain the spatial coordination of planning across different parts of Britain. The first study (Wong, Turner and Ravetz, 2000) used employment and housing as the case study sectors to illustrate the fragmentation of policy frameworks at all spatial levels and the second study (Wong, Schulze-Baing and Rae, 2006) examined the spatial structure and spatial connections across the UK. The latest study, *A Map for England* (Wong et al., 2012), aims to bring the two together by examining the interplay between policy needs and spatial contexts via mapping the government's policy and programme delivery. Since there is not a single place or data source within government that makes all of these maps available to view, 'A Map for England' is designed to help policymakers make better judgments about how individual policy proposals interact with, and affect, the development of the country as a whole through better spatial planning. Despite the advancement of GIS technology, there has been a long acknowledged gap between its research development and its application in planning practice (Vonk et al. 2005), which Batty (2004, p. 327) regards as 'the tragedy of the field'. The comments made by Gilfoyle and Wong (1998) in the late 1990s remain true today that the large majority of GIS applications in British planning remain lower order activities of cartographic outputs, rather than for analysis and decision making. The RTPI initiative has, however, been greeted with an enthusiastic response and associated debate from planners and the media. Interestingly, some politicians have subsequently drawn upon the findings of the study to argue their points in public government committee hearings (see RTPI, 2014).

Against this context, this chapter aims to serve two purposes. First of all, it considers whether there has been any concerted effort to improve the spatial coordination of policy during the two decades or more of policy experiments since the 'patchwork quilt' and 'spaghetti bowl' analogies. Secondly, rather than using traditional quantitative data or qualitative texts, this chapter demonstrates the analytical and visual impact values of employing GIS mapping overlays to identify the patterns of spatial synergies and conflicts arising from policies and programmes to serve as a robust monitoring tool for spatial policy coordination. To do so, the key planning challenge of housing delivery in England is used to illustrate how such a methodology can usefully be employed.

The Appraisal of Government Policies and Programmes

Before the commencement of any spatial mapping work, the study started with a systematic scanning exercise of policy documents and web sites of different government departments and their agencies and NDPBs (non-departmental public bodies). This exercise was undertaken in January 2012 and covered a total of 95 relevant sources (see Wong et al., 2012). These included documents and websites from the Department for Business, Innovation and Skills; the Department for Communities and Local Government; the Department for Environment, the Department for Environment, Food and Rural Affairs; the Department for Transport; the Department for Culture, Media and Sport; HM Treasury; Home Office; Cabinet Office; British Waterways; Environment Agency; and Natural England. The scanning exercise first sought to identify which of these national policy documents had a spatial content and which were essentially 'aspatial' in nature. Such an analysis is useful as it 'captures the representation of space in language, and reveals some of the power relations which contest these representations' (Jensen and Richardson, 2001, p. 704).

After appraising the 95 sources, just over a third (37) are found to have an explicit spatial expression and/or spatial consequences in terms of including maps, spatial diagrams, or with clear spatially-referenced data/text. The remaining two-thirds, nonetheless, consist of policies/programmes that have clear spatial consequences and outcomes but do not articulate such characteristics explicitly. Given the spatial consequences, explicit or otherwise, of a wide range of government policy, it is perhaps rather surprising that there is no single national strategy document that tries to provide any kind of integrated spatial framework, drawing on all these sectoral policies and programmes and how they cumulatively interact and affect the development of the country. Among the documents that shy away from making any spatial articulation, and can thus be categorized as 'aspatial' in nature, it is something of a concern to find not only the Coalition government's recent National Planning Policy Framework (NPPF) (DCLG, 2012), but also the majority of the earlier national Planning Policy Statements/Guidance (PPS/PPGs), largely lack spatial expressions. Yet these national policy documents are supposed to be an integral part of the spatial planning system, providing an overarching framework for spatial planning across England. This begs the question of how local planning can be expected to perform the coordination and management role of sustainable development when there is no obvious spatial framework to join up different sectoral policies across different regions in England, let alone across the UK as a whole.

Despite the fact that the NPPF is an 'aspatial' document, it is clear that government policies and programmes do have spatial implications. In some instances these are made explicit, such as in the case of the high speed rail network proposals and other projects set out in the National Infrastructure Plan (HM Treasury, 2011). Other mainstream government policies (e.g. on supply-side measures for tackling non-employment, on investment in flood risk prevention projects, etc.) also have implications for the spatial distribution of economic activity and development opportunities. Yet a clear understanding of these differential spatial operating contexts and potential outcomes is particularly important if local authorities are to successfully contribute to the UK Government's national objectives and planning policies (see DCLG, 2012) for sustainable development and securing economic growth, via their preparation of local plans and strategies.

Moreover, administrative boundaries at regional and local levels do not define functional entities. Economic, social and spatial processes do not stop at an administrative boundary. Rather, there are 'spill-over effects' reflecting inter-area linkages. This is perhaps most clear in the case of London, the South East and East of England regions, where all three regions are influenced by the role of London as a 'World City' and where a key policy area, the Thames Gateway, straddles regional boundaries. Elsewhere, there are also important functional (commuting and migration) links, for example, between parts of Derbyshire (in the East Midlands) and Sheffield (in Yorkshire and the Humber), and southern Milton Keynes. Certain sub-regions, such as High Peak and northern Cheshire, have strong linkages with the Greater Manchester Local Enterprise Partnership. Relevant information with regard to these spatial functional connections, taking account of issues such as strategic housing market areas and flood risk, can help to inform local authorities in developing more robust local spatial planning and growth strategies.

Mapping the Spatial Contexts for Future Planning

Planning is a future oriented activity and development across different local authority and partnership areas is very much affected by a combination of the physical constraints of landscape designation and emerging development trends, as well as the government's own policies and programmes that can exacerbate or reduce such trends. The designation of national parks and areas of outstanding natural beauty, as protected environments and landscapes within England, sets out physical restrictions on the development of land (see Figure 10.1). The longstanding planning tool of national green belt policy, which limits growth of the conurbations and major towns into their surrounding countryside, imposes further restrictions on development, often in areas exhibiting the greatest development pressures. This has led to controversial debates over whether there is a need to revise current green belt policy in order to release land for house building to address the pressurised housing markets, especially in the wider South East region that surrounds the Greater London conurbation. These restrictions also mean that the country's main urban centres are often tightly defined by the green belt and the scope for development in some regions is very limited. For example, potential development in the South West region is limited by environmental and green belt designations and, as a consequence, mainly focused within particular parts of the counties of Somerset and Devon. In contrast, the areas with less national land use restrictions tend to be located on the eastern side of England, such as within the East Midlands and Eastern regions and parts of Yorkshire and the Humber.

Besides landscape designations, the general socio-economic development patterns reveal very important contextual information that underpins the rationale and vision for spatial planning across the country. Figure 10.2 shows the distribution of gross value added per head as well as highlighting areas containing England's 20 per cent most deprived population.[1] This map clearly illustrates the long standing 'North-South' divide by drawing a line between the Severn and the Wash Estuary. This line encapsulates the socio-economic and political differences between different regions in Britain, mainly the southern regions against the rest. This broad generalization of an affluent south and a less well-off north does, however, mask major variations in these regions. While southern England, on the whole, enjoys higher GVA per capita, some of the southern coastal areas do have lower GVA and exhibit significant pockets of deprivation. It is also interesting to note the complex spatial phenomenon of London where affluence and deprivation co-exist in the same space.

Meanwhile, the situation in northern England shows that deprivation tends to go hand in hand with low levels of GVA in the major urban conurbations, especially in the Mersey belt (the urban areas running across the North West region from Liverpool in the west to Manchester in the east), the West Midlands, Tyne and Wear, Teesside and West and South Yorkshire. Many of these areas have suffered from the decline of their industrial legacy and many are previous coal-mining areas. There are, however, pockets of very affluent areas, such as

1 The 20 per cent most deprived areas are based on the value of the 2010 Index of Multiple Deprivation. The spatial unit of analysis is the Lower Super Output Areas (LSOAs), which are socially homogeneous areas. Each have an average of around 1,500 residents and 650 households.

Figure 10.1 Key landscape designations

Note: © Natural England, 2010, reproduced with the permission of Natural England. Data accessed from GIS Digital Boundary Datasets of Natural England, http://www.gis.naturalengland.org.uk.

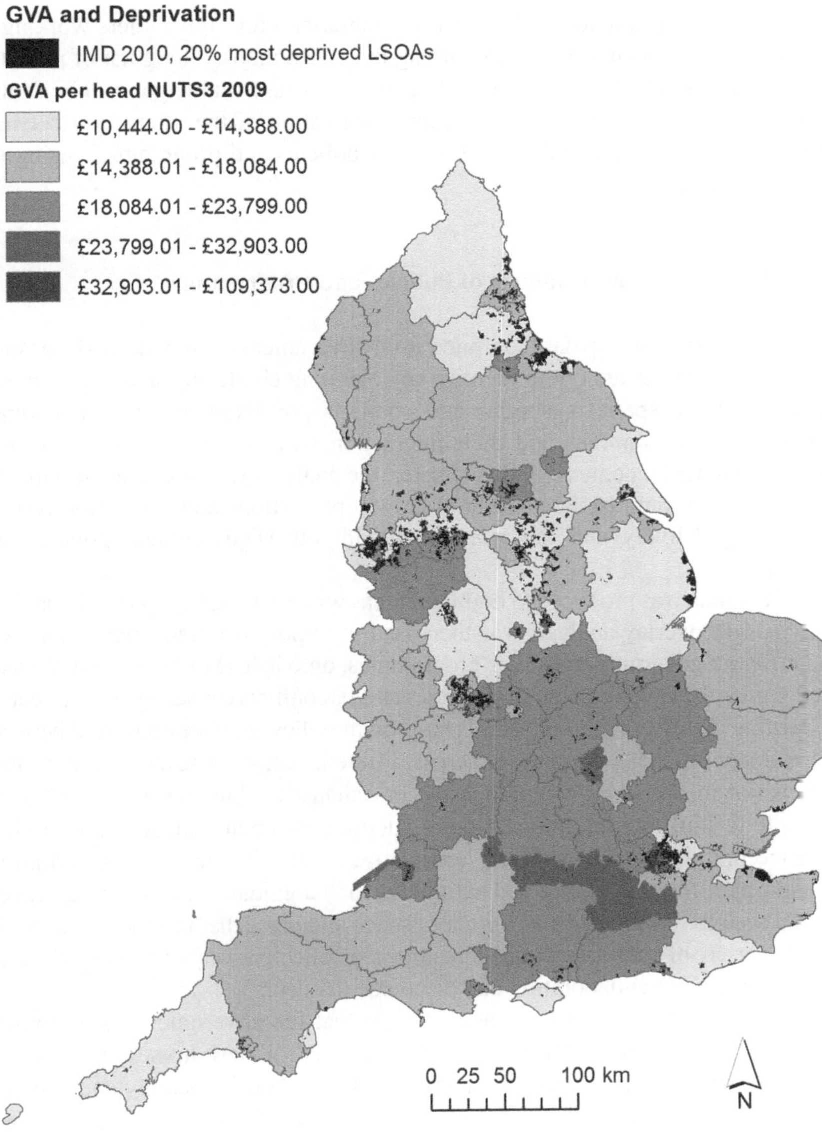

Figure 10.2 Gross value added and Index of Multiple Deprivation

Note: For the English Indices of Deprivation 2010, Department for Communities and Local Government, http://www.communities.gov.uk/publications/corporate/statistics/indices2010. For GVA, Office for National Statistics, http://data.gov.uk/dataset/gross_value_added_gva_per_head, Table 3.6 GVA per head (constrained to headline NUTS2) at current basic prices.

Cheshire which is a renowned commuting hinterland for high earners working in Liverpool and Manchester. By combining these two maps, a long list of potential spatial planning challenges has already emerged. These challenges will be further analysed in the next section through an examination of the interaction between these spatial contexts and the government's policies and programmes, using the case of housing delivery.

Spatial Synergies and Conflicts of Policies and Programmes

Based on the spatial map data compiled from government documents and websites, an analysis was carried out to identify key planning challenges and opportunities in terms of the spatial synergies and conflicts presented in different national policies and programmes and their interaction with the wider socio-economic and environmental context. In this chapter, the analysis focuses on examining the relationship of trend based household growth projections and the spatial context of housing delivery and the interaction with other government policies and programmes.

A core map was produced to highlight areas with very high projected household growth. GIS overlay analysis was then used to impose different spatial contexts, and different government policies/programmes, onto it. The map overlays are used predominantly to display spatial synergies and/or conflicts caused by the interaction of existing development trends and pro-growth policy interventions, and between government policies/programmes across different sectors. The analysis presented here is not meant to be comprehensive or exhaustive, but rather to serve as an example to illustrate the importance of adopting an open, guidance approach to enhance the understanding of the cumulative spatial impacts and opportunities brought by different policies and activities. Such a guidance or visioning process should seek to formulate a conception of space and place that is understandable by a wide variety of actors, such that it results in a collective way of 'seeing' (Healey, 2007) to facilitate collaboration and discourse to occur.

The NPPF (DCLG, 2012) makes it clear that the government's key housing objective is to significantly increase the delivery of new homes. However, it does not articulate the spatial distribution of future housing supply and demand. Instead, it requires local planning authorities (with neighbouring authorities where housing market areas cross administrative boundaries) to prepare a Strategic Housing Market Assessment (SHMA) to assess their full housing requirements in terms of the scale and mix of housing and the range of tenures. They are also required to prepare a Strategic Housing Land Availability Assessment (SHLAA) to establish realistic assumptions about the availability, suitability and the likely economic viability of land to meet the identified requirement for housing over the plan period. The outcomes of conducting SHMA and SHLAA are to ascertain the spatial requirements of different types of housing within the local plan. The National Infrastructure Plan (HM Treasury, 2011), however, goes further to articulate the inter-sector connections between new housing delivery and infrastructure provision. It argues that new housing has to be supported by infrastructure, and that infrastructure providers therefore need to have certainty over new housing delivery before making investment decisions. The National Infrastructure Plan makes reference to the land supply proposals of emerging local plans and the need for local authorities to work together across boundaries to properly plan for infrastructure provision (para. 4.24) as well as potential funding sources (para. 4.5). However, in spite of the emphasis on policy coordination of housing delivery and the provision of services, employment and infrastructure, there is not any overview of the spatial patterns of delivery at the strategic, national level across different government departments. The analysis here aims to illustrate this knowledge gap.

Areas with the highest projected household growth (above the 27 per cent average) during 2008–2033 are mainly found in commuting areas to major townships and urban areas. In Eastern England, high growth areas are found around the major townships of Ipswich, Colchester, King's Lynn, Cambridge and Northampton. High growth is also expected in the Ashford/Kent area and in part of London in the South East. Significant growth is projected in the South West Region: the M5 corridor near Bristol; and areas around Exeter and North Dartmoor. In the Midlands, major increase is projected around South Derbyshire, Warwickshire, and the High Peak area. When moving further north, there is a growth belt running from the Yorkshire Dales, Bradford, Leeds, Selby, York, West Lindsey, East Riding and Hull.

By simply overlaying the maps of physical constraints and projected household growth together in Figure 10.3, the potential conflicts in locations with high landscape value and with high projected household growth are clearly highlighted. Obvious clashes are found in the Yorkshire Dales, the coastal areas of Norfolk and Suffolk, and in Devon. Similarly, Figure 10.4 shows that certain high growth areas may have conflicts of interest over areas with low levels of habitat fragmentation such as in Dartmoor and the Ashford and Kent area. It is interesting in this context to note the recent thoughts of the Coalition government to introduce a seismic change to the long-standing policy of restricting housing developments areas with high landscape and ecological values. The proposal is to relax planning regulations that allow for up to three dwellings to replace or convert existing farm buildings in national parks and other areas of outstanding natural beauty (see Wollaston, 2014). This has set off a contentious debate over the shortage of affordable housing in national parks and the devastating effect of adding development pressures to already fragile rural landscapes and ecosystems. There is a major worry that, rather than meeting the housing needs of the low income rural population, these additional developments will simply cater for the second, holiday and luxury home markets. The GIS mapping analysis that we demonstrate here could offer a useful toolkit for the public and key stakeholders to be better informed on proposals like this.

With the growing concerns of climate change and the unpredictable natural events associated with the temperature change, large scale housing developments have to take account of flood risks and water supply issues. High levels of flood risk are projected in the high growth areas near the coastal areas of North Somerset and Hull and East Riding (see Figure 10.5). At the opposite end of the problem is relative water stress. As clearly shown in Figure 10.6, areas on the eastern side of England (with the exception of the Yorkshire growth areas) tend to suffer from more serious levels of water stress (partly related to their higher levels of drought risk). Unfortunately, these eastern locations also incorporate most of the projected high household growth areas. Our map overlays clearly indicate that any major housing developments in these sensitive areas will need to seriously take flood risk and water supply into account.

Another key concern of housing policy and delivery is to meet market demands and ensure housing affordability. Figure 10.7 and Figure 10.8 provide mapping overlays between GVA and deprivation with the projected high household growth areas. The projected high household areas tend to be found in southern England – in areas with very high GVA per capita such as London, Cambridgeshire, Oxfordshire, Bristol and North Somerset. However, very high growth is also projected in some areas with moderate GVA per capita levels, such as in the Eastern region and Devon. High growth areas in northern England exhibit varied levels of GVA, with a mix of affluent and rural localities. As might be expected, with the exception of London and parts of West Yorkshire, there are no areas of high projected household growth which are also in areas with concentrated deprivation. Figure 10.9 shows the relationship between areas projected to have very high household growth rates and their respective housing affordability ratios (normalised house price versus average household income). It is interesting to discover that there is a very neat spatial divide in terms of the affordability ratio of different high growth areas. The 'North-South' divide neatly summarises the situation, revealing the high growth areas north of the dividing line are much more affordable than is the case on the southern side of the divide. If we overlay this map on top of Figure 10.3, it can be seen that the high growth areas that are least affordable are those located in national parks and areas of outstanding natural beauty. This justifies the concerns expressed by the media that the government's proposal to allow for more house building in these areas will mainly fuel more second homes rather than addressing local housing needs.

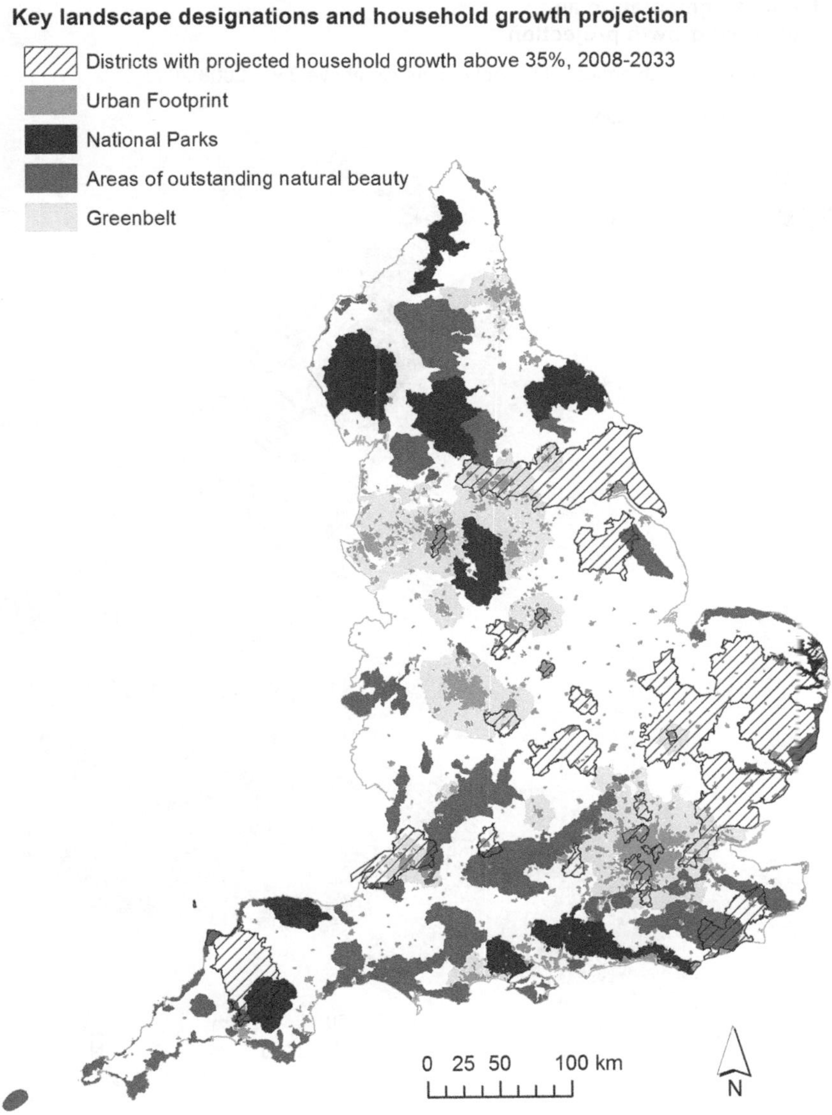

Figure 10.3 Key landscape designations and household growth projections

Note: For National Parks and Areas of Outstanding Natural Beauty see Figure 10.1. For 2008-based household projections to 2033, Department for Communities and Local Government Table 406, Household projections by district http://www.communities.gov.uk/publications/corporate/statistics/2033household1110.

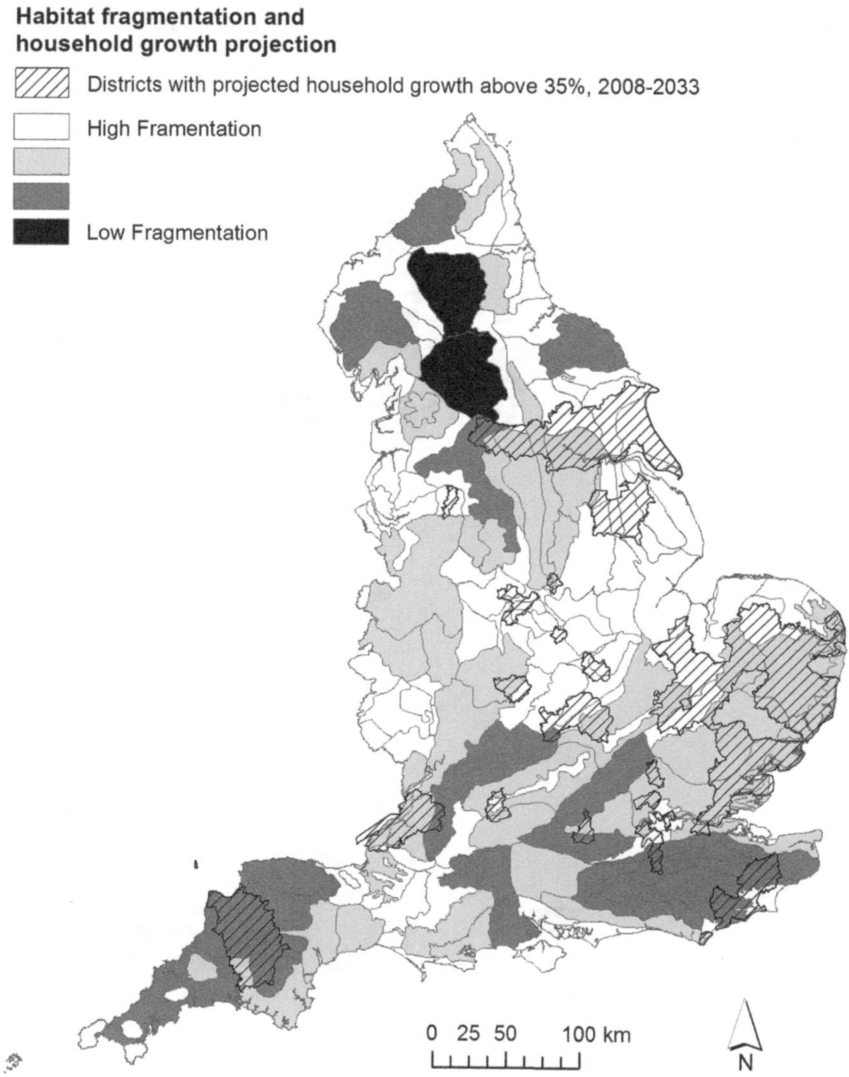

Figure 10.4 Habitat fragmentation and high household growth projection

Note: Habitat Fragmentation: Lawton, J.H. Brotherton, P.N.M. and Brown, V.K. et al., 2010. 'Making Space for Nature: A Review of England's Wildlife Sites and Ecological Network'. Report to Defra, p. 51. http://www.defra.gov.uk/news/2010/09/24/nature-news/. For household growth data see Figure 10.3.

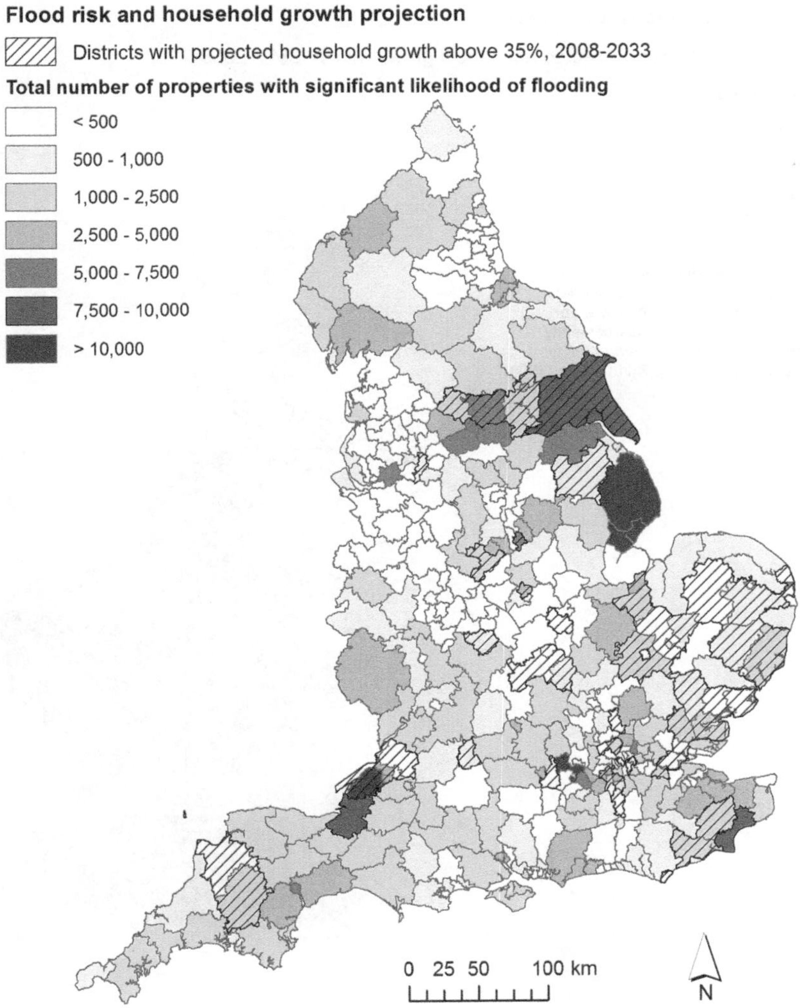

Figure 10.5 Flood risk and high household growth projection

Note: Environment Agency. 2009. 'Flooding in England – A National Assessment of Flood Risk', p. 27. http://publications.environment-agency.gov.uk/pdf/GEHO0609BQDS-E-E.pdf. For household growth data see Figure 10.3.

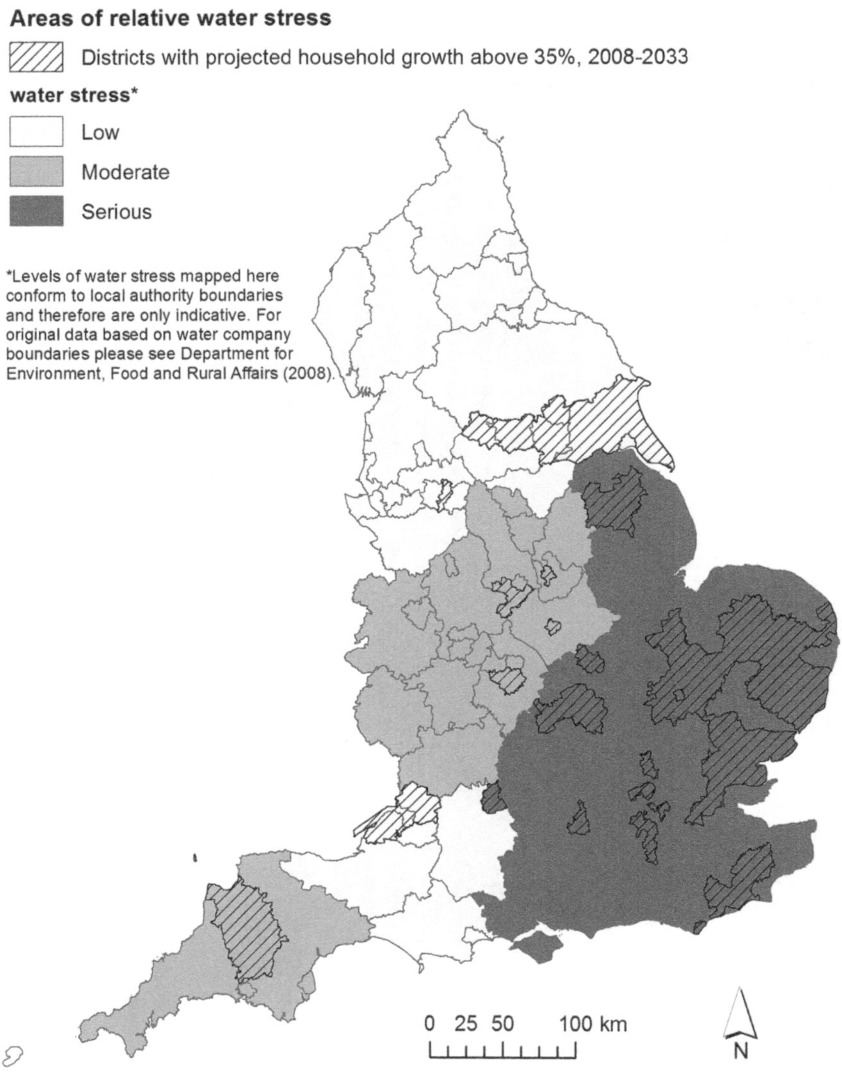

Figure 10.6 Relative water stress level and high household growth projection

Note: For Areas of relative water stress, see: Department for Environment, Food and Rural Affairs. 2008. *Future Water: The Government's Water Strategy for England*. London, Stationary Office, p. 22. For household growth data see Figure 10.3.

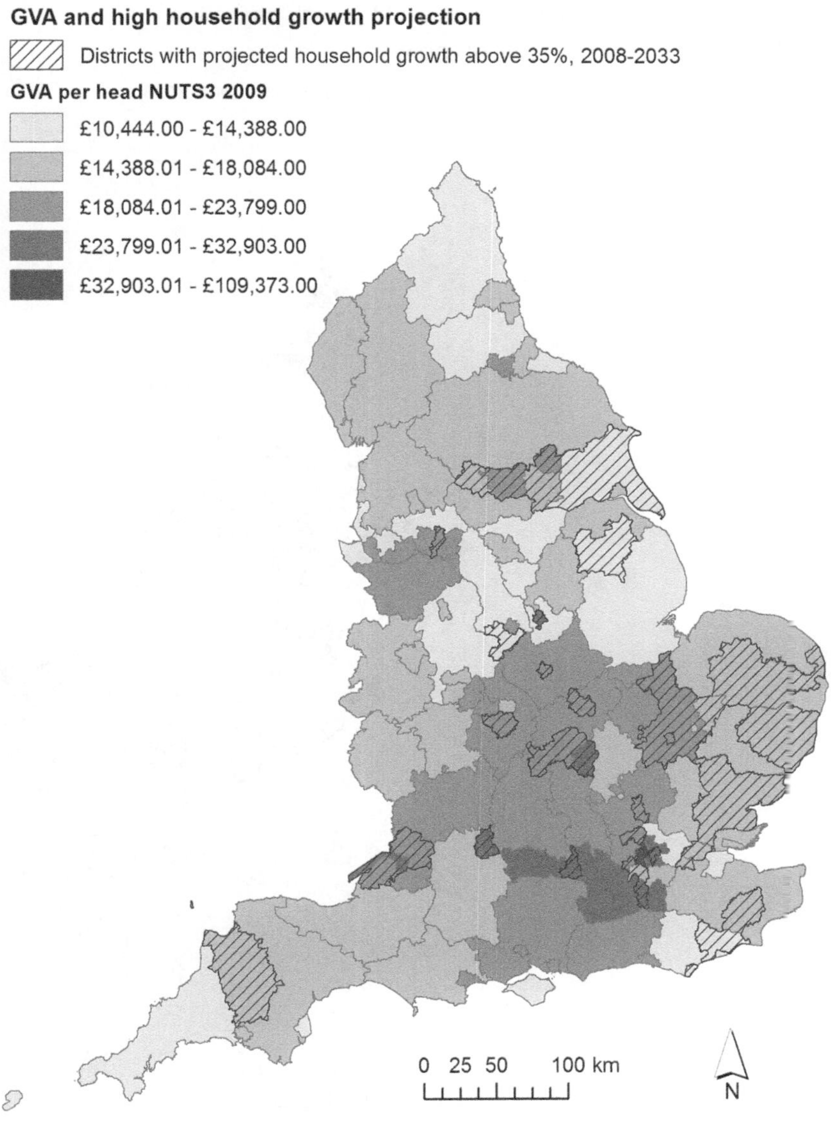

Figure 10.7 GVA and high household growth projection

Note: For household growth data see Figure 3. For GVA see Figure 10.2.

Figure 10.8 IMD and high household growth projection

Note: For the English Indices of Deprivation 2010 see Figure 10.2. For household growth data see Figure 10.3.

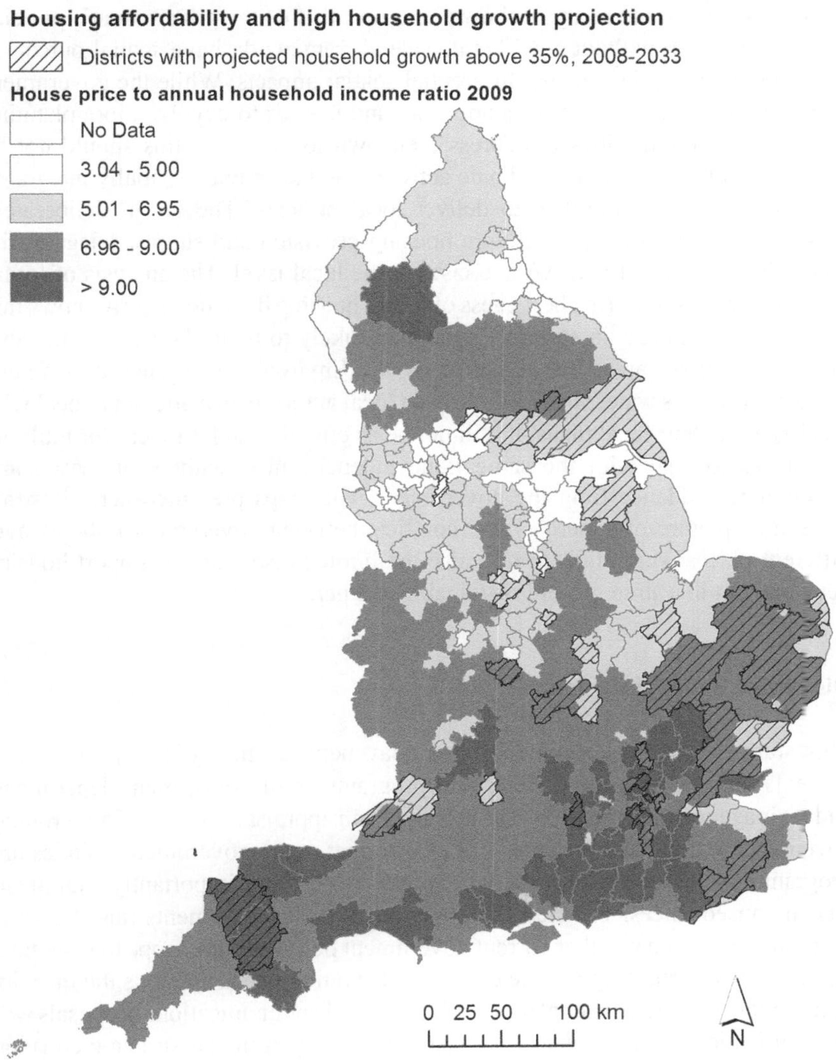

Figure 10.9 Housing affordability and high household growth projection

Note: For housing affordability, House Price Data: DCLG. Household Income Data: ONS Neighbourhood Statistics. For household growth data see Figure 10.1.

The housing trend analysis shown here reveals that the government's policies and actions, even without a deliberate spatial framework, have spatial outcomes and, cumulatively, they create differential spatial impacts. While the government acknowledges that one size does not fit all and intends to devolve more planning power to local authorities to address their own local issues, this should not be confused with the need to coordinate activities and to provide spatially integrated guidance to local authorities to deliver local policies. The NPPF deliberately avoids any spatial steering of future housing provision and simply delegates the job to the SHMA and SHLAA exercises at the local level. The analysis of future household projections (the best guess of future housing demand) clearly shows that the high growth areas in eastern England are likely to be in the least sustainable locations as these areas are classified by the Environment Agency as amongst those with serious water stress. While each local authority can attempt to deal with this issue via their local plans, it will not be as effective and efficient for multiple authorities to deal with the same issue independently, without at least some overarching guidance from the government. The maps presented here illustrate a sporadic picture of synergies and conflicts between government policies over different places which fails to optimise the limited resources to support housing development in a strategic and sustainable manner.

Discussion and Conclusion

This study set out to examine the interplay between policy needs and spatial contexts via mapping the policies and programmes of government departments and their agencies and NDPBs. The scanning and appraisal exercise of government documents, reports and websites has proved that many government policies and programmes do have strong spatial expression or, more importantly, significant spatial consequences. The study of government policy documents raised a series of questions around whether current government policy has the scope to respond to strategic spatial challenges. One example, remaining unanswered, is the question of the extent to which national road, rail and digital communications proposals will, in combination, simply concentrate resources along particular strategic corridors and/or complement the potential concentration of resources on eight identified core cities. Similarly, the relationships between these and other issues (such as the location of new power stations, or future airport capacity) and the spatial context of environmental constraints (for example, National Parks and nature conservation sites) and risks (for example, flooding) are given limited consideration at national level.

This review, and the associated mapping exercise, suggested that there was apparently little understanding in government policy of how these sectoral issues might complement or conflict with one another, thus creating synergies or tensions that might lead to future successes or failures in delivery and implementation. Ultimately this situation will contribute to, or frustrate, the future levels of economic growth and sustainable development that the current Coalition Government has committed itself to achieving. Indeed, the delivery of policy priorities in one local authority does not happen in isolation of the delivery of policy priorities in other local authorities – adjacent or otherwise. The strategic level of spatial planning, to address the bigger picture ('larger-than-local') is as important now as it was before the current Coalition government abolished the regional strategies and associated regional institutions such as the regional development agencies. In response, there is a new statutory requirement for neighbouring local authorities to cooperate over addressing a range of strategic priorities, as set out in the NPPF (DCLG, 2012). Although this so-called 'duty-to-cooperate' has some potential to bring policy makers in different local authorities together, the capacity for effective strategic policy development through this means alone remains doubtful. Yet strategic spatial thinking is a skill that policy makers can ill-afford to neglect in an era of localism. Without a strategic spatial perspective, there are risks that strategic policy blindness will further undermine efforts to deliver more effective spatial policy co-ordination and will limit the capacity for local policy to respond strategically to spatial outcomes.

The analysis of the household growth and housing delivery presented here further demonstrates that, by placing these spatial expressions together, a spatial reference framework starts to reveal some of the implicit assumptions, or random decisions, of government policy making. Making these spatial challenges and opportunities more explicit would help to inform policy debate and encourage partnership working to better coordinate and manage the delivery of what are often very complex spatial planning policies. On the contrary, by not articulating the spatial relationships of planning issues, such as future household growth and the spatial strategy of housing delivery, can lead to disjointed and ad-hoc management of infrastructure and service provisions at the local level. Only when such spatial relationships of policies and programmes are clearly presented and understood, can different spatial scenarios and metaphors, for future spatial opportunities and challenges in England, be developed in a way that can inform long-range, strategic planning and development in the wider European and global contexts.

In summary, there is a general political consensus that policy implementation should be contextual, and that more power should be devolved to local authorities in order to allow them to address their own policy priorities. However, it is important that the co-ordination of activities across boundaries is not lost, and that our understanding of spatial synergies and conflicts is sensitized to the long-term strategic priorities that lie beyond the policy agendas of individual local authorities. The analysis here illustrates that government policies and actions – even without a deliberate spatial framework – create spatial outcomes that, cumulatively, create very stark spatial impacts. In addition, the responsibility for co-ordinating the provision of future housing rests squarely on the shoulders of local authorities, with the Government's own national planning policy framework (NPPF) deliberately avoiding giving any strategic spatial direction. While each local authority can attempt to deal with a wide range of spatial issues through their

Local Plans, it will not be effective or efficient for multiple authorities to deal with these issues independently without some overarching guidance from central government. It is these kinds of acute tensions, synergies and conflicts, emerging from different policy agendas and coupled with the asymmetric consequences of devolved political responsibilities, that need to be more clearly understood and monitored, not only for England but across the UK as a whole.

In contrast to this strategic spatial policy vacuum, the analysis in this study does demonstrate the value and potential of applying a GIS analytical methodology to a whole array of planning and sectoral policy issues to identify spatial synergies, conflicts, challenges and opportunities across different spatial scales to inform policy-thinking. With more resources, more sophisticated mapping packages and datasets could be utilized to refine and develop the analysis further and extend the coverage to all the key topics of the NPPF. The potential value of such GIS mapping lies in its powerful visualization functions and its associated role in enhancing shared cognition and values to improve decision making in groups, enhancing trust and facilitating collaboration and consensus during policy-making processes and stakeholder consultations. But even without focusing on the development of perfect technologies and high-end software, the 'A Map for England' study demonstrates that a simple mapping overlay and associated analysis of specific planning issues can catch the attention of planners, politicians and the media (RTPI, 2014). The development of a flexible, guidance approach, with the aide of GIS to facilitate 'planning-as-learning', is therefore advocated as a way forward to achieve better spatial policy coordination (Rae and Wong, 2012; Wong et al., 2008). This collective way of 'seeing' can facilitate debate and learning among stakeholders and help them to 'think spatially, act spatially'.

References

Albrechts, L. 2004. Strategic (spatial) planning re-examined. *Environment and Planning B: Planning and Design*, 31, pp. 743–58.

Audit Commission. 1989. *Urban Regeneration and Economic Development: The Local Government Dimension.* London: HMSO.

Baker, M. and Wong, C. 2012. The delusion of strategic spatial planning: What's left after the Labour Government's English regional experiment? *Planning Practice and Research*, 28(1), pp. 83–103.

Batty, M. 2004. Editorial. *Environment and Planning B: Planning and Design*, 31(1), pp. 326–30.

Department for Communities and Local Government [DCLG]. 2007. *Taking Forward the Review of Sub-national Economic Development and Regeneration.* London: DCLG.

Department for Communities and Local Government [DCLG]. 2012. *National Planning Policy Framework.* London: DCLG.

Gilfoyle, I. and Wong, C. 1998. Computer applications in planning: Twenty years' experience of Cheshire County Council. *Planning Practice and Research*, 13(2), pp. 191–7.

HM Government. 2004. *Planning and Compulsory Purchase Act 2004*. London: HMSO.

HM Government. 2011. *Unlocking Growth in Cities*. London: Cabinet Office.

HM Treasury and UK Infrastructure. 2011. *National Infrastructure Plan 2011*. London: TSO.

Jensen, O. and Richardson, T. 2001. Nested visions: New rationalities of space in European spatial planning. *Regional Studies*, 35(8), pp. 703–17.

Krugman, P. 1991. Increasing returns and economic geography. *Journal of Political Economy*, 99, pp. 483–99.

Krugman, P. 1998. What's new about the new economic geography? *Oxford Review of Economic Policy*, 14(2), pp. 7–17.

Lords Hansard. 2003. Lords Hansard text for 1 April 2003 (230401-01).

Rittel, H. and Webber, M. 1973. Dilemmas in a general theory of planning. *Policy Sciences*, 4, pp. 155–69.

Rae, A. and Wong, C. 2012. Monitoring spatial planning policies: Towards an analytical, adaptive and spatial approach to a 'wicked problem'. *Environment and Planning B: Planning and Design*, 39, pp. 880–96.

Royal Town Planning Institute (RTPI). 2014. *A Map for England*. London: RTPI. http://www.rtpi.org.uk/knowledge/core-issues/map-for-england/ [accessed 8 March 2014].

Vonk, G., Geertman, S. and Schot, P.P. 2005. Bottlenecks blocking widespread usage of planning support systems. *Environment and Planning A*, 37(5), pp. 909–24.

Wollaston, S. 2014. Plans to relax building regulations in our national parks will spell disaster. *The Guardian*, 23 February 2014. http://www.theguardian.com/commentisfree/2014/feb/23/building-regulations-national-parks-disaster-planning [accessed 8 March 2014].

Wong, C., Ravetz, J. and Turner, J. 2000. *The United Kingdom Spatial Planning Framework*. London: The Royal Town Planning Institute.

Wong, C., Schulze Bäing, A. and Rae, A. 2006. *Uniting Britain – the Evidence Base: Spatial Structure and Key Drivers*. London: The Royal Town Planning Institute.

Wong, C., Rae, A., Baker, M., et al. 2008. *Measuring the Outcomes of Spatial Planning in England*. London: The Royal Town Planning Institute and the Department for Communities and Local Government, RTPI.

Wong, C., Baker, M., Hincks, S., et al. 2012. *A Map for England: Spatial Expression of Government Policies and Programmes*. London: The Royal Town Planning Institute. http://www.rtpi.org.uk/knowledge/core-issues/map-for-england/ [accessed 8 March 2014].

Wong, C. and Watkins, C. 2009. Conceptualising spatial planning outcomes: Towards an integrative measurement framework. *Town Planning Review*, 80 (4/5), pp. 481–516.

Chapter 11
Spatial Patterns of the 'Mestre Through Highway' within the Venetian Metropolitan Area (Italy)

Domenico Patassini, Matteo Basso and Giorgio Borghelot

Introduction

This chapter evaluates spatial changes of location patterns of economic activities, which have been generated by a large piece of infrastructure: the 'Mestre Through Highway' (known in Italian as 'Passante di Mestre', hereafter referred to as the MTH). These changes become a component of a wider 'spatial discourse' stimulated by the MTH project. Generally, spatial location patterns are empirically observable, as for instance shopping malls, hotels, restaurants, business and leisure centres which lie along the transport infrastructures.

From a theoretical viewpoint, infrastructures are an important pull factor for economic activities and consequently a key factor of regional competiveness: Biehl (1986), Hirschman (1988) and Aschauer (1989) consider infrastructures as a basic instrument to develop policies at different levels. Furthermore, in the field of regional economics, they have been theorized as being part of 'agglomeration economies', influencing overall business and enterprise location decisions.

The MTH has been operating since 2009, after a gestation period of more than 40 years. It passes through the central Venetian Metropolitan Area and is approximately 32.3 km long, running south-west to north-east. This infrastructure was supposed to reduce the through traffic in congested areas around the lagoon basin and it strategically links the TEN network and the Corridor V in the north-eastern part of Italy (Figure 11.1).

In addition to the above-mentioned objectives evaluated with traffic simulation models and supported by spatial planning tools, many implicit objectives are not as popular in the feasibility studies and impact analysis. They generate direct and indirect effects that qualify the infrastructure as a 'public good' in a broad sense, but also as an economic multiplier and accelerator.

From a comparative evaluation, the effects that qualify the infrastructure as a 'public good' concern the dynamics and the location of economic activities, the geography of land use patterns, the changes in territorial structures shaped by the highway and by its complementary works, and the adaptation of the local spatial planning system.

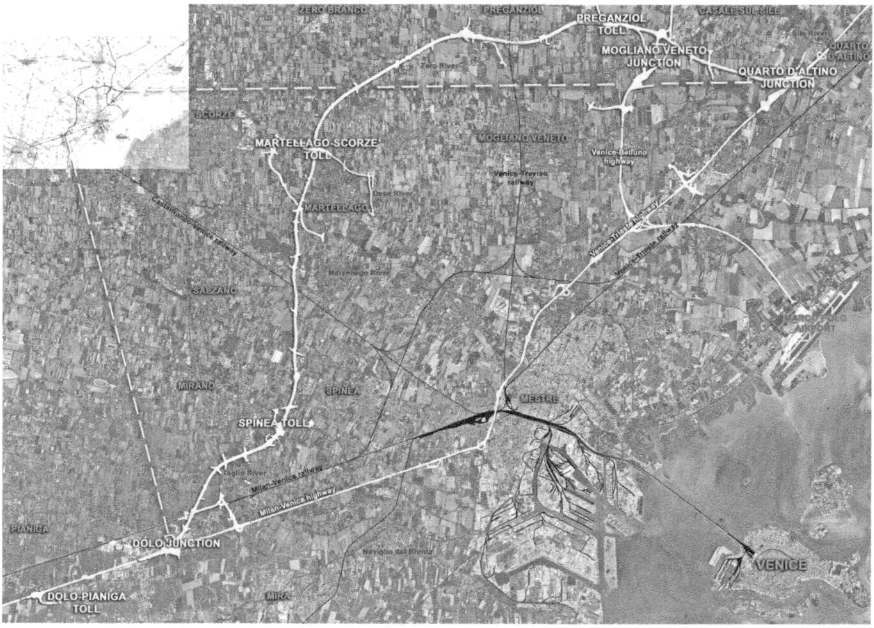

Figure 11.1 Territorial framework

Note: Figures 11.1, 11.4 and 11.6–11.12 use the following data sets: orthophoto Province of Venice 2006, orthophoto Veneto Region 2003, mosaic of master plans within the Province of Treviso 1991–2004, mosaic of master plans within the Province of Venice 1993–2007, current master and operational plans in January 2011, land cover map of the Veneto Region 2009.

The infrastructure has triggered a heated 'public discourse' which goes well beyond the theme of 'crossing Mestre'. It has also helped to introduce, although with some delay, the concept of a 'Green Through Highway' (the exact opposite of a supposed metropolitan 'city wall'), a large-scale landscape project that features ecological corridors. This perspective strives to mitigate, compensate and redesign ex-post this particular territory. New projects, often based on conflicting rationales, have been blossoming alongside it. All of these effects put together have helped to shape a new 'relational capital', casting a competitive space.

The estimate of the 'net effect' of the infrastructure was initially assumed as a crucial evaluation issue despite the complexity of the metropolitan context and the transition towards a new regional planning system. To estimate the net effect, an empirical analysis was carried out over a period of 17 years, i.e. from 1993, the year in which the project was presented, to 2010, the year of partial work completion. This period is long enough to capture the 'announcement' and 'anticipation' effects and to understand the adaptive behaviour of companies and public institutions, but not for an ex-post evaluation.

Using available statistics on population, employment, businesses, projects and plans, the effects were evaluated in an area that comprises 12 municipalities crossed by the highway, and 7 municipalities affected by complementary works (tolls, access roads, and so forth). Two aggregate scenarios could thus be designed on the geography of production areas and on spatial morphology: the first according to plan provisions, the second as a possible map of an adaptive strategy.

Evaluation Demand within Public Discourse

As stated in the introduction, the main objective was to evaluate the extent to which the MTH changes the geography of the main economic activities and influences the metropolitan area of Venice.[1] Since this geography presumably changes the metropolitan structure, it becomes crucial to identify the evaluation questions and 'spatial devices' it generates.

A first set of questions (set A) arose from the fact that the MTH can be regarded as a strategic infrastructure for the local landscape and for the entire North-East of Italy. A second set of questions (set B) came from statutory evaluations made during the design, implementation and monitoring of the project's initial operational stages. The last and most significant set of questions (set C) matured into what we shall call a 'public discourse'.

Set A

The MTH seeks to respond to serious congestion problems related to the Mestre-Venice node, a convergence (but also *cul de sac*) of axes and flows of regional, interregional and international relevance.

For many years, congestion has produced negative externalities locally and reduced the competitiveness of Venetian territories as a hub connecting Central

1 The current structure of the metropolitan area is the outcome of a long-term process spanning about a century and characterized by five cycles. The first begins in 1908 with the master plan of the passenger port ('Marittima') and continues in 1917 with the approval of the first master plan of the new commercial port of Venice. The second master plan provided the area for a commercial port, for a small oil port with an internal basin and areas for factories and warehouses, an industrial area with service channels and a garden city for some 30.000 inhabitants. The first industrialization phase dates back to the 20s, was consolidated in the 50s and 60s with the development of the chemical hub and the preparation of the second plan of the port in 1965. The development of the industrial district reached its peak in the 70s, followed by a rapid decline, the end of the business cycle of the chemical industry, restructuring and regeneration of the site. From 1979 to 2012, port traffic recorded a decrease of industrial components and oils, and an increase in commercial traffic, especially containers. Currently industrial traffic, of oils and commercial commodities, has reached 15 per cent, 39 per cent and 46 per cent respectively. Passenger traffic is booming with 2.3 million passengers in 2012, of which 1.8 million were cruise passengers.

Europe and the Balkans. The territory has also experienced a loss of potential comparative advantages, especially in terms of access to the Mediterranean and to central and northern Europe.

The location of the MTH is strategic not only from an international perspective, because it is the crossing point of the TEN corridors, but also from a regional and local perspective. It may, in fact, be considered the 'backbone' of the metropolitan area as well as an anchor for the port and the airport.

The projects under discussion to adapt and develop port and airport have acquired a different significance after the completion of the MTH, giving rise to optimistic scenarios in terms of flows of people and goods.[2]

At the local level, the MTH acts as a 'support arch' and at the same time as a 'sequence of doors' for the reorganization of the road network, the land, the lagoon and for sea access. With such characteristics, two main effects deserve special mention: on the one hand, enhancing urban hierarchy in a quasi-isotropic environment and, on the other, diversifying access in order to avoid bottleneck created by the only bridge connecting the mainland to the historic centre. Hierarchisation and diversification can increase the urban footprint within the neighbouring areas of the lagoon.

Strengthening the two poles (the airport and port) could further accelerate urbanization with marked effects on lagoon morphology and flood risk, due to the exceptionally high waters, even after the MOSE dam starts operating.[3] Its activation requires updating the lagoon's ordinary maintenance in tandem with portfolio investments of the Morphological Plan of the Lagoon of Venice (2013–2023).

But the local effects of the MTH do not stop here. The route's geometric and building characteristics cause significant impacts on the environment and on current and expected land use patterns. The first ones to be influenced will be the ecological capacity of territories which have a rich habitat and hydraulic system, characterized by a dense water network in a context of prevailing sprawl.

Even the road network partition in question records changes in morphology and efficiency. This, combined with emerging land use patterns, reinforces the multiplier and acceleration effects on investment and consumption activities. In fact, the generative capacity of the MTH, in terms of economic activity, employment, public and private projects, seems remarkable.

---

2 See for instance the Three-year Operational Plan (2014–2017) of the Venice Port Authority and the proposal to build a new airstrip with related ground services and facilities.

3 MOSE (*Experimental Electromechanical Module*) is a defence system at the inlets designed to protect the lagoon area, its inhabitants, towns and cities of inestimable historical, artistic and environmental heritage from all high waters, including extreme events. The system is expected to be operative in 2016.

Set B

Several procedures were applied to answer specific evaluation questions, especially in terms of feasibility: EIA to assess environmental and landscape impacts,[4] MEPLAN to estimate the overall transportation costs, CBA to estimate the discounted net economic benefits related to a given set of operational goals. The TRIPOD model was applied in similar contexts to capture the spatial effects generated by a significant infrastructure.

Yet, despite the variety of assessment procedures used, only direct spatial effects were considered, paying little attention to the relationship between infrastructure and socio-economic context. Field research identified 'indirect' effects.[5]

The 'evaluation domain' tends to change significantly with the interaction between project and context, in particular when facing situations of mutual adaptation. During both the design and implementation phase or operational management, adaptation proves to be strategic. It is defined by several factors as for instance the adjustment, which is multi-actor and conflictual but also sensitive to cumulative impacts, and helps to update the spatial agenda and redefine the terms and conditions of the negotiation.

Set C

The spatial discourse created by the MTH deviates from a predictable project cycle to form a process where the infrastructure becomes 'space' in an incremental manner, a factor that characterizes the planning process and outlines a contingent valuation domain.

It is therefore evident that the evaluation issues of the first two sets A and B are unable to capture three important processes. The first concerns strengthening the Venetian metropolitan 'voice', the second adapting the planning frame, and the third considering the new important projects that affect the early rationale of the local planning. These three processes update the empirical evaluation domain by basing it on what is called spatial discourse.

The first process shows the extent to which the MTH was capable of strengthening the metropolitan city structure.[6]

Starting from the 60s, the area began consolidating its metropolitan structure with new production and service hubs encouraging densification and changing the

4 The available documentation on preliminary project and impact assessment (2002) helps to better understand the limitations of the approach.

5 Among the indirect spatial effects generated by a new transport infrastructure the following should be considered: the dynamics of economic activities and their location, the geography of land use patterns, the changes in territorial structures, the adaptation of local planning rules and strategies.

6 The documentation is available in Italian. For more details see the publication of the Gianni Pellicani Foundation (2008, 2011).

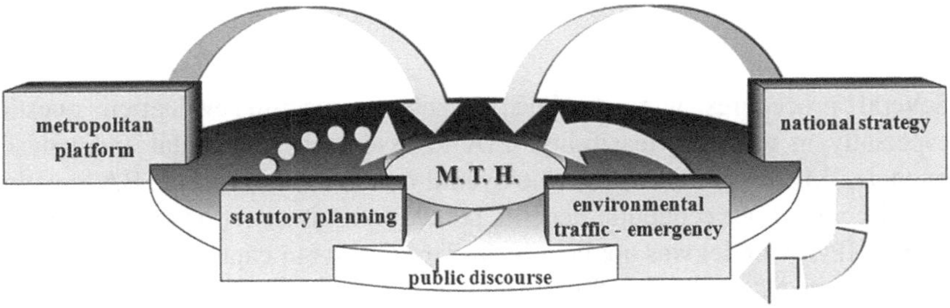

Figure 11.2 Public discourse on MTH

land use patterns that are typical of sprawl. The MTH contributes to a qualitative leap in the structure, distinguishing the metropolitan area from neighbouring areas and redefining new connections.

The MTH reinforces the metropolitan structure by creating a new economic geography with activities and services as well as new territorial scenarios. These have given rise to a much richer and more documented public discourse (Figure 11.2) compared to what occurred during the design stage: a process that lasted nearly 50 years, from 1958 to 2009.[7] The second process involves a significant change of the local planning framework.

The MTH's spatial buffers can easily be recognized as a public good because they update the rules of the game and the institutional competences in the area. The idea of a 'green road infrastructure' (GTH) has stayed in the backdrop. Yet it could have served for landscape management encouraging collaborative behaviour.

A continuous adaptation of the planning systems and practices is shown to change the rules of the game as well as the power and duties of the actors involved. Such adaptation often generates either announcement or response effects. It is also shown to strengthen negotiations that are often in conflict with statutory planning and current environmental compensation strategies. The effects on the real estate market are significant.

The third process shows significant changes in the economic geography and the location of services. These changes were analysed using an ex-post assessment of some of the measurable phenomena in the 1993–2011 period (Figure 11.3). Its duration allows detecting some of the announcement as well as anticipatory effects stemming from the implementation of the infrastructure. The idea of the MTH in fact arose within the decision-making process at the beginning of the 1990s.

Spatial discourse arises from the way the MTH comes into the public agenda and then is being implemented. It includes the design phase together with the relations of mutual adaptation in time and in space. It is believed to be a sort of 'collective actor' that has not been legitimated by a defined cluster of subjects. The

7 See also Vanni (2009).

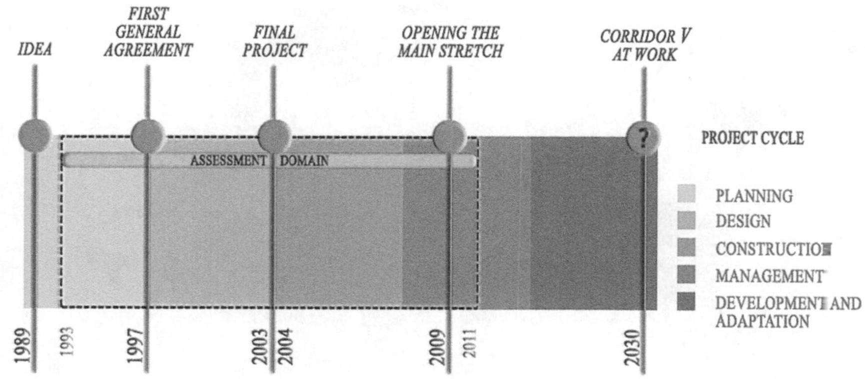

Figure 11.3 Project cycle and assessment domain

strategic role of MTH depends greatly on the fact that the spatial discourse tends to guide the life cycle of the MTH and distinctively characterizes it: in particular, because it gives the MTH the role of strategic infrastructure.

In conclusion, the public discourse seems to have drawn three metropolitan conflicting visions: a wider one, which treats the entire region as a diffuse metropolis, a second one, smaller and selective, which recognizes the main features of the Venetian metropolitan area, while a third one counteracts it by means of a hypothetical metropolitan area that should include additional cities as Vicenza, Verona and Rovigo. As shown in the conclusions, the second and third vision can support either an incremental or a planning scenario.

Empirical Evidence and Evaluation Domain

The third set of questions (Set C) helps to expand the size of the evaluation domain as they raise the issue of evaluability of the MTH process (that is, screening).

With the aid of empirical evidence, the analysis seeks to identify the generative forces produced by the contact between the project and the context. The analysis is thus closely related to the approach known as 'evaluating the complex'.[8] Due to the complexity of the MTH process, its evaluation is equally complex.

The empirical analysis highlighted the difficulty in estimating the MTH 'net effect', i.e. the MTH specific contribution to the metropolitan growth. It was instead easier to recognize the main features of the 'spatial discourse' which considers the symptoms of a number of synapses within a MTH fuzzy influence

8 'The evaluation process seeks a balance between order and chaos. The interaction of four elements – simplicity, inventiveness, flexibility and specificity – allows complex pattern to emerge. On this point, see Forss et al. (2011).

Figure 11.4 Area of influence

area as empirical evidence. The area includes 12 municipalities crossed by the MTH in addition to another 7 municipalities affected by complementary works (Figure 11.4).

Maps and time series have been employed striving to distinguish between direct and indirect effects. Empirical evidence comprises the demographic dynamics, the economic geography of activities and the changes in the land uses and of the infrastructure stock, but also of the interpretation of MTH given by plans, projects and programs.

Population dynamics can be considered one of the main components. The population (on December 31, 2009 was roughly 590,000) was higher in all the municipalities crossed by the MTH, whereas the municipality of Venice experienced a drop. The increase indicates densification processes of the metropolitan belt (Figure 11.5) which comprises small to medium size municipalities, with some booming growth rates as in the case of Casale sul Sile (+70 per cent), Casier and Santa Maria di Sala (+50 per cent), and Marcon (+40 per cent). The municipalities with the highest rates are mapped in Figure 11.5 and are numbered 1, 2, 14 and 4, respectively.

There were about 60.000 economic units in the area (handicraft excluded) on 30 April 2010. Although 51 per cent of the economic activities were located in

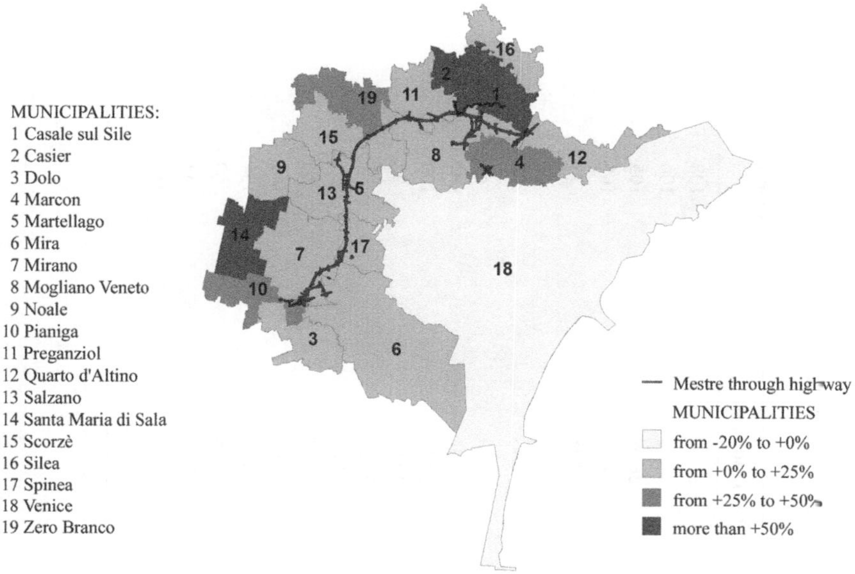

Figure 11.5 1993–2010 population growth rate
Source: Central Statistics Office, ISTAT.

the municipality of Venice, the index of entrepreneurial density had its highest values in the municipalities crossed by MTH (Table 11.1). Between 1993 and 2010 there were also significant location patterns and dynamics. In fact, the largest contributions came from the services sector (business services +184 per cent in 2008), building and construction (+166 per cent in 2008) and transport activities (+116 per cent, always in 2008). The main clusters were located within the areas served by toll booths, where the variation in location quotient was greatest, in particular in the transport sector, and around the junctions between MTH and A4 highway.

Land use changes of residential and productive zones (by location, size, and accessibility) were assessed, comparing 1993 to 2007 statutory planning schemes. The changes were estimated, net of intermediate plan variations, illustrating the influence of toll junctions, but above all complementary works.

Concerning the residential function, there was a noticeable increase of nearly 23 per cent, from 5.198 to 6.400 hectares, with rates over 60 per cent for municipalities such as Santa Maria di Sala, Pianiga, Zero Branco and Dolo With regard to the industry (Figure 11.6), if the whole area showed a 7 per cent increase, i.e. from 4.099 to 4.388 hectares (Table 11.2), that is to say roughly 29 per cent, excluding the negative trend of Venice, there were municipalities like Noale, Pianiga, Zero Branco and Salzano that had increases which exceeded 80 per cent.

Table 11.1 Economic units in the area on 30 April 2010

Municipalities	Economic units at 30 April 2010				Index of entrepreneurial density
	Total	%	Handicraft	% Handicraft	
Casale Sul Sile	1,363	2.31	432	31.69	108
Casier	1,112	1.88	358	32.19	100
Dolo	1,582	2.68	422	26.68	105
Marcon	1,615	2.73	460	28.48	105
Martellago	1,653	2.80	636	38.48	78
Mira	2,592	4.39	900	34.72	67
Mirano	2,615	4.42	757	28.95	98
Mogliano Veneto	2,442	4.13	599	24.53	87
Noale	1,456	2.46	458	31.46	92
Pianiga	1,297	2.16	411	32.13	109
Preganziol	1,331	2.25	400	30.05	78
Quarto d'Altino	787	1.33	246	31.26	97
Salzano	1,020	1.73	433	42.45	81
Santa Maria di Sala	1,541	2.61	562	36.47	89
Scorzè	2,130	3.60	704	33.05	112
Silea	1,320	2.23	354	26.82	131
Spinea	1,839	3.11	581	31.59	69
Venice	30,152	51.01	6,918	22.94	111
Zero Branco	1,280	2.17	503	39.30	119
Total area	59,109				

Source: Chamber of Commerce.

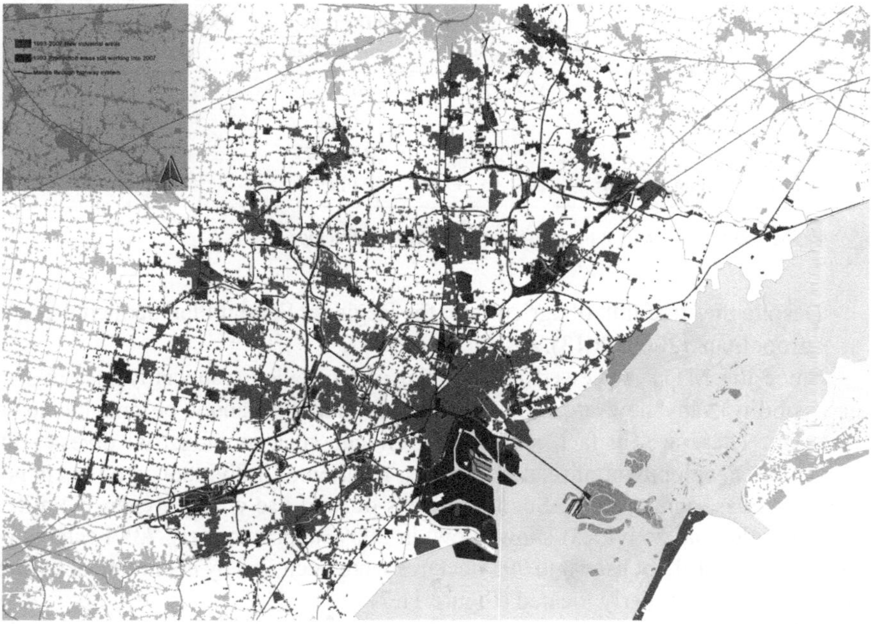

Figure 11.6 1993–2007 new industrial areas

Table 11.2 1993–2010 growth rate of productive areas

	1993 industrial areas (has)	2007 industrial areas (has)	1993-2010 growth rate of productive areas
Casale Sul Sile	109.49	188.30	+ 71.98
Casier	105.21	152.30	+ 44.74
Dolo	62.51	85.18	+ 36.26
Marcon	171.61	172.06	+ 0.26
Martellago	69.60	81.04	+ 16.43
Mira	283.10	301.57	+ 6.53
Mirano	87.09	92.03	+ 5.67
Mogliano Veneto	145.73	229.89	+ 57.75
Noale	46.25	135.97	+ 194.00
Pianiga	82.81	161.72	+ 95.29
Preganziol	59.04	74.18	+ 25.63
Quarto d'Altino	156.86	128.41	- 18.14
Salzano	26.58	48.53	+ 82.59
Santa Maria di Sala	165.33	138.71	- 16.10
Scorzè	132.92	175.07	+ 31.71
Silea	99.53	126.32	+ 26.92
Spinea	49.30	54.43	+ 10.41

	1993 industrial areas (has)	2007 industrial areas (has)	1993-2010 growth rate of productive areas
Venice	2182.16	1917.93	- 12.11
Zero Branco	64.10	124.24	+ 93.84
Total area	4099.22	4387.87	+ 7.04
Total area (excluding Venice)	1917.05	2469.94	+ 28.84

Source: Master plan mosaic of the provinces of Treviso and Venice.

Despite the 'free fall' of agriculture, the Regional Spatial Plan defined the area as 'agropolitan' (2009, 2013).

Since the MTH is viewed as a spatial order device (especially in terms of accessibility), any new areas destined for production cannot be considered in isolation. There are, in fact, substitution effects, such as industrial divestment, relocation, development of related services, location of individual activities in not suitable areas, and so forth. Such a process of land use change (even if limited only to industrial activities) shows how, in 2009, approximately 735 hectares of the area were still not used and 905 hectares were occupied by economic activities which were not properly located (Figure 11.7).

It is worth noting that the dynamics of the metropolitan environment results from several factors. The MTH, although it acts as the infrastructure catalyst, it is certainly not the only one (Figure 11.8). In fact, other dynamics need to be considered as for instance those related to road, railway, water supply, and network changes.

The local and regional planning system 'adopts' the MTH with a variable interpretation and adaptation process. In some cases, the MTH requires schedule adjustments, in other cases it challenges established value systems or priorities or even requires changing the operational planning device: land use schemes, standard, rules and regulations. Not embodying the features of a green infrastructure, environmental innovations, landscape protection, the development of urban agriculture and urbanity become less likely.

The areas served by the new infrastructure are affected by 'mega-projects' (Flyvbjerg et al., 2003) and 'edge-cities' (Garreau, 1991). They are located particularly close to motorway exits and junctions, but also in agricultural areas. They involve real estate projects initiated mostly by private investors, entailing partial or total agreements with the local authorities.

A series of examples serve here illustrate the point above (Figure 11.9). 'Veneto City' is expected to be developed at the junction near the new infrastructure and the A4 motorway. The area covers over 50 hectares designed for industrial, commercial and leisure uses and crosses three municipalities. A theme park, located north of the municipality of Casale sul Sile, is expected to be built for production, trade and services, in an area stretching more than 50 hectares.[9] Move Hotel is a new complex opened in Mogliano Veneto (North of Venice), at the junction between

9 See Municipality of Casale Sul Sile (2005, 2009).

Figure 11.7 Land use substitution effects

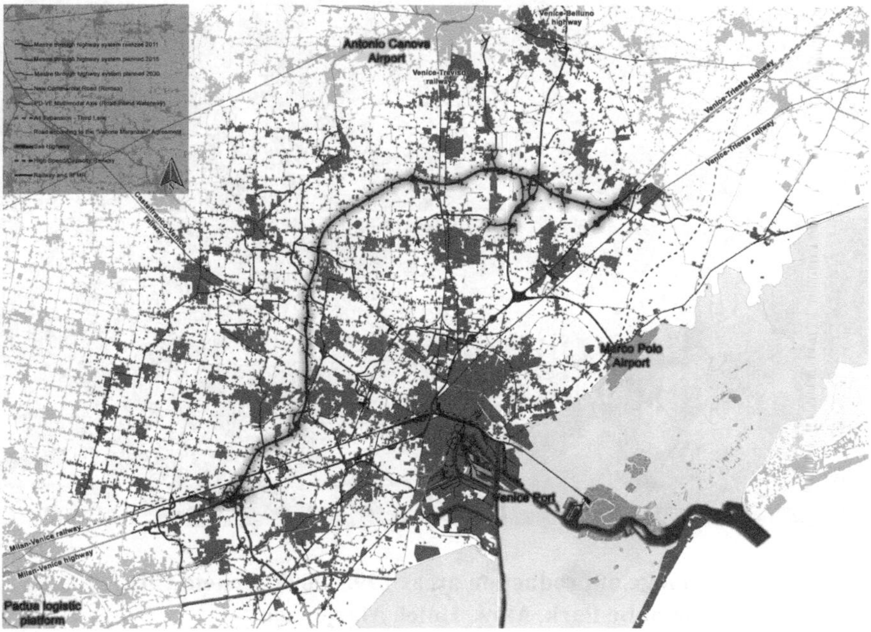

Figure 11.8 Infrastructure projects in the Venetian mainland, 2011

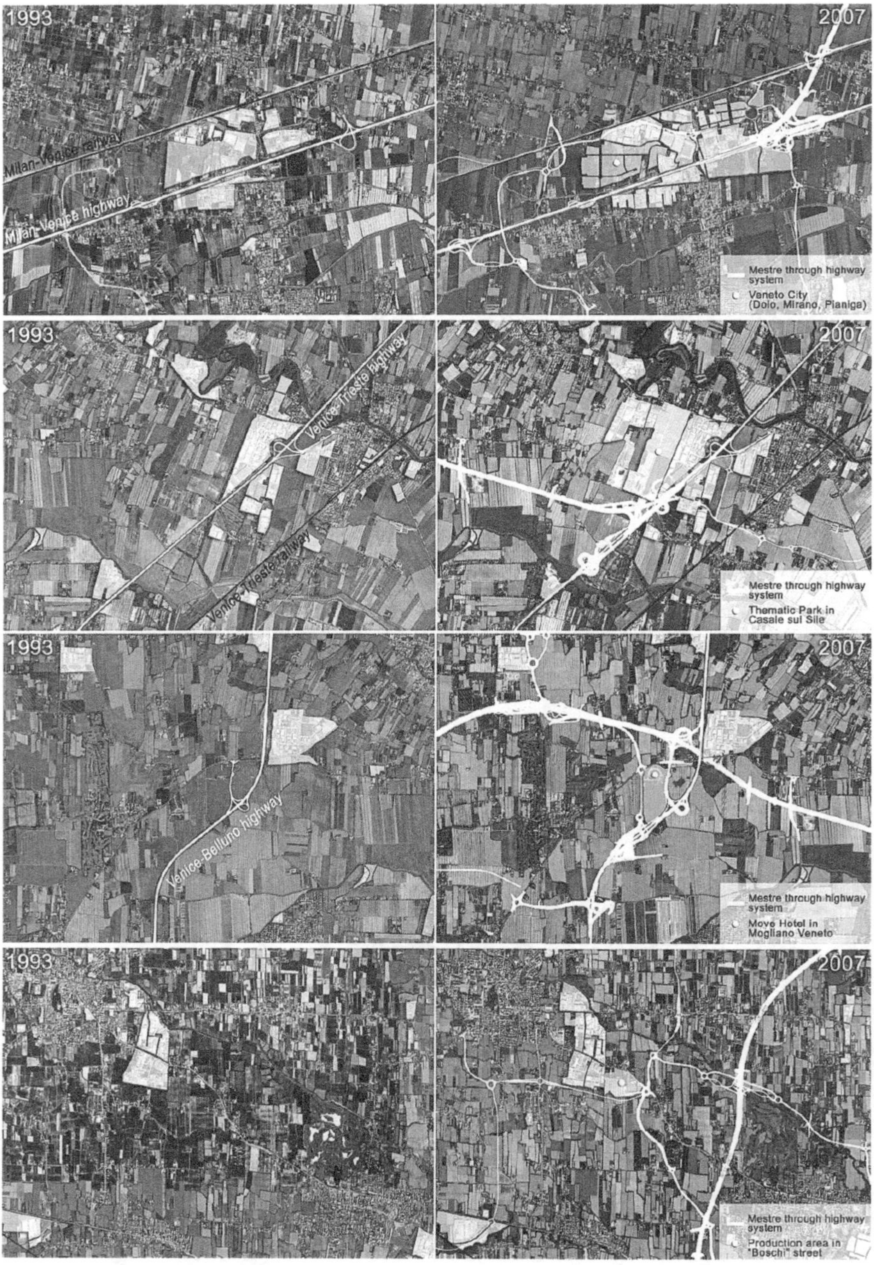

Figure 11.9 Change of production areas 1993–2007: Veneto City, Thematic Park, Move Hotel, Martellago

the A27 motorway and the MTH. The new complex is designed for mixed-uses which include a business centre, hotels and leisure. Another example is in the municipality of Martellago, which close to the MTH exit sees a new area set aside for production, logistics and complementary works.[10]

Spatial Discourse as an Added Value: Identifying Relevant Criteria for Monitoring and Ongoing Evaluation

Since the 1980s, 'communicative planning theory' has led to the acceptance that planning is not merely an exercise in rational analysis and decision making but rather, a 'stubbornly social and political activity' involving conflicts and negotiations among actors (Ozawa and Seltzer, 1999). The fulcrum of the communicative planning theory is Habermas' 'communicative rationality': the idea that a shared, reasoned understanding of truth may be reached by having actors communicate in an 'ideal speech situation'. Yet, the public discourse process in social changes emphasizes the effects of power, asymmetry and conflict.

The spatial discourse that the MTH introduces privileges conflict and competition over consensus. It derives from an unlikely project-cycle, incorporating a process where the infrastructure itself becomes 'unexpected space'. The capture process cannot be gradual even if, as has been seen, some transformations operate as a pivot.

Empirical evaluation is updated in the domain of project-context interactions, yielding relevant criteria, as indicated below:

a. the efficiency and safety of the transportation network for metropolitan and trans-regional traffic flows (in the wide area network);
b. the effectiveness of land use in terms of consumption, generation of productive activities and location patterns;
c. the urban-rural environmental quality;
d. statutory and strategic capacity of planning tools;
e. adaptation of the local plans to the scenarios defined by incremental large-scale projects.

These criteria can enhance a monitoring system that can serve for an on-going evaluation according to two scenarios.

Alignment of two Scenarios

Monitoring and ongoing evaluation devices can be aligned to two scenarios by generating a wider cognitive space. The evidence gathered so far can be used to

10 See Municipality of Martellago (2010).

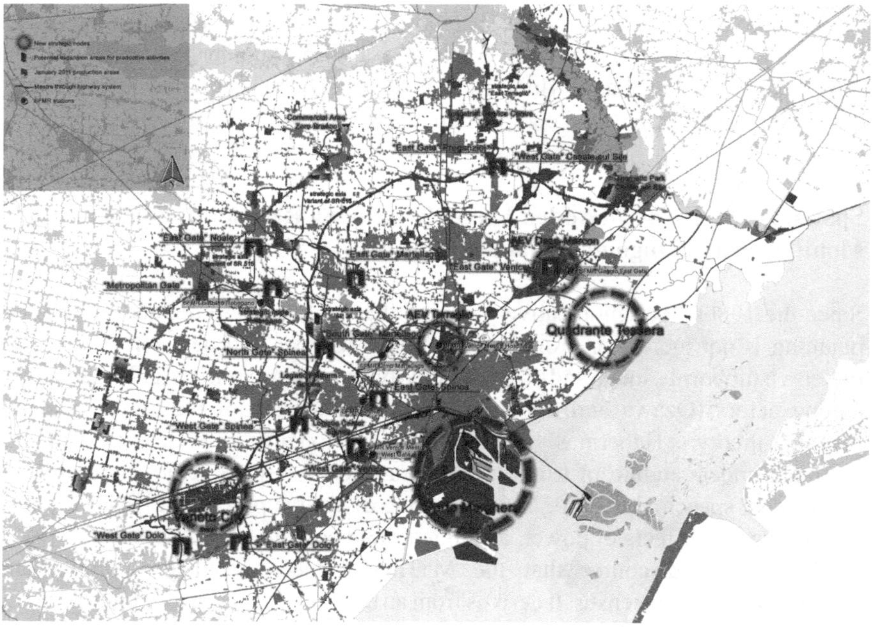

Figure 11.10 First scenario-additive outcome

construct two different scenarios based on the geographic location of production and commercial activities within the metropolitan area of Venice.

The first scenario, the 'incremental' type, represents the simple aggregation of the expected expansion and the projects promoted by opportunistic individual municipalities, without any kind of supra-local coordination. From the spatial perspective, what is obtained in such case is a fragmented development strategy that can be very conflicting, since it is based on the development of new 'edge cities' (Garreau, 1991) localized along the infrastructure, with high impacts in terms of land consumption and with the attraction of new flows of private mobility (Figure 11.10).

In the second scenario instead, the new infrastructure is designed as an organizational element of the territory, along with the rail system. This strategy, which can be called 'adaptive', requires the introduction of large-scale governance. It should identify and prioritize strategic development nodes (to direct investment projects), contain urban sprawl, re-locate economic activities that are not properly located, and help manage urban-rural interactions (Figure 11.11).

Spatial discourse changes with the scenarios and it characterizes relevant evaluative criteria according to the incremental or planned rationale. To confirm the importance of managing the location opportunities generated by the new transport infrastructure, we shall briefly refer to a regulatory device that was recently introduced by the Veneto Region's Administration.

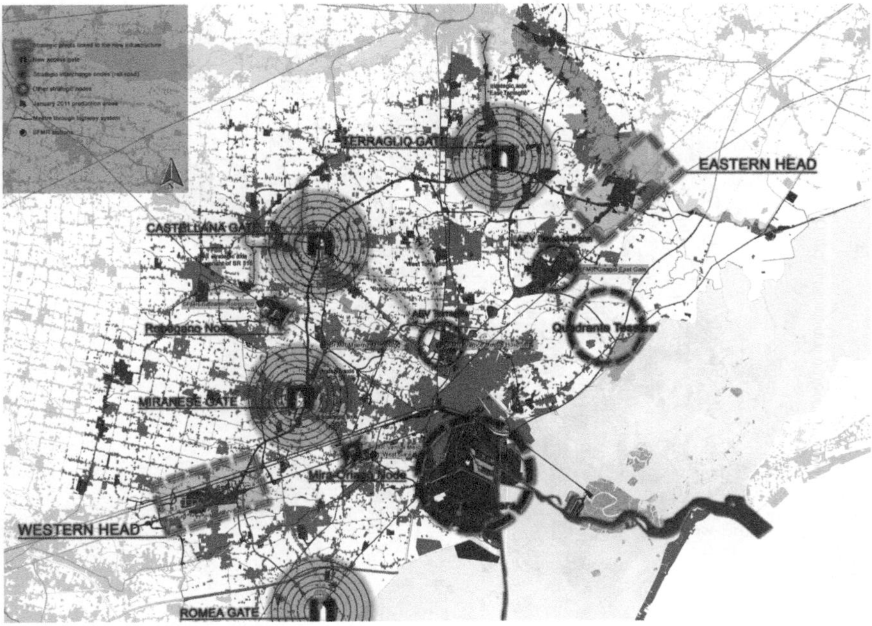

Figure 11.11 Second scenario based on integrated opportunities

According to article 38 of the Regional Spatial Plan (2013), the areas around toll roads, accesses to the primary network and to the regional railway system (SFMR), within a 2 km radius, are considered strategic areas of overriding public interest, that need to be planned and managed with appropriate regional strategic projects.[11]

The above-mentioned article, despite its vagueness, focuses on supra-management (directly coordinated by the Regional Administration) of the potential location effects and of the real estate pressures around the main nodes of the infrastructure system. Figure 11.12 shows an outline of potential areas. Greater gravity towards the bipolar city of Mestre-Venice can easily be noted, together with the amount of land for which planning should be primarily up to the Region.

11 To some extent, this regulatory device can be compared to a specific planning policy provided both within the 2004 and the 2011 London Plan (for further details, see Mayor of London 2004, 2011), the spatial development strategy for the whole Greater London Area. As a result of this policy, the Mayor of London identifies a set of areas close to the major nodes of transport infrastructures (railways, underground and motorways) where to directly promote and manage strategic development projects, as well as to coordinate local authorities and private developers. This sort of 're-scaling' of planning powers from the boroughs to the Mayor of the Greater London is absolutely important for an effective management of the so-called opportunity or intensification areas.

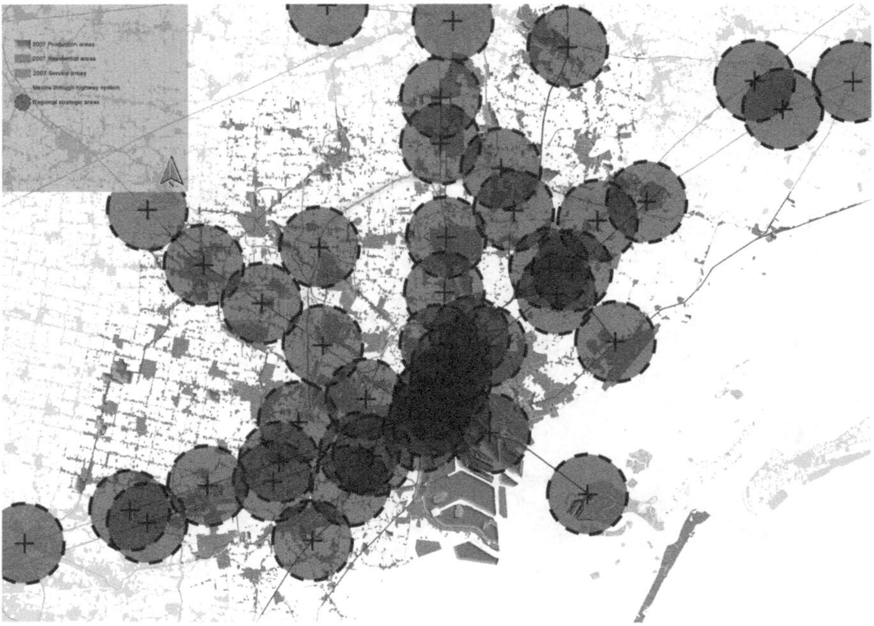

Figure 11.12 Regional strategic areas of overriding public interest

A two-fold interpretation of the results arises from applying such an article. On the one hand, the spread of productive activities, which is mainly due to local planning practices within the MTH, should be administered and, perhaps, limited. On the other hand, potential diffused logistics may not be properly exploited. As known from similar sprawl contexts, a low density logistics model might achieve satisfactory levels of efficiency, even with mini-hub and small road-rail interchange points. For these reasons, there may be conflicts between local and regional administrations in planning and land management.

Conclusions

The empirical evidence from this case-study may be a subject of further investigation and reflection.

The first point concerns the way the project relates to the spatial context. The synaptic nature is typical of reticular infrastructure as well as of any action carried out within an interconnected environment. Its nature may be recognized in at least two ways: as a sort of graft or as a mere outcome. A graft can potentially shape a territory as well as practices in an unpredictable, non-linear manner. Only ex-post can the logic of the change be grasped as a verifiable proof of what has happened. Looking at other similar projects, in terms of size, physical and functional features, inertia or context profile, can be helpful, but it is not decisive. Although similar cases may contribute to enrich the logics of change repertoire, use of this repertoire will necessarily be effective in terms of forecasting or simulation. Even a statistical Bayesian approach can prove to be weak in this regard, mainly due to vagueness and circularity of the relationship between project and context. Circularity tends to turn the project into context and the context into the project, questioning any assumption of causality. The 'mechanisms' of planning, rules and regulations, location and urbanization patterns can affect the relationships between context and project outcomes.

The project considered in this study does not only produce territorial, environmental, economic, social or institutional effects. It also prompts different responses from politicians, enterprises and communities that, in turn, contribute to update the spatial agenda with constant changes of direction. In the first place, such a situation weakens the concept of project cycle as well as some of the evaluation approaches (either process, performance or impact oriented). This happens even when the so-called spatial integrated evaluation is applied. The 'announcement effect', for example, influenced the project's implementation. It changed the geography of land use and location opportunities, but especially it redefined the mutual relations among the stakeholders.

Some evaluation approaches become less plausible. The assumptions of benefit-cost analysis (BCA), including the community impact assessment/evaluation (CIA/CIE) are no longer valid; neither experimental nor theory-driven evaluation are directly applicable.[12] Given the circumstances, the realistic evaluation is in big trouble, while adaptations of the pragmatist approach seem to keep their polish: namely, those designed as user-focused or oriented towards pluralistic values. Scenario writing approaches can no doubt benefit from these adaptations, mostly when they are considered not only as a series of 'scenes', but rather as a sort of 'social theatre' that defines and updates the political game.

A second point of reflection concerns the difficulty to measure the net effect of a synaptic project in a dynamic environment. Currently, there is no stable term of comparison for an area-based project. The net effect becomes a concept of little use.

A third point of reflection, related to the previous two, is the project's ability to activate a public discourse. While implementing the project, certain engineering features have been redesigned (path sections or tollbooth location, for example) and some planning rules were also changed. In particular, public discourse failed

---

12 Theory-based (or driven) evaluation (TDE) is an approach in which attention is paid to theories of policy makers, program managers and other stakeholders through the collection of assumptions and hypotheses. Both assumptions and hypotheses are empirically testable and logically linked. Over the last 50 years a number of evaluation experts have contributed to such development, among all Suchman (1967), Chen (1990), Pawson and Tilley (1997), Weiss (1998), Donaldson (2007).

to reshape the local government system, revealing a strong inertia at various administrative levels.

At the local level, the municipalities served by the new infrastructure did not hesitate to act as free riders, incrementally revising their spatial strategies in the light of perceived opportunities. This has led to a substantial real estate investment around the track, at least until the beginning of the economic crisis.

At a larger scale, the spatial changes due to the MTH implementation have been of little strategic value, and except for some sporadic cases, discussions on good spatial governance have proved limited and inconclusive. Even the local production system, whose behaviour was the object of specific assessment, was confined by business logic and traditional spatial aggregation patterns. The institutionalization of the metropolitan area of Venice has also been slow, despite the many factors that seemed to be moving in that direction.

Conversely, in contexts such as London an effective management of the location opportunities generated by transport infrastructures (underground lines, railway lines, motorways) is supported by an overall re-scaling process of planning-powers. The 2004 and 2011 London Plan, for instance, identifies a set of areas close to major transport nodes where the Mayor of London (and not the mayors of the single local municipalities) directly promotes and manages strategic development projects, as well as the overall coordination between metropolitan government, local government and private investors. Such a measure aims at fostering a supra-local spatial strategy which is then supposed to lead to a better definition of uses and functions for each single area served by the new transport infrastructures.

Acknowledgement

The chapter draws on and elaborates on the results of a post-graduate thesis entitled 'Tracce di passante? Nuove geografie localizzative delle attività economiche e delle aree produttive' ('Traces of a through-highway? New patterns and location of economic activities and productive areas'), by M. Basso and G. Borghelot, tutors M.E. Turvani, D. Patassini and S. Tonin. The thesis is part of wider research on Venetian territories.

References

Aschauer, D.A. 1989. Is public expenditure productive? *Journal of Monetary Economics*, 23, pp. 177–200.

Biehl, D. 1986. *The Contribution of Infrastructure to Regional Development: Final Report*. Office for Official Publications of the European Communities.

Chen, H.T. 1990. *Theory-Driven Evaluations*. Thousand Oaks: Sage.

Donaldson, S.I. 2007. *Program Theory-Driven Evaluation Science*. New York, NY: Lawrence Erlbaum.

Flyvbjerg, B., Bruzelius, N. and Rothengatter, W. 2003. *Megaprojects and Risk. An Anatomy of Ambition*. Cambridge: Cambridge University Press.

Fondazione Gianni Pellicani. 2008. *La Metropoli del Passante. Una fotografia della 'metropoli' delimitata dalla nuova infrastruttura*. Venezia: rapporto di ricerca Idee per Mestre.

Fondazione Gianni Pellicani. 2011. *La Metropoli del Passante due anni dopo*. Venezia: Quaderni di Idee per Mestre.

Forss, K., Marra, M. and Schwartz, R. (eds) 2011. *Evaluating the Complex. Attribution, Contribution, and Beyond*. Comparative Policy Evaluation, Vol. 18. London: Transaction Publishers.

Garreau, J. 1991. *Edge City: Life on the New Frontier*. New York: Doubleday.

Hirschman, A.O. 1988. *The Strategy of Economic Development*. Boulder: Westview Press.

Mayor of London. 2004. London Plan.

Mayor of London. 2011. London Plan.

Municipality of Casale Sul Sile. 2005. Local Masterplan.

Municipality of Casale Sul Sile. 2009. Operational Masterplan.

Municipality of Martellago. 2010. Local Masterplan.

Ozawa, C.P. and Seltzer, E.P. 1999. Taking Our Bearings: Mapping a Relationship among Planning Practice, Theory and Education. *Journal of Planning Education and Research*, 18, pp. 257–66.

Passante di Mestre S.C.p.A (a cura di). 2008. *Il Passante di Mestre. Una grande opera al lavoro*. Venezia.

Pawson, R. and Tilley, N. 1997. *Realistic Evaluation*. London: Sage.

Suchman, E. 1967. *Evaluative Research*. New York, NY: Russell Sage Foundation.

Vanni, V. 2009. *Un passante per il Nordest*. Una storia lunga cinquant'anni. Venezia: Marsilio.

Veneto Region. 2009. Regional Spatial Plan.

Veneto Region. 2013. Regional Spatial Plan.

Venice Port Authority, Three-year Operational Plan (2014–2017).

Weiss, C.H. 1998. *Evaluation: Methods for Studying Programs and Policies*. Upper Saddle River, NJ: Prentice Hall.

PART III:
Spatial Analysis for Integrated Projects

Chapter 12
Overcoming Lock-In: Instruments for Value Creation and Assessment Early in the Infrastructure Planning Process

Niels Heeres, Taede Tillema and Jos Arts

Introduction

Transport infrastructure plays a vital role in the socio-economic development of regions as it facilitates accessibility between spatial functions, within and between regions (Van Wee et al., 2013; Wegener and Fürst, 2004). However, the development of (new) infrastructure proves to be difficult because of conflicting interests, scarce space, complex environmental issues, dynamics in economy and land-use, changing roles of (national) government and (local) public resistance. Because of the considerable impacts the development of transport infrastructure may have, it is traditionally subject to extensive ex ante evaluation of environmental, social and economic impacts by instruments such as Environmental Impact Assessment (EIA), Social Impact Assessment (SIA), Cost-Benefit Analysis (CBA). These evaluation instruments play an important role in delivering sustainable outcomes in infrastructure planning. In practice, however, there is often much criticism on assessments for infrastructure development being too lengthy, too costly and the quality of resulting project studies is often poor (Arts, 2007; Arts and Niekerk, 2010; Arts et al., 2012; Runhaar et al., 2013). Causes are the content of impact assessment reports (too much detail, info-overload), procedure (complex regulations) and process (financing, decision making and management of projects), all relating to the issue of dealing with uncertainty intrinsic to planning (a culture of hedging risks). More specifically, an important reason for this cumbersome practice of infrastructure planning is so-called 'lock-in', which refers to the over-commitment to suboptimal policies as a consequence of path dependency (Cantarelli et al., 2010; Priemus and Van Wee, 2013; Elverding, 2008). Currently, transport infrastructure and spatial development are usually planned in different silos, by different authorities in different institutional settings. In infrastructure planning, government agencies are usually responsible for only a certain infrastructure mode – road, water, rail, etc. – therefore they usually develop also projects with a limited, locked-in scope. However, development of transport infrastructure is usually done in situations characterized by strongly interrelated land use functions and interdependent, but fragmented public actors, which calls for more integrated planning.

This chapter aims to explore to what extent current planning instruments are equipped for overcoming lock-in and achieving integration of major transport infrastructure (i.e. national highways) and other land use functions in planning. As a case we use the recent introduction of a new planning instrument in the Dutch national transport infrastructure planning sector. In the Netherlands attention to integrated approaches is abundant, including a recent discussion on the need and role assessment and evaluation of integrated plans and ambitions. This has been analysed in a recent inventory study of Rijkswaterstaat (the executive agency of the Dutch Ministry of Infrastructure and Environment) into planning and evaluation instruments used in national infrastructure planning in the Netherlands (Arcadis, 2012). We especially focus on a newly developed instrument called the 'Omgevingswijzer' – which might be translated as 'context appraiser' or 'sustainability check' – that aims at gaining insight into the potential for sustainable area developments around transport infrastructure initiatives (RWS, 2013; see also Heeres et al., 2012c and Burger, 2014). Two of the authors have been involved in the development of this instrument and the mentioned inventory study. Our experiences are complemented with data derived from interviews with experts in the field and document analysis. Illustration is provided by several case examples. In this chapter the term assessment is used to refer especially to the appraisal of impacts and the term evaluation to denote the broader set of tools for assessing impacts, creating values and the related processes. In this chapter the focus converges from broad to specific, taking into account the system for assessment and evaluation as a whole and highlighting a specific qualitative instrument ('Omgevingswijzer') by comparing it with quantitative instruments (such as CBA and BREEAM), but for an in-depth discussion of other planning and evaluation instruments the reader is referred to other chapters in this volume (e.g. chapters 13 and 15 discuss in more detail CBA).

The following section reviews the academic literature and develops a conceptual understanding of the contemporary function and requirements for evaluation and assessment in integrated transport infrastructure planning, arising from the challenges of functional interrelatedness and fragmentation of responsible actors. After that, we present the findings of our empirical exploration, based on studies on and involvement in infrastructure evaluation and assessment in the Netherlands. The third section explores the general functioning of instruments in infrastructure-land use integration. Subsequently, the focus shifts towards an in-depth empirical analysis of the instrument 'Omgevingswijzer' in order to understand the potential contribution of this new tool to evaluation and planning. In the final section general and specific findings are combined in an overall analysis to gain insights into the current state of evaluation of area-oriented infrastructure ambitions and potential improvements. The chapter concludes with a discussion about the question of how the process of evaluation can better support assessment and value creation for preventing lock-in and achieving more integrated plan-making in a complex context of interrelated functions and fragmented public actors.

Challenges for Evaluation and Assessment in Infrastructure Planning

In academic literature as well as in recent policy documents in various countries (in e.g. the Netherlands, Finland, USA, Japan) it is acknowledged that overcoming lock-in and achieving a more inclusive integrated scope in the planning of transport planning is crucial for more resilient and sustainable planning outcomes; however, such integration is scarcely present in practice (Te Brömmelstroet and Bertolini, 2009, Heeres et al., 2012b; Priemus and Van Wee, 2013). In response to this cumbersome planning and decision making on infrastructure projects, for instance, the Dutch minister of Transport and the minister of Environment appointed an external committee on the 'faster decision making of infrastructure projects' (Elverding, 2008). This Committee analysed problems thoroughly and provided influential recommendations, including: early and broad participation, a staged inclusive planning process, and an integrated 'area-oriented' approach to infrastructure planning. Other Dutch authors (e.g. Arts, 2007; De Zeeuw and Licher, 2008) have also indicated that it is crucial to make infrastructure projects much more inclusive in their scope both regarding content (scope of issues, impacts) and process (parties, stakeholders involved). Infrastructure planning needs careful consideration of the area surrounding projects, as well as the establishment of integrated planning and evaluation processes and corresponding procedural arrangements. Also the need for more integration is seen as an essential element of a more inclusive, sustainable transport planning (see Arts, 2007; Banister, 2008; Cervero, 2009; Van Wee et al., 2013).

Integrated Approaches

As a consequence, increasing attention for planning approaches that take an integrated approach of transport infrastructure and other issues is observable in several countries. While these approaches strongly differ in motivation and practical implementation – e.g. 'area-oriented' approaches in the Netherlands, 'context-sensitive design' and SAFETEA-LU-regulations in the USA, 'regional packages' in Sweden and a 'transport revolution' in Finland (Heeres et al., 2012a and b) – there is an underlying common rationale. In order for transport infrastructure development to be a viable activity in situations characterized by strongly interrelated land use functions and interdependent but fragmented public actors there is need for a transformation from transport infrastructure planning as a technical effort towards a more socio-technical approach (De Roo et al., 2001; Amekudzi and Meyer, 2006; Klieverik, 2010; Runhaar and Driessen, 2010). The essence of these approaches is that the functional and spatial scopes of traditionally separate planning sectors are widened and are considered coherently to address the functional interrelatedness and fragmentation of responsible public actors. Regarding the governance of these approaches this asks for exploration of a shared development direction as an addition to the optimization of specific spatial issues (Heeres et al., 2012b).

Coping with Interrelatedness

For major transport infrastructure development the emergence of integrated strategies implies engaging with other land use developments (see Figure 12.1). Engaging with other land uses means that the functional and spatial scope of infrastructure development are broadening and that boundaries between infrastructure and other land uses are becoming fuzzy in order to include (relevant) interrelated land uses into the processes of infrastructure development (Graham and Marvin, 2001; Graham, 2009). Consequently, development of transport infrastructure networks and local area quality are becoming intertwined activities (in the Netherlands often referred to as area-oriented planning; see e.g. Van der Heijden, 1996; Priemus, 2004; Handy, 2008). Exploration of potential solutions is not only concerned with optimization of infrastructural issues, but explicitly takes into account the interrelationships of transport infrastructure with other land use functions. Through synergy among individual interests, integration of infrastructure and other land use functions may be expected to lead to higher quality outcomes than traditional planning. Coherence among different spatial development aspects in a location is expected to create higher value outcomes. This added value can be experienced as higher financial-economic value or other values (social, environmental) – see e.g. Peek (2006), Ruth et al. (2014), Hamers et al. (2014).

Addressing Fragmentation

In addition to higher local spatial quality of the functional-spatial outcomes of planning, integrated planning is also incentivized by a need for more dense cooperation of fragmented public actors on different government levels, representing different spatial interests. All actors base their ideas about plan-making and discussion of spatial development on historically developed referential frames. These frames differ strongly between various types of planners and make it difficult to come to a shared understanding of problems, scope and potential solutions: transport infrastructure planners often have civil engineering orientations, while land use planners have a social science background. The creation of shared development visions, exploiting potentially fruitful linkages between transport infrastructure and other land uses (see Figure 12.1), is thus also dependent on the ability to overcome these differences and adjustment of disputed or conflicting frames is required – ‘reframing’ see e.g. Kaufman and Smith (1999) and Van den Brink (2009).

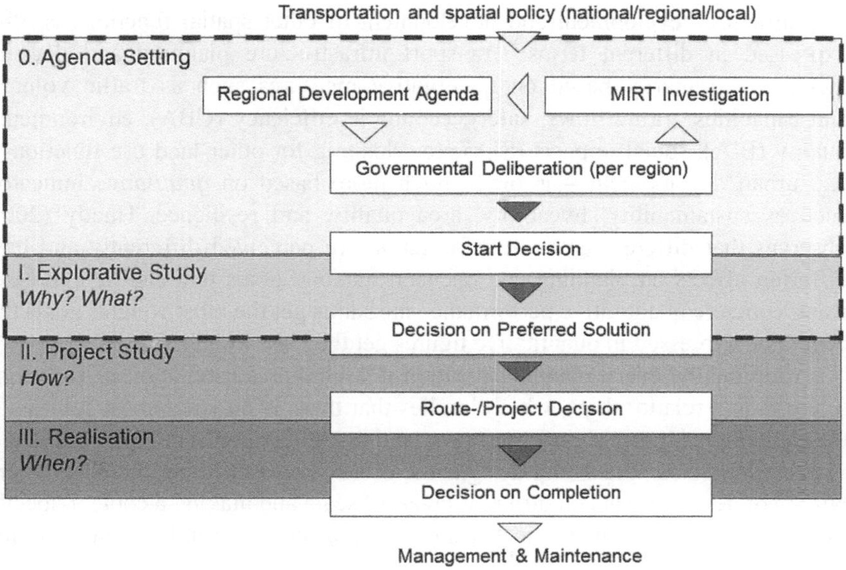

Figure 12.1 Dealing with interrelatedness and fragmentation infrastructure planning

Note: Broadening of functional and spatial scope of plans and designs (left) and shifting from conflicting interests to forming shared development directions (right).

Implications for Evaluation

The dual character of area-oriented infrastructure planning – emerging from thoughts about the co-existence of technical rationality and a communicative rationality in planning (Woltjer, 2000) – asks for evaluation instruments that are capable of serving this hybrid purpose (Runhaar and Driessen, 2010). To combine the logical scientific attitudes of infrastructure planning with the narrative argumentation of spatial development hybrid forms of evaluation, taking a position in between formal and informal rationalities, are required (Barbanente and Borri in Khakee et al., 2008; Van der Knaap, 2004). Next to providing an insight into the effects of integrated planning proposals, instruments also have a potential role in stimulating collective operations and facilitating cooperation among actors of different backgrounds (Beukers and Heeres, 2012).

Content of Assessment

The viability of the integration of transport infrastructure and other land use functions by means of area-oriented approaches strongly depends on the ability to make clear the added value of integrated plans to planners and decision makers (Meyer and Miller, 2001). This is problematic as information about transport infrastructure development and development of other spatial functions is often expressed in different terms. Transport infrastructure planning and decision making is primarily based on quantitative measures such as traffic volumes and capacities, traffic flows, safety, economic efficiency (CBA), environmental quality (EIA), social impacts (SIA) etc. Planning for other land use functions – e.g. urban development – is often much more based on *qualitative* indicators such as sustainability, liveability, area quality, and resilience. Handy (2008) observes that different types of information are perceived differently and have differing effects on planning and decision making: goals that can be expressed using concrete/quantitative performance measures get the most weight; goals that cannot be expressed in quantitative figures get the least weight.

Additionally, every planning location is a unique constellation of functions, interests and relationships, which implies that there is no single best solution to the challenge of transport infrastructure-land use integration. The establishment of viable *multi-sector* plans and designs therefore requires linking the benefits and burden of developments on different levels of scale and making a context-specific trade-off between potential effects expressed in quantitative and qualitative terms.

Assessment Instruments in Collaborative Processes

Traditionally, assessment instruments should be able to make accessible the relevant information in which the added value of integration is expressed and made understandable for planners with different backgrounds. Additionally, in line with the turn from restrictive to development-oriented planning (Hartman et al. 2012, Salet and Woltjer, 2009), instruments could also take an expanded role, a move from *testing* against rules and limiting development to *generating* alternatives and supporting interactive processes (see also Hamers et al., 2014).

The existence of 'multiple truths' and the need for 'reframing' further complicate mutual understanding and communication in integrated developments on the infrastructure-land use interface (Wildawsky in Van der Knaap, 2004). Research by Beukers (2012) on the use of cost-benefit analysis by groups of planners indicates that different backgrounds potentially leads to misunderstanding between representatives of different interests, a lack of communication, mistrust and resistance towards each other's approaches and interests. Different scholars (Handy, 2008; Beukers et al., 2012; see also Chapter 15 of this volume) subsequently indicate that difficulties with respect to the communication of potential impacts lead to the loss of important roles for assessment instruments, such as opportunities for learning and better plan-making. Therefore, in addition to the improvement of technical-substantial aspects of the planning process authors such as Handy (2008) and Runhaar (2010) also stress the need for attention to the effect of planning instruments on the promotion of a more open and participatory way of plan-making.

Importance of Early Stages

In addition to implications for the content and process of evaluation instruments, the position in the planning process is important for integrating infrastructure and land-use in planning. For the creation of added value the early stages of integrated planning processes are essential (Holland, 1998; Ruth, 2011; Peek, 2006). In these pre-consent stages the planner's freedom for experimentation is largest and the chances of successfully exploring and developing shared ambitions are higher than after official decisions on future directions have been taken and details are elaborated (Elverding, 2008). In order to optimize the effects of infrastructure-land use integration, the focus of the integrated planning processes is first of all on the early stages. Samset (Williams, Samset and Sunnevåg, 2009) refers to these stages as 'front-end governance' in projects, when degrees of freedom are large and allow for strategic and conceptual choices to be made. The front-end phase – between initial idea and the final decision to finance the project – allows for conceptual thinking in which 'the focus is on economic and societal aspects, rather than technical aspects' and where 'the anticipated effects of a perceived project' can be leading 'rather than the present undesired situation' (Williams and Samset, 2010, p. 39). In the Dutch planning context this would be the agenda setting phase and the stage of early strategic explorations, before the preferred

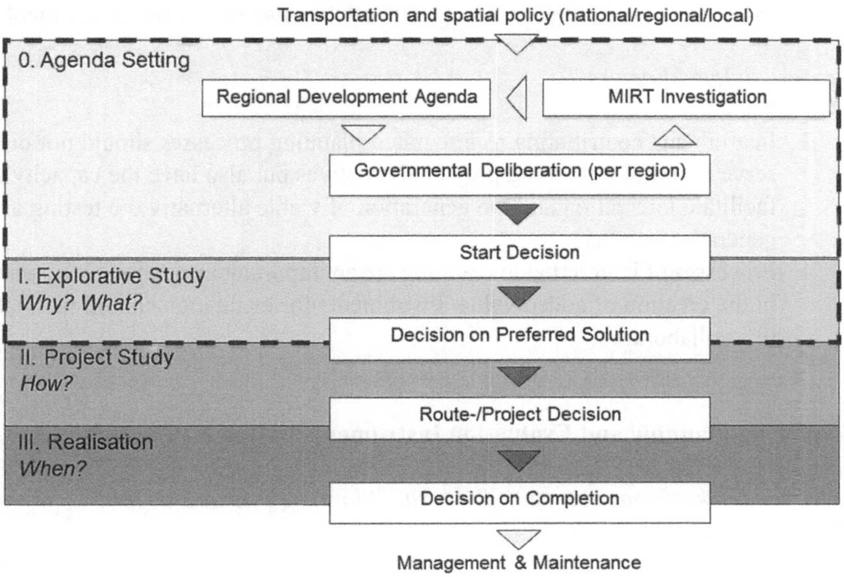

Figure 12.2 The Dutch infrastructure planning process: Strategic, operational and realisation stages

Source: Based on I&M (2011).

alternative decision and complementary budgetary allocations (see Figure 12.2; Heeres et al., 2012c). Implementation of innovative instruments for planning and evaluation in these phases allows for the necessary strategic freedom to seriously consider different conceptual solutions and overcoming lock-in.

Synthesis

Based on the discussion above, we conclude that for evaluation instruments to remain a platform, which supports infrastructure development appropriately by helping to avoid lock-in situations and promoting the creation of added value on the infrastructure-land use interface, the following hybrid factors can be identified:

Substantial Factors

1. Regarding the content of plans and designs, evaluation instruments must pay attention to the creation of coherence among different functions involved as well as the individual functioning of the infrastructure development à mono-functional and multi-functional scope;
2. Regarding interactive processes qualitative measures may be more useful than quantitative measures for expressing and discussing potential effects collectively. However, impacts such as traffic and transport effects are best expressed in quantitative terms à quantitative and qualitative assessment;

Process-Related Factors

3. Instruments contributing to integrated planning processes should not only serve as a tool for testing project alternatives but also have the capacity to facilitate interaction and the generation of viable alternatives à testing and generation of alternatives;
4. Processes of interaction in planning are an important aspect for instruments in the creation of added value. Instruments for evaluation should facilitate this collaboration.

Exploring Planning and Evaluation Instruments in Dutch Practice

'Meerwaardeketenmanagement': Exploring Planning Instruments for the Creation of Added Value

Recently Rijkswaterstaat made an inventory of the various planning and evaluation instruments that are used frequently in the planning of national infrastructure in the Netherlands (see Arcadis, 2012). This study was part of an action within Rijkswaterstaat – called 'Meerwaardeketenmanagement' – to revise and restructure the current framework of planning instruments that are used for the evaluation of potential added value throughout the transport infrastructure planning process. The objective was to create an overview of the available instruments, matching project roles for the instruments, planning phases for suitable application, what added values are to be created and where in the planning process this is located. Regarding this, added value for transport infrastructure projects had been defined as all qualitative, economic and organizational aspects of infrastructure projects that are higher than primary functional specification; 'beyond primary functional specifications' and 'with less effort to get more results'. (Arcadis, 2012). A wide array of tools proved to be available as some 21 evaluative instruments were found to being used frequently in the planning process for transport infrastructure. Instruments related to assessing impacts (CBA, SEA, EIA, value engineering, Life-cycle costing), procurement and contracting (market scan, best value procurement, economically most favourable bid, integrated contracting such DBFM), rating systems (BREEAM, DuboCalc, CO_2-performance ladder) – for a more detailed discussion of similar (international) instruments see Arts and Faith-Ell (2012). They deal more with environmental, social and/or economic impacts. Also some of the instruments are long-standing, others such as Omgevingswijzer have been developed rather recently in response to the discussion about more integration in Dutch planning (see previous discussion).

The Meerwaardeketenmanagement study reveals that instruments seem to be overlapping and having a strong focus on technical issues, whereas the relationship of transport infrastructure with other issues seems to receive far less attention. For instance, social-interactive issues are largely left unaddressed. However specific process-related values can be reached. Creating sufficient public support for plans and proposals may be regarded as a primary process expectation: aiming for acceptance of project plans as can be assumed a core value in planning processes. To rise above this acceptability level planning instruments should initiate efforts for creating support to stimulate additional functionality (e.g. integrated area) or lowering barriers between actors to reduce the efforts required for this – see also Ruth et al., 2014.

The inventory of the Meerwaardeketenmanagement study seems to indicate that cooperation and collective action are primarily stimulated by instruments with a broad functional scope, applied early in the process. Such, often qualitative, instruments offer the possibility to explore the value of possible future combinations of functions (co-production) from a qualitative perspective, in addition to creating an understanding of the mutual interests of parties involved. In that way parties involved increase their understanding of interdependence and can, by getting to know each other, even better understand how they can mutually reinforce one another. Such an inclusive process could eventually lead to the generation or design of alternatives that provide higher societal value. Regarding this especially the Omgevingswijzer instrument is considered potentially interesting (Arcadis, 2012; RWS, 2014).

Tools Compared

Before discussing Omgevingswijzer in more detail in the next section, we compare this instrument briefly with some other instruments available to give more insight in the nature of this instrument and more in general in the instruments used in Dutch infrastructure planning and discussed in the Meerwaardeketenmanagement study (Arcadis, 2012). Because of limitations of space, we focus especially cost-benefit analysis, EIA and BREEAM – as these instruments are also widely applied in other countries (see e.g. Arts and Faith-Ell, 2012). Table 12.1 shows a comparison of the four instruments. More in-depth discussion about CBA, EIA and BREAAM can be found in extensive literature (see e.g. Beukers et al., 2012; Sijtsma et al., 2012; Runhaar and Driessen, 2012, Runhaar et al., 2013; Arts and Faith-Ell, 2012).

Table 12.1 Comparison of Omgevingswijzer with some well-known instruments used in infrastructure planning

	Omgevingswijzer	Cost-benefit analysis	EIA (Environmental Impact Assessment)	BREEAM (infra or area)
Functional scope (content of assessment)	Broad, 12 themes clustered around economic, social and environmental issues	Primarily economic benefits and costs. To certain extent also issues such as energy and martials used, ecology and wellbeing	Primarily environmental themes, although often also social and sometimes economic issues included	Broad range of environmental and social themes, minor attention to project finances
Quantitative or qualitative measures	Qualitative indication	Quantitative	Usually primarily quantitative, sometimes also qualitative	Quantitative
Testing and/ or generating alternatives	Testing and generation of alternatives	Primarily testing alternatives	Primarily testing alternatives	Primarily testing, more specifically rating. Applicable in early stages for comparing alternatives
Attention to process in assessment	Explicitly developed as a tool for facilitating discussion and collective decision making (to be developed further)	Until now very little (process influence is currently being researched)	Originally mostly content, but last decade much attention for process	The instrument is a means for discussing sustainability in a common language

Together with the review of Meerwaardeketenmanagement, the above comparison of four instruments reveals several issues about added value creation in transport infrastructure planning. First of all, there are instruments with both broad and dedicated functional scope. Secondly, the planner's toolbox includes qualitative as well as quantitative instruments. Thirdly, explicit attention to the planning process, more than general attention to creation public support, is largely absent. Finally, while most instruments are equipped for testing, generation of viable alternatives seems to be far less present. In general this implies that a wide range of instruments is available for infrastructure planners. Specific attention to processes of interaction and collective design is however only included in a few instruments.

Omgevingswijzer: A Tool for Integrating Infrastructure and Land-Use Planning

The short exploration above reveals that there are only a few tools that link in with the instrumental requirements for the support of infrastructure-land use integration (broad range of issues, qualitative nature, testing and generation of alternatives). However the so-called Omgevingswijzer seems to have potential. This tool has been developed over the past years as part of the sustainability strategy of Rijkswaterstaat. Rijkswaterstaat has identified integrated area development as one of its core themes regarding sustainability (RWS, 2011, 2013, 2014). Since 2012 also other organizations within the Dutch infrastructure sector, such as provinces, municipalities, water boards as well as private developers and construction companies, use the Omgevingswijzer for their projects – RWS (2013), for a more extensive discussion of the development and application of this instrument see Heeres et al. (2012c).

The purpose of the instrument Omgevingswijzer is to gain concrete insight into the potential for sustainable area development around transport infrastructure planning initiatives. Therefore the instrument explicitly considers: (a) what potential for infrastructure-land use integration exists and may lead to the creation of additional value; (b) which actors would be involved in these processes of integration; and (c) what the role of the developer (i.e. Rijkswaterstaat) would be to allow for such integration. For this purpose the instrument consists of two main elements: firstly, a tool that visualizes spatial-functional effects and potentials for synergy in sustainable area developments and, secondly, a coupling between project organization and the relevant stakeholder environment of the integrated project. Especially the first element has been intensively developed over the past two years in a series of pilot applications. The second element is currently being further developed through new pilot applications with specific attention to interaction between the stakeholders involved (Heeres et al., 2012; RWS, 2012; Burger, 2014).

The main purpose of the instrument is not to provide quantitative insights, but rather to start a discussion on possible infrastructure-land use integration. Therefore, the indicators represent a broad perspective on sustainable development and can be linked to the People-Planet-Profit (PPP) pillars of the triple bottom line (Elkington, 1997). To this end the instrument contains a checklist to assess whether all sustainability aspects are addressed and to what extent. In addition, the instrument can be used to compare alternatives options in a project or alternative projects. Furthermore, by comparing early strategic principles and aims with concrete development proposals the instruments also allows for a consistency check on projects as they progress towards establishment of final proposals (RWS, 2013; see Figure 12.3).

The instrument works by exploring a project alternative or comparing different alternatives on 12 different indicators of sustainability. These indicators, linking in to either the people, planet or profit dimension of sustainability, are operationalized for transport infrastructure development through the formulation of principles. Table 12.2 provides an overview of the 12 indicators and the relevant principles. In practice, project alternatives are rated on sustainability characteristics by means of available policy and project documentation and by using standardized interviews questions. The outcomes of this examination are visualized in a 'synergy wheel' (see example below), in which the 12 sustainability indicators are included. A lighter shade above the zero-line indicates a positive effect on a specific theme, a darker shade below zero indicates negative effects to be expected for an alternative. The scoring mechanism also allows for both positive and negative effects to be shown for the same theme. This way a nuanced insight into the expected effects of an alternative is presented. The use of this wheel to visualize the sustainability of proposals seems to resemble other recent initiatives regarding the assessment

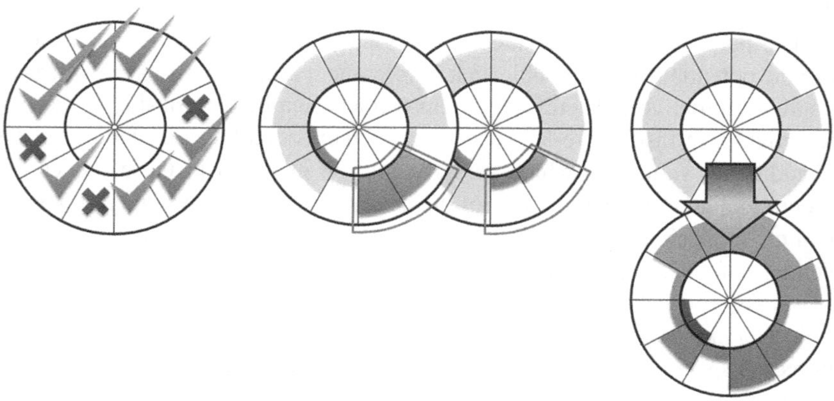

Figure 12.3 Three different purposes of Omgevingswijzer: a) checklist; b) comparing alternatives; c) consistency between stages

Source: Based on RWS (2013).

of sustainability effects (see also the development of the SPeAR (ARUP) and Adaptive Capacity Wheel: Gupta et al., 2010). The use of a wheel allows for an accessible and comprehensive overview of sustainability effects and provides insight into the relative positive and negative effects of different alternatives.

Table 12.2 Twelve indicators of sustainability and relevant principles

Sustainability theme	Indicator	Relevant principles relate to … (e.g.)
Ecology/planet	Water	Water safety/water flooding/water quality/water shortage/climate proofing
	Underground/soil	Soil quality/diversity of soil types/soil biodiversity /archaeology/soil subsidence
	Energy and materials	Reduction of energy consumption/use of renewable energy/use of fossil fuels/consequences of extraction of resources/exchange of energy/robustness of energy networks
	Ecology and biodiversity	Room for flora and fauna/biodiversity/ecological structures
Social/people	Use of space	Linkage with existing (regional) construction needs/restructuring/expansion/multiple land-use
	Public space and functions	Spatial quality values: experience, use and future/integrated design
	Social relevance	Social well-being/demographical composition and trends/public support/local expertise
	Well-being	Health: positive contribution and limitation risks/physical and social safety/prevention of nuisance
Economy/Profit	Mobility and accessibility	Robustness of transport system (within/between modalities)/efficient infrastructure use/accessibility and connectivity/adaptiveness for mobility policies
	Investments	Cost-benefit ratio/area potential/value capturing
	Economic benefits for companies	Business climate/economic policy/capacity for innovation and adaptation
	Economic benefits for citizens	Employment/accessibility of job market/development of labour force/available services

Source: RWS (2013).

If applied in practice, the Omgevingswijzer can take any of three particular roles in the planning process. First of all, the instrument can serve to measure the sustainability potential of a specific plan or proposal. Secondly, the Omgevingswijzer can serve as a tool for comparison of alternative options in a project on the overall sustainability picture and on 12 specific sustainability facets. Thirdly, the Omgevingswijzer can indicate the consistency of the planning progress by comparing consecutive proposals (see Figure 12.3, RWS 2013; Hanekamp and Tromp, 2012). Two examples of application are described below in order to illustrate the potential of this instrument for broadening of the scope of infrastructure projects.

Two Pilot Applications of Omgevingswijzer

To assess and improve the application of the developed instrument, the Omgevingswijzer has been tested in pilot projects by the Dutch department of public works and water management (Rijkswaterstaat) in 2012. Two pilot projects are selected here to show the results and potential usage of the instrument: one that combines road infrastructure with opportunities for spatial development (A1-zone) and one co-development project of a canal, nature and recreation (Zuid-Willemsvaart) – see also RWS (2012), Heeres et al. (2012c), Burger (2014).

First of all, the A1-zone program aims to enlarge the capacity of one of the major highways in the Netherlands on a stretch of 70km in the eastern part of the country. It is an important connection between the Randstad-area and Germany. The aim of the program is to utilize opportunities for the region that originate from the capacity enlargement of the road. Regional opportunities that have been identified are the enhancement of spatial quality and sustainability. The program's functional scope also comprises better exploitation of the region's economic potential, for which 10 focus issues have been identified. The spatial scope of the proposed developments is first of all determined by the administrative boundaries of the municipalities that are cooperating within the program.

Secondly, the Zuid-Willemsvaart project comprehends the realization of a new 7 km waterway, replacing an older canal with a new bypass that is suitable for larger ships. This project is part of a larger waterway project that aims to increase the capacity of waterway system in the south of the Netherlands. Where the new section of waterway passes between the city of 's-Hertogenbosch and the town of Rosmalen the required construction works are expanded with the interest of local municipalities. The expanded scope comprises the development of an ecological zone, nature and recreational facilities. This way the reconstruction of a part of the national water network is employed to provide for local wishes.

The results of the explorations for these integrated infrastructural developments are visualized in Figure 12.4.

Table 12.3 Two integrated projects compared

Project	A1-zone	Zuid-Willemsvaart
Infrastructure type	Highway	Waterway
Objectives	Combination of road infrastructure and spatial objectives on a regional level of scale to enhance spatial quality, sustainability and economic potentials	Construction of a new waterway in combination with the creation of an ecological zone, including nature and recreation facilities
Main motivation	Capacity increase	Capacity increase
Functional scope	A bundling of existing projects on infrastructure, housing, business, water and recreation	Water transport infrastructure, nature and recreation
Spatial scope	Zone around highway A1 between Apeldoorn and the border with Germany (ca. 70 km × 5 km)	A 7 km stretch of new waterway and the area between the city of 's-Hertogenbosch and the town of Rosmalen (ca. 7 km × 1 km)
Moment of application	Explorative studies (2 applications)	Preparation of realization
Purpose of application	1. checklist and consistency; 2. alternative comparison	Emphasis on consistency check, but also as checklist and alternative comparison

The initiative for the A1-zone shows a general improvement on all themes, with a strong emphasis on the economic aspects. This seems to be in line with the intention of the program to stimulate the economic position of the region, accompanied by spatial quality and sustainability. The information provided by the completed synergy wheels has affected this pilot project also in different ways. In the case of the A1-zone the application of the Omgevingswijzer has confirmed that the selected project alternative has the broadest potential for sustainable regional development. By covering ambitions regarding the road infrastructure issues, as well as other regional land use ambitions, the synergy wheel has identified the broad potential of this regional vision. Additionally, application of the Omgevingswijzer has been a wakeup call for the project organization. Following the discussion during application it appeared that for many of the identified strengths the vision no responsible actor had been appointed in the early explorative stages. To avoid risks of inconsistency between the initial visions and detailed project development in later phases, and consequent implementation problems, this has been identified as a point of attention in later phases.

Comparable to the A1-zone, the application of Omgevingswijzer in the Zuid-Willemsvaart project also shows a broad distribution of positive effects. This implies that the realization of the new waterway is a driver for quality of the area as a whole. The pilot application in the Zuid-Willemsvaart has however taken place during the final preparations of realization. Therefore, the synergy wheel mainly is a review of the planning process that had been undertaken. Application of the Omgevingswijzer has made clear the synergy effects between improvement

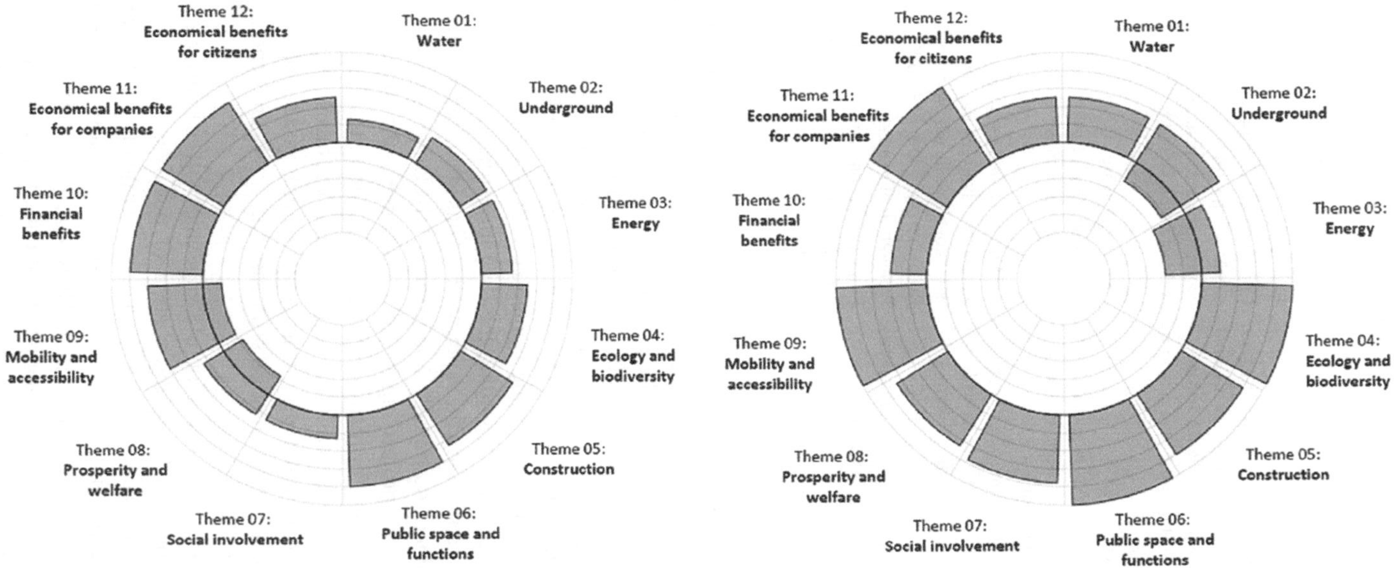

Figure 12.4 The 'synergy wheel' for (a) A1-zone, (b) Zuid-Willemsvaart

of the network of national waterways and nature and recreational facilities with a more local or regional character. Despite the synergies that are created in this project, the project has experienced a troublesome process with struggles between the various involved layers of government. One of the reasons is that it has been difficult to sufficiently communicate the synergies to the stakeholders. The project team commends the instruments straightforward visualization of effects and indicates that it would have been primarily useful as a means for communication between stakeholders. Interviewees suggest that the Omgevingswijzer could have helped in avoiding the experienced struggles between various governmental layers. Additionally the project team mentions the instrument's capacity to better analyse the problems behind the initial assignment. This capacity promotes the instruments function of identifying and creating synergies from an early stage onwards.

Analysis: Potential of Omgevingswijzer for Integrating Infrastructure and Land-Use Planning

If the suggestions of such authors as Handy (2008) and Beukers (2012), for improving evaluation instruments in early planning phases are projected on the Omgevingswijzer, several issues can be noticed – see discussion in the previous sections and Heeres and Beukers (2012). The development of Omgevingswijzer is an attempt to measure and express the various effects of planning proposals in comparable terms. A single evaluation tool has been developed to express the impact of different kinds of effects planning proposals may have. By developing one single measure the effects of a planning proposal on various aspects has been made visible in a qualitative way. This allows for a qualitative comparison between different aspects of a proposal. There have been several reasons for the choice to develop an instrument that creates a qualitative insight into the sustainability potential of plans, most important are:

- Providing an accessible and transparent way of comparing development proposals and using it collectively to facilitate a spatial discussion between fragmented stakeholders.
- It has proven to be not easy to develop an instrument that is able to provide a trustable quantitative insight into developments, which includes such a comprehensive range of themes. Therefore, a qualitative instrument has been chosen to measure effects that are yet difficult to forecast adequately.

During the pilot applications and workshop discussions, participating project managers of integrated infrastructure-land use projects have indicated that the Omgevingswijzer is a useful tool in providing structure to their assignment. First of all, as it shows the interrelatedness among different spatial functions in integrated projects, results from application of Omgevingswijzer reveal the rationale for engaging in an integrated approach and actively addressing stakeholder fragmentation. Secondly, project managers stress that a further advantage is that the Omgevingswijzer facilitates communicating the purpose and importance of integrated planning – from a cross-disciplinary perspective to a multidisciplinary teams of planners. Additionally, making different types of knowledge accessible for larger groups promotes the sense of interdependency within multidisciplinary project teams. Application of Omgevingswijzer seems to be helpful in establishing sufficient understanding of, and respect for, the mutual interests among teams of planners to facilitate appropriate collaboration. Thirdly, project managers experience Omgevingswijzer as a valuable tool for also communicating the broader implications of a proposal to decision makers. With its qualitative nature the Omgevingswijzer can sketch a more comprehensive picture around a few focus points (e.g. traffic capacity, project finances, nuisance and safety) on which decision makers tend to focus (Beukers and Heeres, 2012). As such, the visual strength of Omgevingswijzer's synergy wheel provides a perspective that serves as an addition to the information provided by more traditional, specific and quantitative planning instruments.

Conclusion and Discussion: Towards Transport Infrastructure – Land Use Integration?

In this chapter we have discussed the issue of lock-in and dealing appropriately with the challenges of functional interrelatedness and fragmentation of responsible actors on the infrastructure-land use interface in area-oriented planning approaches. On the basis of a literature review we conclude that a more hybrid role for evaluation instruments seems desirable. In addition to the traditional purpose of assessing and communicating the effects of integration, evaluation instruments therefore also have a potential function in facilitating processes of stakeholder interaction and generating integrated alternatives. This potential is assumed to be highest in the early phases of integrated planning processes.

In order to explore the application of such hybrid planning instruments we have discussed the recent introduction of a hybrid instrument called 'Omgevingswijzer' and its application in Dutch planning practice. On the basis of this discussion we conclude that the evaluation instrument Omgevingswijzer has many characteristics that might address the challenges of lock-in and integrated planning on the infrastructure-land use interface. The instrument applies a straightforward approach to assessment – based on the people-planet-profit concept of sustainability – by sketching a comprehensive picture of the impacts of the planning proposal at hand, providing a common measure to (qualitatively) express the impact of different kinds of effects and visualizes different sustainability aspects in comparable terms. This measurement should however not be expected to provide definitive judgment about the overall sustainability score and the findings of the instrument should be interpreted from a qualitative perspective. However, this is consistent with the aim of the Omgevingswijzer to be applied in the early stages of the planning process in order to stimulate integrative thinking and overcoming lock-in. Usually in these early stages there is very little or no detailed, quantitative data about the planning proposal (and its impacts) available. Moreover, the qualitative results provided by the Omgevingswijzer prove to form a platform for individual or collective interpretation and discussion. Due to its accessible character it is capable of creating awareness of interdependency and also stimulating collective group reasoning. Furthermore, its transparent nature and easy application make this type of instrument particularly useful in the early stages of making plans that integrate major transport infrastructure development and other local land use functions.

As such this instrument is clearly process-oriented and should be considered as a helpful addition to the wider range of available evaluative instruments in infrastructure planning (including EIA, CBA, and BREEAM). These instruments are strong in estimating impacts in detailed information and take a usually quantitative perspective on specific issues (a relatively narrow scope) but seem to be less suitable for the process-related challenges of infrastructure-land integration. An important issue is the lack of common language and understanding on the problem and the need of much expert knowledge which does not facilitate discussion between stakeholders and collective design. Regarding instruments as EIA and CBA, however, also trends towards including a broader scope of impacts (both quantitative and qualitative) as well as process-oriented elements can be seen – see e.g. Runhaar and Driessen (2012), Sijtsma et al. (2012), Chapter by Beukers in this volume). Moreover, in other sectors outside infrastructure planning such as spatial planning, similar instruments as the Omgevingswijzer are developed for instance the Plan Review tool (chapter by Hamers et al. in this volume).

Although the Omgevingswijzer is still being further developed, we think that on the basis of the discussion in the previous sections we may draw some preliminary conclusions about the potential of planning instruments to overcome lock-in and support infrastructure-land use integration:

- The ambition of integrated planning (especially integration of national transport infrastructure networks and local place quality ambitions) requires instruments that address the content of plans and designs as well as processes of interaction.
- Specific attention must be paid to added value creation: rising above primary, sectoral interests and functionality.
- A specific type of instrument seems useful: qualitative measures, easy to use, outcomes easy to comprehend, suitable for collective application.
- A specific strength of the Omgevingswijzer is that through its assessment on the PPP aspects of sustainability it is able not only to appraise the impacts of given project (alternatives) but also to stimulate awareness of interrelations and thereby the sustainability potential of a proposal. It may

challenge the original scope of proposals, point at relevant stakeholders that are not (yet) at the table and thereby it may stimulate also creativity regarding the generation of new alternatives.

The instrument Omgevingswijzer is seen by practitioners as a useful addition in the planner's toolbox by addressing the requirements of early, integrated plan-making that strives to overcome lock-in situations. However, tools such as the Omgevingswijzer, seem to be a useful as a, necessary but not sufficient, planning and evaluation instrument for overcoming lock-in situations in planning. This type of instrument seems especially useful in early plan stages of strategic exploration, to support collective design and choice; however, more detailed and specific instruments remain needed for detailed elaboration and assessment – especially in subsequent stages. When observed as a comprehensive set of supporting instruments for evaluation and assessment, the combined instruments should support the complete cycle of intelligence collection about the problem, collective design of multiple alternatives, and choice for one alternative. Moreover in general, there is need for further investigating the role and position of evaluative instruments such as the Omgevingswijzer in relation to the broader array of evaluation and planning instruments available during the planning process.

References

Amekudzi, A., Meyer, M.D. 2006. Considering the environment in transportation planning: Review of emerging paradigms and practice in the United States. *Journal of Urban Planning and Development*, 132(1), pp. 42–52.

Arcadis. 2012. *Inventarisatie meerwaarde-instrumentarium ketenmanagement Rijkswaterstaat* (Inventory study added-value instruments chain management Rijkswaterstaat). The Hague: Rijkswaterstaat.

Arts, J. 2007. *Nieuwe Wegen? Planningsbenaderingen Voor Duurzame Infrastructuur* (New Roads? Planning approaches for Sustainable Infrastructure). Groningen: University of Groningen.

Arts, J. and Faith-Ell, C. 2012. New governance approaches for sustainable project delivery. *Procedia-Social and Behavioral Sciences*, 48, pp. 3239–50.

Arts, J. and Niekerk, F. 2010. 'Impact assessment and management of infrastructure projects: From risk source to risk control'. In A. Hulll, E. Alexander, A. Khakee and J. Woltjer (eds) 2010. *Evaluating Sustainability and Participation: Fitting Evaluation to Planning Contexts*. London: Routledge.

Arts, J., Runhaar, H. Fischer, T., et al. 2012. The Effectiveness of EIA as an Instrument for Environmental Governance – A Comparison of 25 Years of EIA Practice in the Netherlands and the UK. *Journal of Environmental Assessment and Policy Management*, 14(4), pp. 1–40.

Banister, D. 2008. The sustainable mobility paradigm. *Transport Policy*, 15(2), pp. 73–80.

Beukers, E., Bertolini, L. and te Brömmelstroet, M. 2012. Why Cost Benefit Analysis is perceived as a problematic tool for assessment of transport plans: A process perspective. *Transportation Research Part A: Policy and Practice*, 46(1), pp. 68–78.

Cantarelli, C.C., Flyvbjerg, B., van Wee, B. and Molin, E.J.E. 2010. Lock-in and its influence on the project performance of large-scale transportation infrastructure projects: Investigating the way in which lock-in can emerge and affect cost overruns. *Environment and Planning B: Planning and Design*, 37(5), pp. 792–807.

Cervero, R. 2009. 'Public transport and sustainable urbanism: Global lessons'. In C. Curtis, J.L. Renne and L. Bertolini (eds) *Transit Oriented Development: Making it Happen*. Farnham: Ashgate.

De Roo, G., Schwartz, M.J.C., van der Wal, H. and Oosterhoff, H.A. 2001. *Omgevingsplanning in Nederland* (Environmental Planning in the Netherlands). The Hague: VROM.

De Zeeuw, F. and Licher, H. 2008. *Essay: De actualiteit van infrastructuur en gebiedsontwikkeling*. Delft.

Elkington, J. 1997. *Cannibals with Forks*. Oxford: Capstone.

Elverding, P. 2008. *Sneller en Beter; Advies Commissie Versnelling Besluitvorming Infrastructurele Projecten* (Faster and better, Advice Committee on Faster Decision-making about Infrastucture Projects). The Hague: Ministry of Infrastructure and Environment.

Graham, S. 2009. 'Networked Infrastructure and the Urban Condition'. In T. Rieniets, J. Sigler and K. Christiaanse, *Open City: Designing Coexistence*. Amsterdam: SUN, pp. 157–66.

Graham, S. and Marvin, S. 2001. *Splintering Urbanism, Networked Infrastructures, Technological Mobilities and the Urban Condition*. Abingdon: Taylor and Francis.

Gupta, J., Termeer, C., Klostermann, J., et al. 2010. The Adaptive Capacity Wheel: A method to assess the inherent characteristics of institutions to enable the adaptive capacity of society. *Environmental Science and Policy*, 13(6), pp. 459–71.

Handy, S. 2008. Regional transportation planning in the US: An examination of changes in technical aspects of the planning process in response to changing goals. *Transport Policy*, 15(2), pp. 113–26.

Hanekamp, T. and Tromp, A. 2012. *Presentatie Omgevingswijzer bij Platform Duurzaam GWW.* http://www.duurzaamgww.nl/wp-content/uploads/2012/06/OmgevingswijzerTH18062012.pdf (accessed 18 July 2012).

Hartman, S., Rauws, W., Beeftink, M. and de Roo, G. 2012. 'The Capacity to Adapt: Regional Development in an Age of Quality and Dynamism'. In H. Ovink and E. Wierenga (eds) *Regions in Transition: Designing for Adaptivity*. Rotterdam: Uitgeverij 010.

Heeres, N., Tillema, T. and Arts, J. 2011. Integration in road infrastructure planning: An overall analysis of Dutch experiences. Paper presented at *World Planning Congress*, July 2011, Perth.
Heeres, N., Tillema, T. and Arts, J. 2012a. *Sustainable Planning of Road Infrastructure: An International Perspective*. Report written for Rijkswaterstaat. Groningen: University of Groningen.
Heeres, N., Tillema, T. and Arts, J. 2012b. Integration in Dutch planning of motorways: From 'line' towards 'area-oriented' approaches. *Transport Policy*, 24, pp. 148–58.
Heeres, N., Tillema, T. and Arts, J. 2012c. Functional-Spatial Sustainability Potentials of Integrated Infrastructure Planning. *Procedia-Social and Behavioral Sciences*, 48, pp. 2533–44.
Holland, J.H. 1998. *Emergence: From Order to Chaos*. Oxford: Oxford University Press.
I&M. 2011. *Spelregels van het MIRT* (MIRT Rules of the Game). The Hague: Ministry of Infrastructure and Environment.
Kaufman, S. and Smith, J. 1999. Framing and Reframing in Land Use Change. *Journal of Architectural and Planning Research*, 16(2), p. 165.
Khakee, A. 2008. *New Principles in Planning Evaluation*. Aldershot: Ashgate.
Meyer, M.D. and Miller, E.J. 2001. *Urban Transportation Planning: A Decision-Oriented Approach*. New York: McGraw-Hill.
Peek, G.-J. 2006. *Locatiesynergie: een participatieve start van de herontwikkeling van binnenstedelijke stationslocaties*. Delft: Eburon.
Priemus, B. and van Wee, B. 2013. *International Handbook on Mega-Projects*. Cheltenham: Edgar Elgar.
Priemus, H. 2004. Spatial Memorandum 2004: A Turning Point in the Netherlands' Spatial Development Policy. *Tijdschrift voor economische en sociale geografie*, 95(5), pp. 578–83.
Runhaar, H. and Driessen, P. 2010. 'Planning tools for environmental policy integration – Exploring the balance between supply and demand'. In A. Hull, E. Alexander, A. Khakee and J. Woltjer (eds) *Evaluating Sustainability and Participation: Fitting Evaluation to Planning Contexts*. London: Routledge.
Runhaar, H., van Laerhoven, F., Driessen, P. and Arts, J. 2013. Environmental Assessment in the Netherlands: Effectively governing environmental protection? A discourse analysis. *Environmental Impact Assessment Review*, 39, pp. 13–25.
Ruth, M. 2011. 'Managing regional climate mitigation and adaptation co-benefits and co-costs'. In K. Otto-Zimmerman (ed.) 2011. *Resilient Cities*. Dordrecht: Springer, pp. 205–12.
RWS, Rijkswaterstaat. 2011. *Plan van aanpak 'OmgevingsWijzer'*. Delft/Lelystad: Rijkswaterstaat.
RWS, Rijkswaterstaat. 2012. *Rapportage Omgevingswijzer*. Delft/Lelystad: Rijkswaterstaat.
RWS, Rijkswaterstaat. 2013. *Handleiding Omgevingswijzer*. Delft/Lelystad: Rijkswaterstaat. www.omgevingswijzer.org.

RWS, Rijkswaterstaat. 2014. *Verankering Duurzaamheid in werkwijze RWS – van hulpmiddelen naar een gestructureerd proces*. Delft/Utrecht: Rijkswaterstaat.

Salet, W. and Woltjer, J. 2009. New concepts of strategic spatial planning dilemmas in the Dutch Randstad region. *International Journal of Public Sector Management*, 22(3), pp. 235–48.

Sijtsma, F.J., van der Heide, M. and van Hinsberg, A. 2012. 'Biodiversity and decision-support: Integrating CBA and MCA'. In A. Hull, E. Alexander, A. Khakee and J. Woltjer (eds) 2012. *Evaluating Sustainability and Participation: Fitting Evaluation to Planning Contexts*. London: Routledge.

Te Brömmelstroet, M. and Bertolini, L. 2009. *Integrating Land Use and Transport Knowledge in Strategy-Making*. Berlin: Springer.

Te Brömmelstroet, M. and Schrijnen, P.M. 2010. From planning support systems to mediated planning support: A structured dialogue to overcome the implementation gap. *Environment and Planning B: Planning and Design*, 37(1), pp. 3–20.

Van den Brink, M.A. 2009. *Rijkswaterstaat on the Horns of a Dilemma*. Delft: Eburon.

Van der Heijden, R.E.C.M. 1996. Planning large infrastructure projects: Seeking a new balance between engineering and societal support. *DISP*, 125(2), p. 18.

Van der Knaap, P. 2004. Theory-based evaluation and learning: Possibilities and challenges. *Evaluation*, 10(1), pp. 16–34.

Van Wee, B., Annema, J.A. and Bannister, D. (eds) 2013. *The Transport System and Transport Policy – An Introduction*. Cheltenham: Edward Elgar.

Wegener, M. and Fürst, F. 2004. *Land-Use Transport Interaction: State of the Art*. Dortmund.

Williams, T. and Samset, K. 2010. Issues in front-end decision making on projects. *Project Management Journal*, 41, pp. 38–49.

Williams, T.M., Samset, K. and Sunnevåg, K.J. 2009. *Making Essential Choices with Scant Information: Front-End Decision Making in Major Projects*. Basingstoke: Palgrave Macmillan.

Woltjer, J. 2000. *Consensus Planning: The Relevance of Communicative Planning Theory in Dutch Infrastructure Development*. Aldershot: Ashgate.

Chapter 13

Use and Understanding of CBA in the Evaluation of Infrastructure and Spatial Projects

Emile Dopheide

Introduction

In the search for new discourses and evaluation practices in the development and planning of infrastructure and spatial projects, it is worthwhile to explore the actual role and use of existing and prevailing evaluation frameworks and tools. The use of cost-benefit analysis (CBA) in the evaluation of infrastructure and spatial projects is a case in point.

On the one hand CBA can be considered as an old-school method: CBA is in place and has been widely used for many decades; standard handbooks and guidelines on CBA abound (Boardman et al., 2010; Zerbe and Bellas, 2006; European Commission, 2008; HEATCO, 2006; Eijgenraam et al., 2000); not too many challenges remain from a methodological point of view (Dopheide, Sijtsma and van der Veen, 2011); and the same dilemmas always persist like for example the monetization of non-tangibles; discounting; and indirect effects (see also Mouter, Annema and van Wee, 2012).

On the other hand, at least in the Netherlands, the application of CBA has never been as popular as over the last decade, particularly in the assessment of infrastructure and spatial projects. The application of standard CBA guidelines is mandatory in the Netherlands for larger infrastructure and spatial projects, and in the assessment of smaller projects the tool is being applied at a rate as never before (Annema, Koopmans and van Wee, 2007; de Jong and Geerlings, 2003). So what remains? What is the problem?

Polemics About the Role of CBA

Despite the long history and the institutionalization of CBA as a standard tool in policy-and project evaluation, the polemics about the role and outcome of social cost-benefit analysis continue (e.g. Adler and Posner, 1999; Sen, 2000; Frank, 2000; Naess, 2006; Baer and Spash, 2008; Beukers, Bertolini and Te Brömmelstroet, 2012). Some of the controversial issues include:

- Regional and local authorities perceive social cost-benefit analyses as an obstacle in the achievement of their local ambitions, as the latter don't contribute necessarily to an increased national welfare.
- Visionaries and strategic thinkers see their creative ideas frustrated by bookkeepers' mentality of 'number fetishists'.
- The advocates of 'soft' values such as nature, spatial quality and culture don't find their interests sufficiently reflected in the final monetary indicator (Sijtsma, van der Heide and van Hinsberg, 2011).
- The eventual net losers of plans and projects see their interests outbalanced in the aggregate CBA result (Rietveld, Rouwendal and Vlist, 2007).
- Public administrators see the significance of agenda setting in political decision-making processes put aside by the merits of a rational 'technocratic' assessment instrument (De Jong and Geerlings, 2003).
- Some welfare economists point to the too narrow understanding of welfare in the application of objective CBA and the impossibility of aggregating cardinal measures of individual utility.
- Finally a number of critical observers, including fervent CBA-promoters, are worried about the political use of the instrument and regretfully see CBA being misused as a 'search for benefits ' or an 'ex-post justification' of decisions already taken (Flyvbjerg, 2009).

The above problems might become even more evident in view of expected developments around the use of CBA:

- a broader application of CBA: i.e. from single sector investments (e.g. infrastructure) towards more integrated, multi-functional projects
- a growing inclusion of non-market effects into CBA
- a growing use of CBA by local authorities (municipalities and provinces)

In view of these developments, which will imply a growing integration, complexity and uncertainty in the impact assessment, one might argue that the role of CBA will diminish and that other more flexible assessment methodologies will gain in importance (e.g. Heinzerling and Ackerman, 2002). We maintain, however, the position that given its longstanding position and (sometimes undervalued) virtues, CBA has the potential to continue playing a critical role in the evaluation of integrated projects.

Bottlenecks in CBA Practice

Against the background of the on-going debate on the use and value of CBA in the evaluation of projects and infrastructure and integrated land use projects in particular, a survey was conducted among a number of CBA practitioners, scientists and decision makers (Dopheide, Sijtsma and van der Veen, 2011).

The survey consisted of a set of one-hour interviews with a total of 27 experts in the Netherlands. The interviewees were from universities, planning agencies, knowledge institutes, consulting firms and ministries, and were all experienced in the use and application of CBA in project and policy evaluation. The data from the interviews were processed and analysed through an inductive qualitative research approach following principles of grounded theory (Finch, 2002). Although a relatively positive image of CBA emerged out of these surveys, also a wide range of bottlenecks was identified. Other studies provide a similar and an even more detailed picture of these bottlenecks (Beukers, Bertolini and te Brömmelstroet, 2012; Mouter, Annema and van Wee, 2013).

These bottlenecks were distinguished into substantial constraints and more process-related challenges and were classified according to the phases of the evaluation cycle (see Table 13.1). Some of these challenges are typical for CBA (valuation of externalities; indirect effects), but others apply to any type of evaluation or impact assessment (definition of the alternatives; effect measurement; synergetic effects).

Table 13.1 Bottlenecks in CBA

Phase of the evaluation process		Substantive bottlenecks	Process-based bottlenecks
1. Evaluation-basics	Basic principles	• Scope of welfare understanding • Efficiency as a criterion for decision making • Aggregation of individual preferences	• Role of CBA • Timing within the decision-making process
	Problem definition	• Definition of alternatives (incl. zero option) • Scale Level • Impact Population • Time horizon	
2. Impact assessment		• Effect Measurements • Impact studies • Models • Cost estimates	• Available time and resources • Quality • Relationship with EIA
3. Aggregation		• Valuing non-market effects • Indirect effects • Double counting • Time and risk valuation • Distributional issues	• Relationship with other evaluation methods
4. Communication		• Final indicators	• Black-box character

Source: Dopheide et al. (2011).

As noted above, a large part of the substantial bottlenecks have been dilemmas for ages and will be in the future. Despite admirable attempts to continue working on each of these substantial issues, which from a scientific point remain intriguing, we argue that we will have to look more on how we can effectively deploy CBA as an evaluation tool and to pay explicit attention to the process of CBA.

Process Elements

It seems that the question is not if CBA should be applied, but how. It seems that CBA remains a very useful method of organizing and presenting project effects in a systemic way and helps in the decision making on the use of scarce public and private resources.

As a response to the various problems and in the revitalization of CBA growing attention is now being given to the 'process elements' in the application of CBA (see the right-hand column of Table 13.1. Although common for many other evaluation practices and tools (e.g. like EIA and SIA), the attention given to the process is relatively new in the world of CBA and not so common for the average economist. Also the 'CBA Development Agenda' of the Ministry of Infrastructure and Environment pays explicit attention to process elements like increasing the accessibility, user-friendliness and applicability of CBA (I&M, 2013).

Interesting initiatives are being developed to make the CBA more participatory and to get a better involvement of the various stakeholders (Beukers, Bertolini and Te Brömmelstroet, 2014).

In addition, further standardization of effect measurement, valuation and presentation through standardized models; key indicators and guidelines is aimed for. However, the latter is not necessarily the main way to go. Standardization certainly helps to simplify the application of CBA and makes its implementation more efficient and less time consuming. However further standardization also by-passes and hides the various dilemmas that still persist and might affect the validity and a proper understanding of the presented CBA outcomes.

In addition, more ex-post CBAs are being proposed, but these aim in the first instance at learning from the actual impacts and to develop more key indicators. The actual use, impact and particularly the understanding of CBA is less investigated in ex-post evaluation.

In the pursuit of a better deployment of the tool we should not focus too much on the improvement of the tool, but ask ourselves the question how CBA (as being in place and widely applied) can best fulfil its role as an evaluation instrument. Or in other words how can we eventually enhance the learning and information exchange in the decision-making processes around infrastructure and spatial projects. The latter requires a better understanding of the use and the users, and particularly of how users actually understand the content and scope of CBA.

Understanding of Content and Scope of CBA among Users

The variety of identified bottlenecks (Table 13.1) provides a clear picture of the persistent dilemmas around CBA. Throughout the identification of these bottlenecks also an image emerged that the CBA-language is spoken by only a limited number of people and that only a small group of the CBA-experts understand the instrument properly. Some are even wondering if policy makers really understand a CBA (Dopheide, Sijtsma and van der Veen, 2011).

In addition to the limited understanding of the instrument, awareness among end users of the uncertainty and assumptions in the impact assessment seems low. The aggregation into one single final indicator through some kind of 'spreadsheet logic', conceals not only the assumptions and uncertainties, but also underutilizes the potential of CBA to get a deeper insight into the type, nature and distribution of costs and benefits.

At the international level, an evaluation conducted by the World Bank on the practice of CBA as applied to their own projects revealed amongst others that the use and interest of senior staff in CBA results is very limited despite their overall policy to conduct rigorous ways of evaluations (World Bank, 2010).

A minimal understanding among the users of the assumptions, the approach and the results of CBA is required to ensure that CBA plays a full role in the substantive discussions during project assessment. A risk exists if the final outcome of the CBA is just being used as the final verdict and in an uncritical manner. In addition, a better understanding of the instrument and its outcomes would also enhance the public support for the use of the instrument.

Standard CBA guidelines and handbooks used to include explicit discussions on the limitations and criticisms towards CBA (e.g. de Rus, 2010; Zerbe, 2006; Hanley and Spash, 1994; Pearce, Atkinson and Morato, 2006; and Boardman et al. 2008) and often make the disclaimer that if the various users (analysts; interpreters; public officials; general public etc.) keep the limitations and constraints of CBA in mind, CBA remains a useful tool in the planning and evaluation of specific projects. We question, however, to what extent the various users actually are aware of and understand these limitations and constraints.

The understanding of the meaning and scope of CBA by users seems thus limited. In the next section we briefly elaborate on the bottlenecks listed above, which potentially could lead to misunderstandings and misuse among users.

Fundamentals

Laymen are not always aware of the fundamentals behind CBA and the implications of these for a proper interpretation of the outcome. The understanding of the notion of *welfare* in CBA is a welfare that can be made objective, whereas subjective elements and feelings are not considered. At the same time this does not necessarily imply that welfare is considered equal to money. This principle seems often to lead to confusion on the scope of CBA. This also applies to the *efficiency* criterion against which a project is being evaluated with the use of CBA. Issues of justice, legitimacy, ecological sustainability and effectiveness are in principle beyond the scope of CBA.

The question can also be raised to what extent users understand the principle of consumer sovereignty, willingness-to-pay and the (in) comparability *of individual utility* as basic fundamentals in CBA. A problem always will persist if decision makers promote certain public projects for the sake of the public interest (merit-goods) and attempt to justify this investment with the use of CBA. CBA is by definition based on the aggregation of individual preferences which might conflict with the public interest.

Although in evaluation practice questions about the fundamentals of CBA play only in the background, part of the ambiguities and misinterpretations around CBA seem to relate to these fundamental aspects. Questions about the fundamentals of CBA call for keeping the users alert on the (sometimes limited) scope of CBA.

Problem Definition

The CBA approach forces evaluators to develop and apply a rigorous problem definition and analytical framework, which is not necessarily unique for CBA. However, in the enthusiasm for promoting certain alternatives, the attention given to a proper description of the problem to be solved by the project could disappear to the background.

CBA requires a clear definition of the *null-alternative*. The meaning and significance of the null-alternative it is not always clear to the users. In principle the null-alternative encompasses two quite different elements, which are not always properly distinguished. On the one hand it refers to the alternative allocation of the invested resources. On the other hand it refers to the actual development (autonomous or minimal intervention) if the project would not be implemented.

1. The alternative allocation of the invested resources is reflected through the applied discount rate (see also below).
2. The description of the actual situation if the project were not to be implemented is reflected by a concrete description of this situation and by assessing the *incremental* benefits of the project. Users sometimes confuse a description of the *without-situation* with the definition of a concrete alternative project and don't always understand the need to get a clear picture of the no-go alternative. The formulation of a without situation is considered by these users as a waste of time and resources.

The *impact population* and *scale of analysis* are standard elements in any evaluation practice. In CBA this requires particularly a clear definition and is therefore rigorously applied. For users it is not always clear that the CBA results need to be interpreted within the context of the specific scale level and impact population, which is being applied. The CBA could result in different, seemingly opposing, outcomes at different scale levels. Local policy makers might be surprised when the CBA outcome at the national scale turns out to be less positive than at the local level, caused by local benefits being offset by social losses elsewhere in the country.

The *time horizon* seems for most users unrealistically long and requires a good explanation and justification. A realistic determination remains problematic if one applies a time horizon of more than 30 years for example. A user should get a clear idea whether the length of the time horizon is a critical element in the feasibility of an alternative.

Effect and Impact Assessment

Effect and impact assessment can be considered as a major bottleneck in the practical implementation of CBA. In order to identify and assess the effects in a reasonable manner a vast amount of expertise, resources and time is required. The latter is not always properly understood by the user.

This bottleneck is even stronger as the assumptions and uncertainties underlying the effects are often difficult to interpret and the implications for the end user are less clear. Uncertainties remain intimately linked to impact assessment and as such also to CBA. Sensitivity analysis is only a partial solution to this bottleneck.

The aspects related to effect and impact measurement are often highly complicated in their calculation. However, the intuitive understanding of the outcomes in their own unit of measurement (e.g. area of nature lost, time savings, CO_2 reduction) seems to be easier for the user than the same effects expressed in monetary terms (Sijtsma, van der Heide and van Hinsberg, 2011).

As cost estimates seem to be systematically estimated at a too low level (e.g. Flyvbjerg, 2009), a critical look at the cost estimates and the assumptions behind these seems crucial for a realistic overview of the consequences of the project. A user should be able to develop such a critical view.

Aggregation

Project proponents justify investments often by pointing to all kind of direct and *indirect effects* which however cannot always simply be aggregated as they don't necessarily contribute to additional welfare. This search for benefits often leads to a *double counting* and as such to an overestimate of the benefits of the project.

To facilitate and ensure a good understanding of the effects among users and decision makers, a number of experts are hesitant to call for a far-reaching *monetization* of the non-market effects and plea for an expression of the effects in their own dimension. Also if standard key value indicators (benefit transfers) are being applied in a mechanistic manner to areas other than where these were conceived, final estimates require a careful interpretation.

The principle and idea of *discounting* as applied in a CBA can be considered as an element for professionals and CBA experts, and one which is not necessarily needed to be understood by the layman. However, some awareness of the possible sensitivity of the discount rate on the end result of the CBA might be required for a correct interpretation of the outcome of a CBA, particularly if critical effects occur in the long term. In addition, attention should be paid to the understanding among users of the implication of using alternative discount rates, which could be applied to reflect sustainability principles. In addition to discount rates, also a *risk premium* is usually applied in a standard and sometimes uncritical manner. The original idea and concept behind the discount rate and in particular behind the risk premium seems not be understood among the users (Pearce, Atkinson and Morato, 2006).

The problems of incorporating *distributional effects* in cost-benefit analysis are of a various nature and are well described in the literature (e.g. Drèze, 1998; Rietveld, 2007, Thomopoulos et al. 2009; Van Wee, B. 2011): (i) the sum of individual preferences as the yardstick of societal welfare does not reflect the distribution of impacts; (ii) the final outcome of the CBA is influenced by the prevailing income distribution as the individual preferences of the higher income groups will have a relatively strong weight in the sum of individual preferences; (iii) the applied efficiency criterion assumes a potential compensation between winners and losers (Kaldor-Hicks principle), but often this compensation does not take place; and (iv) the willingness to pay in monetary terms is higher for the higher-income individuals than for the lower income groups.

The implication of the influence of the prevailing income distributions on the final outcome of the CBA in the final assessment is often not realized by the users. Moreover, the aggregated outcome of the CBA often hides information on which groups capture the benefits and who bear the costs. Explicit information about the distributional impacts, which is not necessarily too complicated to supply, would deepen the outcome of a CBA and could enrich the policy debate.

Communication

Particular non expert users are not always aware that alternative *final indicators* (like net present value; internal rate of return; benefit-cost ratio; capital-investment ratio) lead to different rankings if different projects are being compared. Although the net present value is the most commonly used decision criterion in CBA, the application of alternative indicators (e.g. benefit-cost ratio) could lead to a different ranking of project alternatives (Pearce, Atkinson and Morato, 2006).

Need for Further Understanding of the Use of CBA

In practice the final user is often faced with a final result with little involvement throughout the CBA process. A CBA in practice is still too much a study conducted by an external consultancy firm and delivered to a client, where the end result is central but where the client hardly has internalized the essence of the analysis and understood the results in its proper context (Dopheide, Sijtsma and van der Veen, 2011).

As the above outlined persisting bottlenecks always remain a dilemma, without an optimal solution; we would rather make them clear and transparent rather than hiding them or, with all perseverance, trying to solve them.

Most CBA analysts find their way in dealing with these dilemmas, but the question remains to what extent they affect the understanding and interpretation of CBA results among the users. Research has be done about the general implications of CBA for decision making (World Bank, 2010) and about the relationship between CBA outcomes and the decision outcome (Rienstra, 2008), but a proper understanding of the actual use and understanding seems lacking among users.

A Proposal

Given the above outline and argumentation we suggest that the following questions are explored in follow-up research:

- Who are actually reading and using the CBA reports?
- What do the users understand of the CBA contents?
- Which phases and elements of the CBA are most critical for understanding among users to allow an optimal use in its deployment as an evaluation instrument?
- To what extent the users are (and should be) aware of the limitations and assumptions of the various CBA results and their possible implication for the further interpretation?

Activities that we propose to further answer these questions:

- In-depth case studies of a limited number of projects on users, use and understanding of CBA (preferable infrastructure/integral land use projects of different scale; starting at the local level).
  - Stakeholder mapping and analysis.
  - Review of project documents and reports.
  - In-depth interviews and narratives.
- Course and exam setting for policy officials dealing with CBA.
  - Existing CBA university courses/training activities.
  - Tailor-made courses/training activities.
- Assessment of training needs in terms of CBA for policy officials.

Finally, the CBA experts, average economists and insiders should be alert to the knowledge and contributions of other disciplines and remain open-minded for the process elements in the evaluation practice. In such a way, CBA could play a role in an interdisciplinary endeavour beyond the mere technocratic evaluation of infrastructure and spatial projects.

References

Adler, M.D. and Posner, E.A. 1999. Rethinking Cost-Benefit Analysis. *Yale Law Journal*, 109(1999), pp. 165–247.

Annema, J.A., Koopmans, C.C. and Wee, G.P. van, 2007. Evaluating transport infrastructure investments: the Dutch experience with a standardised approach. *Transport Reviews*, 27(2), pp. 125–50.

Baer, P. and Spash, C. 2008. *Cost-Benefit Analysis of Climate Change: Stern Revisited.* Working Paper Series, May 2008, Canberra: CSIRO.

Beukers, E., Bertolini, L. and Te Brömmelstroet, M. 2012. Why Cost Benefit Analysis is perceived as a problematic tool for assessment of transport plans: A process perspective. *Transportation Research Part A: Policy and Practice*, 46(2012), pp. 68–78.

Beukers, E., Bertolini, L. and Te Brömmelstroet, M. 2014. Using cost benefit analysis as a learning process: Identifying interventions for improving communication and trust. *Transport Policy*, 31(2014), pp. 61–72.

Boardman, A., Greenberg, D., Vining, A. and Weimer, D. 2010. *Cost-Benefit Analysis: Concepts and Practice.* Upper Saddle River: Prentice Hall.

De Jong, M. and Geerlings, H. 2003. Exposing weaknesses in interactive planning: The remarkable return of comprehensive policy analysis in the Netherlands. *Impact Assessment and Project Appraisal*, 21(4), pp. 281–91.

De Rus, G. 2010. *Introduction to Cost-Benefit Analysis. Looking for Reasonable Shortcuts.* Cheltenham: Edgar Elgar.

Dopheide, E., Sijtsma, F.J. and Veen, A. van der, 2011. Begrip van MKBA is beperkt onder gebruikers. *Economische Statistische Berichten*, 96(4614 and 4615), pp. 460–463.

Drèze, J. 1998. Distribution Matters in Cost-Benefit Analysis: Comment on K.A. Brekke. *Journal of Public Economics*, 70, pp. 485–8.

Eijgenraam, C.J.J., Koopmans, C.C., Tang, P.J.G. and Verster, A.C.P. 2000. *Evaluation of Infrastructural Projects; Guide for Cost-Benefit Analysis, Sections I and II.* The Hague: CPB, Rotterdam, NEI: Rotterdam.

European Commission, DG Regional Policy. 2008. *Guide to Cost-Benefit Analysis of Investment Projects*. Brussels: European Commission.

Finch, J.H. 2002. The role of grounded theory in developing economic theory. *Journal of Economic Methodology*, 9(2), pp. 213–34.

Flyvbjerg, B. 2009. Survival of the unfittest: Why the worst infrastructure gets built – and what we can do about it. *Oxford Review of Economic Policy*, 25(3), pp. 344–67.

Frank, R.H. 2000. Why is Cost-Benefit Analysis so Controversial? *Journal of Legal Studies*, 29(S2), pp. 913–30.

Hanley, N. and Spash, C.L. 1993. *Cost-Benefit Analysis and the Environment.* Aldershot: Edward Elgar.

HEATCO. 2006. *Developing Harmonised European Approaches for Transport Costing and Project Assessment* [online]. University of Stuttgart. Available at: http://heatco.ier.uni-stuttgart.de [accessed 27 February 2014].

Heinzerling, L. and Ackerman, F. 2002. *Pricing the Priceless. Cost-Benefit Analysis of Environmental Protection*. Georgetown Environmental Law and Policy Institute, Georgetown University Law Center.

I&M. 2013. *De ontwikkelagenda MKBA*. The Hague: Ministry of Infrastructure and Environment.

Mouter, N., Annema, J.A. and Wee, B. van, 2012. Ranking the substantive problems in the Dutch Cost-Benefit Analysis practice. *Transportation Research Part A: Policy and Practice*, 49(2013), pp. 241–55.

Naess, P. 2006. Cost-benefit analysis of transportation investments. Neither critical nor realistic. *Journal of Critical Realism*, 5(1), pp. 32–60.

Pearce, D., Atkinson, G. and Morato, S. 2006. *Cost-Benefit Analysis and the Environment. Recent Developments*. Paris: OECD.

Rienstra, S. 2008. *De rol van kosten-batenanalyse in de besluitvorming*. Den Haag: Kennisinstituut voor Mobiliteitsbeleid (KiM).

Rietveld, P., Rouwendal, J. and Vlist, A. 2007. 'Equity issues in the evaluation of transport policies and transport infrastructure projects'. In M. van Geenhuizen, A. Reggiani and P. Rietveld (eds) 2007. *Policy Analysis of Transport Networks*. Aldershot: Ashgate, pp 19–36.

Sen, A.K. 2000. The discipline of cost-benefit analysis. *The Journal of Legal Studies*, 29(2), pp. 931–52.

Sijtsma, F.J., Heide, C.M. van der and Hinsberg, A. van. 2011. 'Biodiversity and decision-support: Integrating CBA and MCA'. In A. Hull, E. Alexander, A. Khakee and J. Woltjer (eds) 2011. *Evaluation for Participation and Sustainability in Planning*. London: Routledge, ch. 9.

Thomopoulos, N., Grant-Muller, S.M. and Tight, M.R. 2009. Incorporating equity considerations in transport infrastructure evaluation: Current practice and a proposed methodology. *Journal of Evaluation and Program Planning* (Special Issue on Evaluating the Impact of Transport Projects: Lessons for Other Disciplines), 32(4), pp. 351–9.

Wee, B. van. 2011. *Transport and Ethics. Ethics and the Evaluation of Transport Policies and Projects*. Cheltenham: Edward Elgar.

Wee, B. van. 2012. How suitable is CBA for the ex-ante evaluation of transport projects and policies? A discussion from the perspective of ethics. *Transport Policy*, 19(1), pp. 1–7.

World Bank. 2010. *Cost-Benefit Analysis in World Bank Projects*. Washington DC: World Bank, Independent Evaluation Group.

Zerbe, R.O. and Bellas, A.S. 2006. *A Primer for Benefit-Cost Analysis*. Northampton, MA: Edward Elgar.

Chapter 14

The Plan Review: A New Approach to Discussing, Assessing and Improving Urban Projects in a Changing Planning Practice

David Hamers, Like Bijlsma and Anton van Hoorn

Introduction

Motivation and Objective: Improving the Spatial Assessment of Urban Development Plans

The decision-making process around spatial investment projects is faced with a number of problems related to the assessment of urban development plans. For example, there are misunderstandings between administrators and assessors about the relationship between policy targets and development plans. In addition, there is little unity amongst the various disciplines involved in such assessments. Economists, for instance, have another view on cities than urban planners and designers. This chapter introduces a new method intended to facilitate the communication between experts with diverging disciplinary backgrounds, in order to improve the spatial quality of urban plans and the decision-making process around possible public investments.[1] It presents an analysis and assessment instrument that may be used by urban planners and local, regional and national administrators to give them more insight into the main spatial strengths and weaknesses of their projects. In this way, they may be able to assess such projects more effectively, while taking into account certain policy objectives. We call this instrument the Plan Review and have shaped it to benefit the dialogue between various experts (in the spatial domain and related fields), as well as between experts and stakeholders. In actual practice, this involves assembling a list of themes (the 'spatial conditions') in aid of objectifying the analysis of the plan and structuring the dialogue. Depending on the phase of the planning and decision-

1 This chapter is based on the study *Planobjectivering* (Hamers, Bijlsma and Van Hoorn, 2013), as part of a joint research by PBL Netherlands Environmental Assessment Agency and CPB Netherlands Bureau for Economic Policy Analysis called Plannen voor de stad (Romijn and Renes, 2013).

making process, the Plan Review may help to explore possible plan improvements early in the planning process, and to facilitate assessment of the project by an independent expert panel, during a later phase.

Demarcation and Set Up

Although urban projects may serve a variety of purposes, this chapter centres on physical spatial projects. The following section provides a brief description of the current debate on the assessment of such projects in the Netherlands, followed by the Plan Review instrument which was developed in response to the issues involved in this debate. The Plan Review assesses a project by estimating both spatial opportunities and related risks. The most relevant and topical spatial conditions have been ordered in a matrix, and are described in this chapter. The chapter also explores the application of the instrument, according to a number of current projects. The chapter closes with a synthesis and a brief discussion of the steps that could be taken to further develop the Plan Review, in order for the instrument to obtain a fitting function and position within the planning process.

Three Issues around Current Planning Assessments

In the Netherlands, the debate on planning assessments, to date, has focused on the role of cost-benefit analyses (CBAs) in the decision-making process. In planning research and practice, the CBA has been promoted as an instrument that makes project assessment more transparent and increases objectivity and accountability in decision making. In fact, in the Netherlands, CBAs are a mandatory component in the decision-making process of large-scale spatial investments, whenever national public funding is involved. However, academics and practitioners across a wide range of disciplines have criticized these claims (e.g. Bebbington et al., 2007; Tribe, 1972); for instance, with regard to efficiency as the basic criterion of acceptability and the over-reliance on monetization. In addition, various parties in the processes around planning and decision making have pointed to problems related to the way CBAs fit in with the planning of spatial projects in contemporary urban regions (e.g. Beukers et al., 2011; Municipality of Almere and the Ministry of Infrastructure and the Environment, 2011). This chapter addresses three of these issues.

The first issue relates to the multitude and multi-interpretability of objectives of urban projects. These projects often have to contribute to very diverse objectives, such as in economic and environmental areas, as well as those relating to culture and landscape. In order to create a support base for such combined policy objectives, in the political arena, often, more general terms are used that could be interpreted in multiple ways. Examples of such terms are ‘attractive cities’ and ‘international business environment’. The very aspects that make these expressions attractive for administrators to use create difficulties in the planning assessment that precedes decision making, as such descriptions of objectives are difficult to operationalize. Translating these expressions for the purpose of a CBA leads to a 'confusion of tongues' between the various parties involved in the planning and decision-making process. This problem is increased even further by the considerable uncertainties that have characterized the planning practice over the last years. Plans that present a concrete, ultimate goal, increasingly more often are being replaced by plans based on more general guidelines and a phased realization. This also poses problems for the traditional assessment of plans that precede the decision-making phases.

The second issue is one that is related to the various spatial scales on which contemporary urbanization occurs and the governance processes involved. Over the past decades, urbanization – in the Netherlands as well as globally – has increasingly become a regional affair, in a morphological, functional as well as a governance sense. This has led to the formation of urban agglomerations and daily urban systems, and has brought urban planning to the level of the metropolitan region. Moreover, central government also looks to urban regions for the realization of their objectives. The large Dutch cities, thus, are considered to be drivers of the economy; the current Cabinet considers a good international business environment a matter of national interest. Policy objectives of various administrative divisions may be in line, but may just as easily have opposing interests. This causes problems in the assessment of urban projects. After all, an assessment of a plan from a national perspective may look entirely different than one that is based on regional or local objectives. CBAs in the Netherlands only apply to the national scale, causing objectives on smaller scales to be ignored in planning assessments. Therefore, local and regional parties have asked for additional assessment methods.

The third issue around the assessment of urban projects concerns the different types of knowledge required in competent decision making. Because of the complexity of urban projects, knowledge is needed in various fields, such as traffic, economy, planning, geography and urban design. These fields of knowledge are sometimes difficult to unite. There may be differences of opinion, for instance, on the relationship between contributions of 'scientific' knowledge and those of 'crafts',[2] and misunderstandings caused by the use of different definitions could easily occur. Currently, the terminology of welfare economics as used in CBAs is leading. Several parties, however, have indicated that they require additions to this vocabulary from various spatial disciplines (Mouter et al., 2012; Renes et al., 2011).

This chapter introduces an instrument of analysis and assessment that would address the issues as described above. This instrument is the Plan Review; it does not follow the dominant logic of CBAs, but places the knowledge developed within spatial disciplines (e.g. on geography, urban planning, and urban design) at centre stage.

2 Sennett (2008) speaks of 'the craft of experience'.

Foundations of the Plan Review Instrument

Plan Review Positioning in Relation to CBA

The Plan Review instrument has a different function than the CBA in the process around planning and decision making. CBAs primarily focus on the effects and efficiency of a project. However, before a project has any effect, it must comply with several spatial conditions. Does it work? Does the project produce the spatial conditions under which it may have the desired effect, thus contributing to the aspired policy objectives.

In its current form, CBAs are not very well-suited to portray these conditions. This may lead to situations whereby the effectiveness of a project cannot be assessed properly. Moreover, in earlier phases of the decision-making process, options for stakeholders to improve on the project are sometimes overlooked. The Plan Review is intended to improve on the flow of information in the decision-making process in both areas (plan optimization and assessment). In doing so, the Plan Review, unlike CBAs, does not focus on the effectiveness and efficiency of urban projects, but analyses the spatial conditions that projects must comply with in order to achieve the desired effects.

A number of questions related to these issues are paramount: What spatial conditions must a project comply with in order to realize the desired objectives? Which developments does a particular project facilitate and which developments are actually hampered by this project, when comparing it to other, related projects? Which unwanted effects may present themselves? And, finally, which conditions could be improved on, in order to achieve the objectives? The opportunities and risks related to spatial plans may be analysed and presented in a relatively simple manner, both in their physical and policy context.

Connecting to Available and Desired Information in the Planning and Decision-Making Process

The objective of the Plan Review is to improve on the decision-making process around urban projects. Therefore, the foundations for the instrument are to fit in with and connect as closely as possible to the informational requirements of policymakers and government administrators in various situations and phases of the planning and decision-making process.

In Dutch policy and planning practice, very many different plans are being assessed, and the relationship between policy objectives and the desired effects of urban projects differs from case to case. Furthermore, the planning practice currently is subject to considerable change. The old planning practice of integral area development according to large-scale and extensive master plans is taking a backseat, increasingly more often. Under the new planning practice – often referred to as organic area development – this focus is shifting. Large projects primarily focus on sectoral objectives, such as accessibility and innovation. Spatial policy objectives recede into the background. On a local scale, partly due to the financial-economic crisis, also smaller projects are being developed that have a longer runtime and a phased realization. A recent PBL study characterized this development as a change from planning in projects to planning according to an open-ended process (Buitelaar et al., 2012). The parties involved are trying to develop urban plans that address the increase in uncertainties while offering continuity. Therefore, many such current plans are a combination of broad guidelines (to ensure proper footing) and resilience; projects with an open-ended approach and options for intermediate adjustments.

This open approach in planning, typically, is problematic when CBAs are applied. The basic requirements of CBAs are not being met, such as the strict demarcations in time and space that define the project. There is a growing need to evaluate projects within a flexible project definition. This is especially the case when (local) governments want to facilitate grassroots planning and are a direct stakeholder, because they offer land, organize plan processes or invest money in the project. In these cases, the chances of success and added value of the project should be well secured. Plans with a high added value not only locally but also for the city as a whole will have a higher priority than other for the city less beneficial projects. In addition, initiatives should be connected to certain locations, in a fitting manner, for them to be successful. This begs the question of how government could compare the various initiatives, relatively quickly, and how they could discuss and assess submitted plans in a transparent and reproducible way.

It is clear that, in contemporary planning practice, projects are assessed that differ in both approach and scale. They vary from large-scale urbanization projects with infrastructural improvements on a regional scale, to inner city projects with issues and opportunities involving the local scale, such as a train station or part of the city centre. Plans vary from comprehensive master plans with a defined time horizon to more organic, flexible plans. The Plan Review is intended to be applicable to all these cases. This implies that it must provide spatial information on any scale (large to small) and measure of detail (generic to specific), suited to the nature of the project and the phase of planning and decision making. These principles, together, form the entry points of the Plan Review matrix (Figure 14.1).

The rows of the matrix distinguish between the various scales of urban projects; from the entire region to the location of an individual project. These scales relate to the policy objectives that a particular project is to contribute to – objectives that, mostly simultaneously, involve various levels of governance. Think of a large-scale infrastructural project that has to achieve both regional and national accessibility objectives and meet local environmental standards, versus a local public space design that fits with regional and national objectives, such as attracting international knowledge workers in the creative industries. The columns of the matrix distinguish between the available and the required type

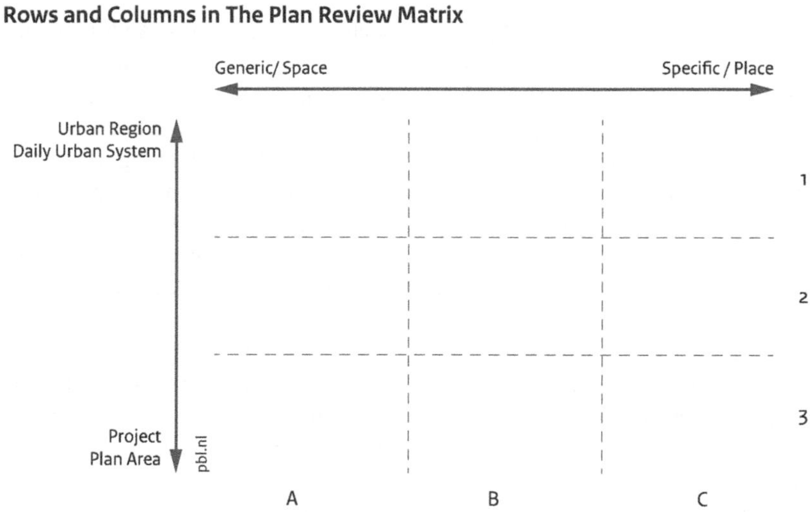

Figure 14.1 Policy objectives and types of information in The Plan Review Matrix

Source: PBL Netherlands Environmental Assessment Agency, 2013.

of information or knowledge, from generic to specific.[3] The type of knowledge required depends on the nature of the plan (e.g. blueprint or open-ended) as well as on the particular phase of the planning and decision-making process.

A number of spatial conditions can be entered into the matrix. These are elaborated in the following section. For a clear presentation of conditions, the matrix has been divided into nine cells (three rows and three columns). The conditions have been distributed over the cells.[4] This enables a clear focus on the relevant conditions in the dialogue between the parties involved. For example, a discussion on a regional project that is still in the early stages of planning is likely to be focused on the cells in the top left-hand corner of the matrix. For projects of a much smaller scale and for those for which more details already have been established, the cells in the lower right-hand corner would feature more predominantly.

3 In geography terms, this would be space and place. Experience has shown that more abstract knowledge often is of a quantitative nature, whereas concrete knowledge can be labelled as qualitative. However, the reverse may also be true.

4 The division in rows and columns is an heuristic, for the purpose of easy applicability in assessments. Because the two axes of the matrix both represent a continuum, this division in rows and columns may seem somewhat contrived. Conditions cannot always be easily allocated to any particular cell. Therefore, cell boundaries are represented by dotted lines.

In addition, if necessary, any of the conditions represented in the other cells also can be discussed. For instance, in situations where urban projects involve various scales (e.g. local business locations near motorway slip roads). Talks between plan assessors may follow a certain 'path' over the matrix.

This allows not only the quantifiable, but also the qualitative aspects of a plan to be analysed and evaluated in a structured way. In addition, it is also possible to compare different plans, placing them in hierarchical order when appropriate. Furthermore, plans may be related to a larger framework through interdependent relationships between cells. Thus, opportunities, uncertainties and risks in spatial projects can be identified effectively and relatively fast.

The Plan Review Matrix

The Cells: Diverging Spatial Conditions

On the basis of literature research (e.g. on urban development, architecture, planning, geography, urban sociology and spatial economy), planning and map analyses, field studies, and expert interviews, we assembled a series of 16 conditions that regularly feature in spatial research and design.[5] This series is not exhaustive. These particular conditions were selected because they are in keeping with contemporary policy goals and the current planning practice. The series was constructed to fit a broad range of projects, while being sufficiently ordered and accessible to function as a checklist in actual practice.

The matrix needs to contain various conditions, as objectives of urban projects also differ, and because projects are realized within cities of different characteristics, on different locations (e.g. think of the increasing importance of spatial transformation and retrofitting, instead of large-scale greenfield development), as well as for different target groups (context sensitivity). This must involve both physical and functional aspects of the urban environment, as well as the manner in which city dwellers experience their surroundings. These separate elements are important in isolation, but also their interconnection is significant; together, they determine how cities are used and experienced.

In addition to these aspects of use and experience, also the time dimension is of great importance in project analysis. After all, analyses of urban investments do

5 There may be considerable differences between traditions directly related to various types of knowledge and know-how. For research that is based more on classical (beta) science, the emphasis is on causes and constants, whereas design research focuses more on options and possibilities within a specific context. In addition to theory, also practical experience may form a valuable source of knowledge. Although some scientific researchers are less inclined to award appropriate value to such physical rather than scientific knowledge, the necessity of 'thinking through making', 'knowing-in-action' and 'reflection-in-action' has a solid basis in scientific theory (e.g. see Schön, 1984, 1987; Sennett, 2008).

not focus on a single moment in time, but must consider long-term developments. They assess the degree to which projects may adapt to change, both the expected and – in the currently changing planning practice – increasingly also the unforeseen. Furthermore, the short term also is of interest in relation to project planning and execution. This includes the circumstances under which a project could be realized and how the pace at which this happens could be increased (e.g. catalysed). In order to address these issues in a planning assessment, the matrix was fitted with a number of variables related to the planning process. These process variables (e.g. flexibility, resilience, catalysts, phasing) can be addressed in all cells instead of being entered into the matrix, separately, because of their importance for risk analysis.

Figure 14.2 shows the series of conditions, as they are distributed over the matrix. The following section provides a characterisation of all conditions, centred on certain questions that are subsequently elaborated on. In addition, attention is given to certain theoretical backgrounds and methods that enable the use of the Plan Review in the planning and decision-making process. The subsequent section provides three examples of how (a number of) conditions may be used in discussions during the planning and assessment process, to objectify the planning analysis and to structure the communication between experts.

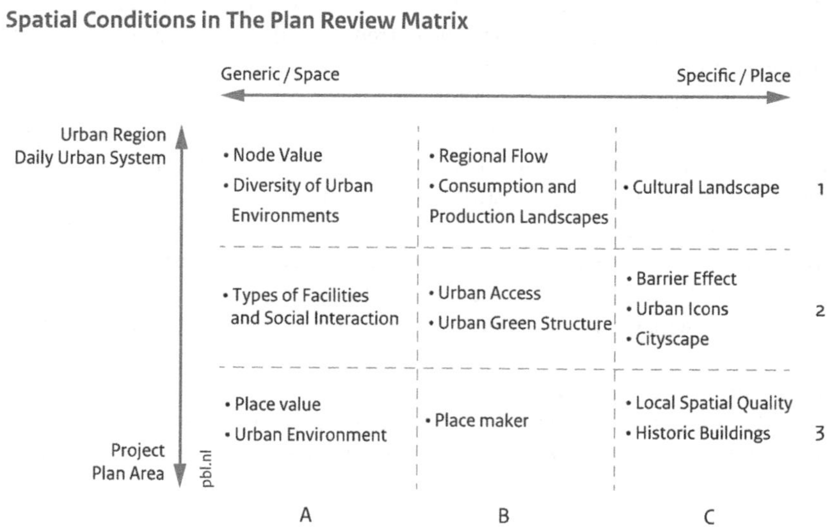

Figure 14.2 Spatial conditions in The Plan Review Matrix

Source: PBL Netherlands Environmental Assessment Agency, 2013.

1. Node and Place Values (1A and 3A)

Are node values and place values sufficiently balanced? The concept of node and place values (Bertolini, 1999) offers starting points for the analysis of development options for mobility in connection with those for the urban functions of an infrastructural node, related to visiting and accommodation. Certain questions are important, such as whether the transport capacity of a node is being fully utilized in view of its position within the network, and whether this node value is balanced with the place value – the facilities in the direct vicinity of the node (e.g. shops, offices and residential homes).

2. Diversity of Urban Environments ('milieus') (1A)

One of the questions is that of whether a project would contribute to the current variety of environments within a particular urban region? In order to meet the needs of a variety of users (dwellers, workers, visitors), urban regions must contain various types of places and functions, together described as 'urban environments' (or 'urban milieus', see Figure 14.3; for elaboration of this concept, see Ritsema van Eck et al., 2009; PBL, 2010; Hamers, 2011). For instance, think of the difference between office towers distributed across the city and a spacious and green campus. Both may contribute to an internationally competitive business climate, but do so on the basis of completely different spatial qualities.

Figure 14.3 Urban environments in the Amsterdam region

Source: PBL Netherlands Environmental Assessment Agency, 2009, 2010.

By considering the regional-urban diversity of environments, arguments can be found for possible changes (additions, restructuring, transformations) that may be required in light of the particular objectives. Here, regional coherence is especially important. Consideration of a project on that scale may indicate that there is a relationship between both competitive and complementary environments (within as well as outside the municipality where the project is being realised).

3. Regional Flow (1B)

Would the regional infrastructure be sufficiently robust and resilient to cope with passenger and freight traffic in the long term, and would it be able to adapt to possible changes in traffic flows, such as increasing cross-traffic in a polycentric region, or decreases due to regional population decline? This includes any parallel infrastructure on connecting routes, connections between suburban centres and choice options for either public transport or passenger vehicles. Another issue is whether solutions could be offered by measures that could be implemented, in the short term, to create flexibility (e.g. changes in public transport timetables, or by the construction of rush-hour lanes).

4. Consumption and Production Landscapes (1B)

Other questions refer to how regional urbanization relates to different landscapes and nature (biodiversity, ecological corridors). Does the landscape in a particular urban region offer any possibilities for water storage, in view of climate change, and is it able to meet urban demands related to regional and local food and energy supply, as well as recreation? In addition to existing and inherent nature qualities, this also relates to planned and still to be designed conditions in consumption and production landscapes.

5. Cultural Landscape (1C)

The extent to which any particular and unique area characteristics have been taken into account must be determined. Will any such characteristics be protected within the project, or maybe used by awarding them a visible position? Depending on the situation, conservation and/or development may offer possibilities. If the cultural landscape stretches over a large area, harmonization between project plans on a regional scale may offer opportunities for making complementary choices in various locations within one region. For example, if historic features become unrecognizable due to the realization of a plan, their conservation elsewhere perhaps could be guaranteed.

6. Types of Facilities and Social Interaction (2A)

A further question is that of whether cities and neighbourhoods have the appropriate facilities, and whether these are sufficiently accessible. Facilities and public space, together, form the social-spatial infrastructure of a city. An effective social and spatial infrastructure is one that connects people, and ensures they feel at ease within that environment. In turn, this may cause them to feel more responsible for the environmental quality. Facilities need to be easily accessible and fit in with the environmental identity (visibility, correct atmosphere), in order for people to feel involved (see Bromm et al., 2011).

7. Urban Access (2B)

Questions that play a role, specifically, for plans related to housing and business development within the existing city and along the urban fringe are those of whether entry and exit roads and light rail or rail connections are sufficiently robust and resilient, on an urban scale, to continue to process passenger and freight traffic, in the long term, and to adapt to possible changes in traffic flows. And whether there are any options for separating through traffic from local traffic on urban ring roads (see also 1B). Will there be sufficient flexibility in the short term? Further urbanization within existing urban areas mostly makes different demands on infrastructure than those related to new housing in expansion areas.

8. Urban Green Structure (2B)

Issues such as how built-up areas relate to green areas in and around the city, and how easy it is for city dwellers to reach these areas of urban green require different solutions in different urban environments (see 3A). Different types of urban dwellers have different preferences with respect to functionality and type of greenery (e.g. recreation, urban agriculture). In addition, the ratio between 'red' and 'green' not only refers to magnitude, but also to characteristics. The size of a park, for example, is not automatically an indication of its quality.

9. Barrier Effect (2C)

The options for clearing spatial (as well as functional, socioeconomic and mental) obstacles within a plan must be considered. Barriers come is all shapes and sizes; for example, in the form of large-scale infrastructure that separates city areas, or a long and closed front of buildings that forces pedestrians to take a detour (e.g. see Lynch, 1960). This may cause locations that are physically very close by to appear very far away, both in travel time and in perception. Such characteristics may have a large impact on the particular meaning of a place, and, thus, on its value. Clearing such barriers can be a project's main objective, but it may also be a side effect of developments that primarily serve a different purpose.

10. Urban Icons (2C)

Iconic features have other things to contribute than a plan of a fairly plain character. Although urban iconic features, by nature, possess a distinct uniqueness, there are ways to analyse them. For example, they are rarely realized within a vacuum, but rather are adapted to suit their surroundings, in which often additional investments take place (e.g. see Lynch, 1960; Verheul, 2012). This connection offers starting points for the analysis of a planned iconic feature, and concerns the convincing logic of a plan, as a whole. In addition to its functional value, especially its perception value needs to be substantiated. For example, will the iconic feature evoke feelings of pride or identification, and/or will it mark the entrance to a certain area or connect two areas?

11. Cityscape (2C)

The degree to which a plan takes into account the presence of a possible historic cityscape must be determined, as well as whether it protects such characteristics or even uses them by awarding them a recognizable position within the project. Local historic characteristics of a city district or neighbourhood may offer starting points for creating a visible identity in a plan. This may enhance attractiveness to residents, businesses and visitors (depending on the planned programme) and may offer possibilities for branding.

12. Urban Environment ('Milieu') (3A)

The spatial and functional conditions that are offered by the planned environment ('urban milieu') in order to achieve related policy objectives must be regarded. When changes to the range of environments within an urban region would contribute to achieving a certain policy objective (see 1A), the focus can be turned to the type of environment that would be needed to realize this change. The main question, here, would be that of which conditions are offered within the planned environment for realizing the policy objective, taking into account possible tensions as well as synergies between objectives in different fields of policy, different municipalities in the region, and different layers of governance.

Following on the example given for point 2 (i.e. the difference between office towers distributed across the city and a spacious campus) planners and assessors could consider which type of plan would contribute most, for instance, to creating so-called knowledge spillovers (a much-debated policy objective in the Netherlands as well as abroad). They could argue that workers of different firms may be expected to share knowledge and experience by making use of shared facilities on a campus. They may be expected to meet more frequently than workers in office towers in different parts of the city.

This then could lead to the relevant question of which type of local spatial quality would be required for knowledge workers to meet (3C, see point 14). By

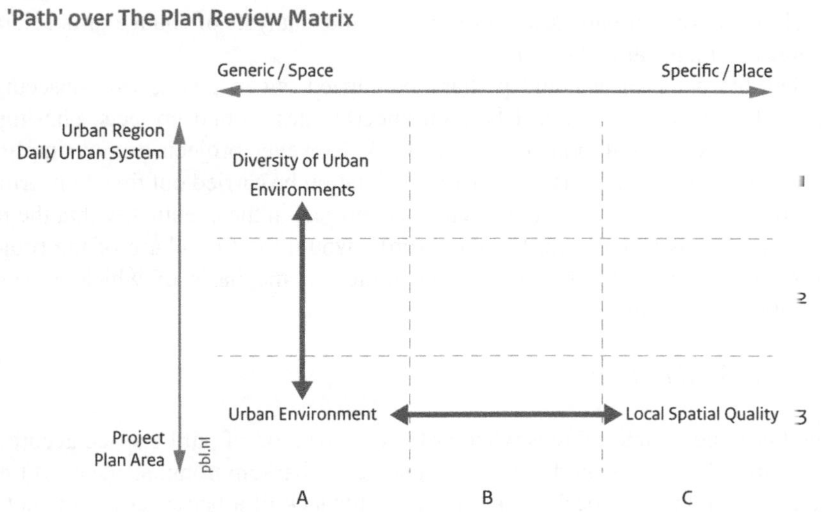

Figure 14.4 'Path' over the Plan Review Matrix
Source: PBL Netherlands Environmental Assessment Agency, 2013.

following a 'path' over the matrix, planners and assessors can analyse, improve and evaluate connected conditions; in the case of the example in Figure 14.4, the range of urban environments (1A), the chosen environment (the campus) (3A), and the fitting local spatial quality (3C).

13. Place Maker (3B)

Another issue is that of how to encourage the transformation of a particular space into a 'place', at locations that are suffering from multiple problems. Here, three options are distinguished. Firstly, the 'golden key': could the plan be altered in such a way that qualities that are difficult to unify may be developed with a certain connection? Changes may affect both the programme and the spatial design. In some cases, a breakthrough in routine thinking can lead to new solutions with a broad support base. In these cases, puzzles are not pieced together again, but the puzzle itself is altered.

Secondly, pioneers may give an extra impulse to the development or redevelopment of an area. Plans may be delayed for various reasons, in certain parts of a city, causing some areas to remain undeveloped for long periods of time, and resulting in the postponement of desired changes. Quite often these 'interim' periods last for many years. Preliminary investments in programmes and buildings (e.g. events, studios and workshops, and meeting places), as well as temporary permissions for unconventional use of space or unconventional urban development (e.g. urban agriculture) may be the catalysts of change or accelerate changes that have already set in.

Thirdly, the project could perhaps be carried out in phases (in consecutive steps). In the current Dutch debate on uncertainties around projects, phasing is considered as a way of reducing financial risk. However, projects cannot simply be cut into several pieces. It is vital that an estimation be carried out first, concerning the practicality and relevance of such a part-project in the eventuality that the rest of the project is abandoned. In other words, would the first phase of the project have an independent functional-spatial logic, on the basis of which it would contribute to the city?

14. Local Spatial Quality (3C)

Another issue to determine is whether the organization of public space according to a particular plan would match the planned urban environment (see 3A) and the target audience? Local spatial quality – for lack of a better term – cannot be described in general terms but must be related to aspects such as location, existing functions, density and building heights. For a square, for example, sun and lighting would be important, in addition to shelter from the wind and a certain degree of enclosure. Furthermore, also information on the main users would be important; after all, they would make different demands on the location. For example, visitors to a certain place would look for entertainment, while residents may seek rest and quiet at the same location. In addition to a pleasant outside space, also local conditions for traffic circulation, as well as social safety and road safety may play a role.

Following on the example given for point 2 (cell 1A) and point 12 (cell 3A), the main question in relation to the local spatial quality on campus could be whether the planned public space is organized in such a way that it meets the needs of the knowledge workers (the campus target group) and achieves the objective of stimulating the exchange of knowledge. Can workers meet in a casual manner and in a comfortable environment? Of course, shared spaces and facilities are no guarantee that people will meet, but they do provide the conditions that encourage such encounters.

Theoretical and empirical knowledge about the spatial quality of public spaces has been documented in handbooks. A number of urban designers and sociologists have studied a variety of squares, parks and streets, documenting their main aspects (e.g. see Cullen, 1961; Gehl, 2001; Jacobs, 1961; Jacobs, 1993; Lynch, 1960, 1988; Whyte, 2001). Notable fact is that these studies name best practices but do not formulate rules for constructing what could be seen as 'good' public spaces. As was mentioned above, each location, function and user target group is different; every situation requires its own approach. However, this does not mean that projects cannot be compared. Although the appropriate spatial quality of a certain area cannot be determined using a general formula, this could be determined from reference projects; a comparison between a planned project and related, already finished projects elsewhere. Questions related to which actions were successful and which were not (ex-post analysis) are important, as well as making certain that the reference projects match the planned project in type of urban environment, magnitude and situation within the urban region.

15. Historic Buildings (3C)

It must be determined, when applicable, whether a plan also deals with cultural heritage, on a building level. Are choices being made solely with respect to preservation of protected historic monuments, or do such historically valuable and characteristic buildings offer starting points for new development? Could they provide a project with some *couleur locale*, or even serve as centrepieces for the development of a recognizable, local identity for the surrounding areas? (see also 2C).

Various Functions of the Matrix

This list, as indicated above, may be used as a checklist for a discussion among experts. Depending on the phase of the planning and decision-making process, this checklist could have various functions. Planners could use the list throughout the development of a project, and subsequently it could also be used in discussions between experts and stakeholders about possible improvement options. Towards the end of the planning and decision-making process, the checklist could become part of the assessment process (an analysis and evaluation by independent experts), as a prelude to the assessment of expected effects and project efficiency; for example (as is current Dutch practice) in the form of a cost-benefit analysis (CBA).

The above brief explanations of the conditions place the emphasis on the desired spatial development within a project. However, in addition, the possibility of unwanted effects must also be considered. For example, the construction of a new exit or entry road may unwittingly create a barrier between two previously well-connected areas, within or outside the planning area. Because such effects may create conditions that could hamper the achievement of policy objectives, they must be brought to light early enough to be considered when improvement options are discussed.

Three Cases in a New Planning Practice

Plan Assessment in Bottom-Up Practices

The search for new ways to deal with the uncertainties in present day planning practice and the real estate market is topical among project developers, investors and municipalities. But governments are also initiating projects around bottom-up development. Although these projects sometimes do focus on a spatial plan, more often goals are defined elsewhere, such as in the health, energy or social sectors.

Even though the spatial dimensions, for many of these initiatives, have become secondary, aligning ideas about spatial development can be beneficial for the city or the region. When the government is playing a coordinating role, including spatial objectives may be a 'quick win'. These objectives may not only refer to the quality of the plan itself, but also to its relationship with the rest of the city. What types of sites are being created: generic places or specific spots with an identity that would complement the existing palette of urban areas? The central question here is whether a project will create the spatial conditions for it to contribute added value to a larger area.

Below, three cases are discussed, according to which we explore how plans that are initiated, bottom up, may also be connected to a broader spatial agenda. This agenda is not determined through rigid top-down regulation, but is created by combining and prioritizing projects and optimizing existing structures. This includes possible connections on and between different scales; of the plan itself, of the city, and/or the Daily Urban System (DUS). And, depending on the type of plan, different methods of analysis may be used; from abstract and numerical to concrete and spatial. The Plan Review matrix connects to these scales and methods of analysis. We indicate which spatial conditions in the matrix may help to structure the analysis and evaluation of plans on a case-by-case basis.

TEEB and the City

TEEB Stad (TEEB City) is a development tool, developed by 11 municipalities, two ministries and the engineering firm of Witteveen + Bos, which calculates the economic amenities of urban green, water and nature in and around cities (KPMG, 2012; Van Wetten, 2012). Results from this tool may be used to support decision-making processes and design choices. The method has a financial approach. The added value related to green and water is determined, those that benefit are identified (e.g. investors, real estate developers, health care institutions, businesses, social institutions, water boards, and municipal land development departments) and may be involved as a co-investors of the project. The financing of public green space in and around the city is no longer just a task of the government.

The green and blue projects are assessed in a numerical way. By adding a more qualitative analysis, such as a categorization of the different types of green and their conditions of use, as well as amenities (such as protected areas, ecosystem services, and functional green), and analysing the connections between urban green and a city's greenbelt, a spatial agenda can be added to the programme. The project may take on a more strategic position, on a local and regional scale, and the level of support may be increased. In addition, innovative location–function combinations may be explored, perhaps offering multiple opportunities to pay off. A striking example of an innovative project which combines blue and urban functions is the 'Watersquare'. It serves all rainwater overflows in cases of extreme rainfall, and can be used as (skate and sports) parks when it's dry (De Urbanisten, 2010). The matrix cells involved in this case are 'urban green structure' (B2), 'consumption and production landscape' (B1), and 'local spatial quality' (C3).

The Heritage Factory

The Dutch province of North Brabant is running a long-term investment programme, called the Heritage Factory (De Erfgoedfabriek). The programme is intended to offer a second life to large heritage sites and structures, such as abandoned cloisters and castles, factories and forts (Provincie Noord-Brabant, 2012). Funding does not take place through subsidies, but rather the province invites other parties to present ideas. Any and all may apply, including municipalities, property owners, designers, enthusiasts, development companies and investors. The province takes on the role of connector, investor, or developer, and it demands value for money. Each project is granted a suitable strategy for development, with a return on investment sometimes taking a long time to come about. The approach of the province is one that favours 'time for quality' rather than speed. The province acknowledges that tailored solutions, good ideas and experiments take more time than any off-the-shelf investment option.

The province must determine, beforehand, which projects will be viable. High returns on investment are evidently preferred, and the province has to decide whether or not that might be achieved by investing in as many development complexes as possible, and whether to do so on the basis of generic key figures (as is done in CBAs). It seems to make sense to first select the most promising projects via a multiple level chance and risk analysis. The projects that score highly on multiple spatial scales potentially have the greatest added value. Providing space for dynamics in the longer term, and connecting local plans in their specific context with targets on a regional level may offer opportunities to create a stronger regional profile. In the Plan Review matrix, the conditions of 'place maker' (cell B3), 'historic buildings' (C3), 'cityscape', 'urban icons' (C2) and 'cultural landscape' (C1) offer points of reference for prioritization.

PLUG Rotterdam

PLUG Rotterdam is a programme initiated by the AIR Foundation (Architecture centre of Rotterdam) in collaboration with the municipality of Rotterdam (Municipality of Rotterdam, 2013) that calls for development projects in and around Rotterdam. PLUG Rotterdam is aimed at future occupants and entrepreneurs. The municipality has opened up several inner city sites for experiments with a new type of project: a user-defined business proposition. A website displays the development of the sites. Larger investors typically are not interested in sites such as these. The sites are too small and the planning situation is too complicated. However, these sites are ideally suited for bottom-up development. This programme aims to make the city more liveable.

The focus of the municipality on the financial feasibility of the propositions provides, however, no guarantee of plans being realized that contribute to the quality of their location or the social climate of the city. This would require a comparison of various designs for any specific location. In this way, the spatial qualities may be explored and plans may be optimized. Such an evaluation would not immediately focus on the financing or 'the bricks', but would first explore the possible public qualities of the intervention.

Depending on the type of project, several of the conditions in the matrix may be relevant. On the basis of building plans for parcels of land of moderate size, plans can be assessed for their contribution to the urban environment ('environment') (A3), 'types of facilities and social interaction' (A2) and 'local spatial quality' (C3). Even before a calculation is made for the project, it can be established if and how the plan would contribute to a city's public space and how it would help this public space to function better as a whole.

Synthesis

This chapter has argued that the introduction of the Plan Review instrument could make communication on urban projects more transparent, and thus contribute to improvements in planning and decision-making processes around urban investments. Insight into the specific environment of a project location is vital to the assessment of urban plans (context sensitivity). In addition, there is a need for a system of analysis that would be widely applicable (scope, repeatability), and that uses clear considerations (transparency) which are comprehensible to a wide variety of experts and stakeholders (communication beyond the boundaries of individual disciplines).

The Plan Review meets the combined needs for context-specific knowledge as well as a system. It features a series of 16 spatial conditions that a project has to meet in order to achieve policy objectives (possibly at various governance levels and in different fields). This series is presented in a matrix, from which relevant conditions may be chosen, per case. This matrix may help to structure the dialogue with stakeholders, and maximize the objectification of plan analysis by independent experts. Depending on the phase of the planning and decision-making process, the Plan Review may improve urban planning and contribute to the related spatial assessment, in light of the particular policy objectives.

The Plan Review has been developed on the basis of knowledge and know-how from various spatial disciplines. It deviates from the welfare economics approach of cost-benefit analyses (CBAs) – the instrument most used in the Dutch assessment process. The Plan Review is not intended as an alternative to CBAs, but rather as a complementary instrument. Because of its deviating theoretic and methodical basis, a successful application of the Plan Review would require an expert panel with a composition of disciplines that are different from those of the teams that currently mostly assess project plans. The application of this instrument requires a multidisciplinary team that represents both knowledge and experience in spatial research and design.

Until recently, Dutch spatial policy and planning practice was dominated by large-scale, straightforward and all-inclusive urban development. Planning practice, however, is currently undergoing considerable changes. The use of top-down, comprehensive and detailed plans is declining – and is giving way to a wider spectrum of more diverse, grassroots-oriented planning. New planning documents (dubbed 'organic development') often include small-scale goals or aim for longer-term objectives in a variety of fields. As a result, CBAs, which require fully fledged, detailed plans, with clear parameters, are less suitable to deal with all the issues in present day plans. In this chapter we have explored the use of a tool that can assess the value of these new plans more effectively than could be done using a CBA.

In the assessment of traditional large-scale plans, the Plan Review can be used alongside the CBA. Whereas CBAs are concerned with overall (financial) efficiency, the Plan Review first establishes a plan's conditions for effectiveness and aims to answer the question of whether the planning goals would be achievable. In the assessment of 'organic planning' projects, where attention shifts to the benefits of multiple stakeholders in an often open-ended process, the public responsibility of municipalities shifts towards aligning multiple goals and finding common ground for development. In cases like these, the Plan Review can help decision makers rank and optimize plans, in relation to higher-level spatial agendas and visions.

The scope of the Plan Review is wide. In addition to an assessment instrument on a national level, it is also a strategic tool for local government (e.g. provinces and municipalities). Local governments increasingly are faced with innovative plans that are not easy to assess using a system of traditional assessment methods. To position the Plan Review not at the tail end of the planning process but at the beginning, the instrument may contribute to a chance and risk analysis in an early stage of the decision-making process, when there is still room for improvement. It may help government to take responsible and transparent decisions, with uncertainties not being ignored, but rather used as input in a search for flexible or robust responses to them. The instrument can connect comprehensive spatial plans with a bottom-up agenda. That agenda is not determined by rigid top-down regulation, but is created by combining and prioritizing projects with the aim of optimizing existing spatial structures. In this way, the Plan Review may support the government to develop a policy that offers opportunities in several respects; to various promoters, to experimental planning and to innovative uses of space.

Discussion

The new instrument presented in this chapter will be developed further over the years to come – it has not reached a definite form. Over the coming period, its usefulness will be tested in practical applications, and further discussions with a wide variety of users may help to fine-tune the instrument. This would also require further research into the weighing of the conditions as entered into the matrix. In its current form, the matrix does not provide any ranking. The experts involved must choose the most relevant conditions, depending on the nature of the project. Further research as well as practical experience could indicate whether the possible incommensurability of various conditions would stand in the way of clear decision making. Something similar applies to the connection of various spatial conditions with policy objectives on different scales and in different fields.

Attention should be paid to the embedding of the instrument in local and self-organizing processes. How would the input of an independent expert panel relate to organic processes in which room is provided for the ideas of various promoters? Finally, the relationship between the Plan Review and the CBA in the planning process should be considered. Because of the focus of CBAs on effectiveness and efficiency of a project plan, and the Plan Review's focus on the conditions under which a plan could be most effective, we recommend that application of the Plan Review would precede the use of a CBA. Testing of the instrument in actual practice will show whether this approach indeed delivers positive results.

References

Bebbington, J., Brown, J. and Frame, B. 2007. Accounting technologies and sustainability assessment models. *Ecological Economics*, 61, pp. 224–36.

Bertolini, L. 1999. Spatial development patterns and public transport: The application of an analytical model in the Netherlands. *Planning Practice & Research*, 14(2), pp. 199–210.

Beukers, E., Bertolini, L. and Te Brömmelstroet, M. 2011. *Knelpunten in het MKBA-proces*. The Hague: Nicis Institute.

Bromm, S., Vreeswijk, E. and Wiggers, R. 2011. Sociaal ruimtelijke ambities: verankering sociaal met fysiek. *Plan Amsterdam*, (3), pp. 4–13.

De Urbanisten. 2010. *De Urbanisten and the Wondrous Water Square*. Rotterdam: NAI Publishers.

Cullen, G. 1961. *The Concise Townscape*. London: Van Nostrand Reinhold.

Gehl, J. 2001. *Life between Buildings: Using Public Space*. 5th Edition. Kopenhagen: Danish Architectural Press.

Hamers, D. 2011. *The Fringe in Focus. A Mosaic of Urban Milieus as the Basis for a Regional Planning Approach to Urban Development*. International Forum on Urbanism (IFoU), Proceedings. 5th Conference of the IFoU, Singapore.

Hamers, D., Bijlsma, L. and Hoorn, A van. 2013. *Planobjectivering. Een nieuw instrument voor de ruimtelijke analyse van stedelijke projecten in aanvulling op. de maatschappelijke kosten-batenanalyse*. The Hague: PBL.

Jacobs, A. 1993. *Great Streets*. Cambridge, MA: MIT Press.

Jacobs, J. 1961. *The Death and Life of Great American Cities*. New York: Modern Library.
KPMG. 2012. *Groen, gezond en productief.* The Economics of Ecosystems & Biodiversity. Rotterdam: KPMG
Lynch, K. 1960. *The Image of the City*. Cambridge, MA: MIT Press.
Lynch, K. 1988. *Good City Form*. Cambridge, MA: MIT Press.
Mouter, N., Annema, J.A. and Wee, B. van. 2012. *Maatschappelijke kosten- en batenanalyse inhoudelijk geëvalueerd*. The Hague: Nicis Institute.
Municipality of Almere and Ministry of Infrastructure and the Environment. 2011. *Verslag MKBA-symposium: tussen methodiek en politiek*. Den Haag/Almere: Ministry of Infrastructure and the Environment and municipality of Almere.
Municipality of Rotterdam. 2013. *Klein en Fijn* [online]. http://www.rotterdam.nl/kleinenfijn, [accessed 11 September 2013].
PBL Netherlands Environmental Assessment Agency. 2010. *De staat van de ruimte 2010: de herschikking van stedelijk Nederland*. The Hague/Bilthoven: PBL.
Provincie Noord-Brabant. 2013. *Verkenning Erfgoedcomplexen*. 's Hertogenbosch: Provincie Noord-Brabant.
Renes, G., Hoorn, A. van and Hamers, D. 2011. *Verbetering van de communicatie en presentatie rondom de MKBA bij verstedelijkingsprojecten*. The Hague/Bilthoven: PBL.
Ritsema van Eck, J., Amsterdam, H. van and Schuit, J. van der. 2009. *Ruimtelijke ontwikkelingen in het stedelijk gebied: dynamiek Stedelijke Milieus 2000–2006*. The Hague/Bilthoven: PBL.
Romijn, G. and Renes, G. 2013. *Plannen voor de stad. Een multidisciplinaire verkenning van de effecten van verstedelijkingsprojecten op het functioneren van een stad*. The Hague: CPB and PBL.
Schön, D. 1984. *The Reflective Practitioner: How Professionals Think in Action*. Aldershot: Ashgate.
Schön, D. 1987. *Educating the Reflective Practitioner: Toward a New Design for Teaching and Learning in the Professions*. San Francisco: Jossey-Bass.
Sennett, R. 2008. *The Craftsman*. London: Allen Lane.
Tribe, L. 1972. Policy science: Analysis or ideology. *Philosophy & Public Affairs*, 2(1), pp. 66–110.
Van Wetten, J. 2012. *Groen loont met TEEB stad*. Deventer: Stichting open boek.
Verheul, W.J. 2012. *Stedelijke iconen. Het ontstaan van beeldbepalende projecten tussen betoog en beton*. The Hague: Boom Lemma.
Whyte, W. 2001. *The Social Life of Small Urban Spaces*. New York: Project for Public Spaces.

Chapter 15

The CBA Dialogue: Early Interaction between Plan Owners and Evaluators for the Assessment of Integrated Transport Plans in a Cost-Benefit Analysis

Els Beukers

Introduction: Problematic CBA Assessments

In Dutch planning, integrated land-use transport plans requiring national funding are assessed in a cost-benefit analysis (CBA), an obligatory step in planning and decision-making processes (Ministry of Transport and Water Management and Ministry of Housing, Spatial Planning and the Environment, 2009). However, the practice of CBA assessments of integrated land-use transport plans seems problematic, both in the Netherlands and beyond. Problems include disputable calculation methods for translating soft variables like quality of nature into monetary terms, or leaving these effects out of the analysis altogether (Mackie and Preston, 1998); missing information about winners and losers (the distribution effects) and ignoring equity issues (Ackerman and Heinzerling, 2002); missing information about expected synergy and agglomeration effects (Wee 2006); poorly constructed reference cases (Annema et al., 2007; De Jong and Geerlings, 2003; Naess, 2006; Van Wee, 2006); poor incorporation of uncertainties (Salling and Banister, 2009; Ševcíková et al. 2011); and too much focus on how infrastructure can help solve traffic bottlenecks (i.e. decreased travel time) and too little on how it can support a vision for spatial economic developments (ECMT, 2004; Van Wee et al., 2006). In addition to many content issues (Annema et al., 2007; De Jong and Geerlings, 2003; Beukers et al., 2012; Mackie, 2010; Naess, 2006; Mackie and Preston, 1998), the process of applying CBA is often hampered (Sager and Ravlum, 2005; Martinsen et al., 2010; Eliasson and Lundberg, 2010; see also the contribution by Emile Dopheide in this volume).

As an analysis of CBA practitioner perceptions of CBA process issues revealed, CBAs have been perceived mainly as a final examination of the plan's fulfilment of CBA criteria, leading some planners to see it as just another obstacle to overcome (ECMT, 2004; Page et al., 2009; Beukers et al., 2012). Local authorities in the United Kingdom perceived appraisal, like the CBA, as a barrier to developing sustainable urban transport and land use strategies (Page, 2009). Dutch planning practitioners and CBA evaluators also perceived several issues when integrated land-use transport plans were assessed in a CBA, for example fearing that easier-to-calculate effects (like travel time savings) would dominate difficult-to-calculate effects (like effects on sustainability); not understanding the assessment method and outcome – perceiving it as a black box; or that the assessment would be used selectively and strategically (Beukers et al., 2012). The CBA process issues in this research were synthesized into three fundamental issues: (1) communication deficits; (2) distrust between plan owners and evaluators; and (3) the use of the CBA as a final assessment.

This third issue was perceived to be problematic because it leaves out the opportunity to utilize CBA assessments for optimizing plans (and assessments) in a process of gradual learning. In the process of developing integrated transport plans, it is important to be able to continuously adapt the plan's goals and means based on emerging insights. Although a process of gradual learning is desired by CBA practitioners – both plan owners and evaluators (Beukers et al. 2012) – the CBA as commonly used in Europe hardly supports this (ECMT, 2004). Furthermore, CBA processes in the Netherlands were characterized by two groups with opposing perspectives towards the CBA: planners and plan supporters on one side, and economists and evaluators on the other (Mouter et al., 2013; Beukers et al., 2012).

Assessing Integrated Land-Use Transport Plans

Many CBA process issues are rooted in the changing (transport) planning context. Issues like quality of life and sustainability have become core subjects of public debates about transport plans. However, such intangible effects are difficult, if not impossible, to translate into quantitative and monetary terms, and therefore are often neglected in CBAs. For example, plans that generate less mobility and more accessibility (and provide desired economic development) often form a sustainable solution. However, CBAs tend to focus on travel time savings and therefore on valuing more mobility instead of less (Ferreira et al., 2012). Furthermore, transport planning is no longer an isolated profession. Integrated approaches that combine various aspects (e.g. local economic development, social equity, environmental preservation) are becoming increasingly relevant (Bertolini et al. 2005; Banister, 2008; Straatemeier and Bertolini, 2008). As described by Heeres et al. (2012), transport planning traditionally focused on finding infrastructural solutions for mobility issues (transport projects), and grew to finding solutions within a broader spatial system (land-use transport plans). These land-use transport plans are regularly the outcome of a communicative planning process in which various stakeholders with different interests and goals participate (Curtis, 2008). CBAs, however, are regularly conducted by experts (CBA evaluators) asked to assess already finalized plans, separately from the planning processes.

These shifts have altered the demands imposed on assessing transport plans (Handy, 2008; Willson, 2001), requiring multiple dimensions and context specificities to be taken into account (Curtis, 2008; Hull et al., 2011). Commonly used, conventional assessment tools that mainly rely on quantitative data appear ill-suited for this task. The central issues of evaluation have expanded from measuring all performances of a plan or program to facilitating an integrative and learning approach (Khakee, 2003, p. 345). Evaluation should not solely be considered as the final outcome of administrative actions with the aim of approving or rejecting a given plan. Instead, evaluation should be considered as a work methodology, a 'process of gradual learning' to increase the awareness of the choices taken, within the context of the decision-making process (Selicato and Maggio, 2011, p. 173). Evaluation, then, may be seen as a cognitive process in which social realities are constructed and actors develop self-reflective learning abilities, find unexpected meanings in their actions, and build up networks of people, actions, and thoughts. Therefore, it is crucial that actors actively engage with each other, appreciate each other, recognize commitments and struggles, and gain insight into their strong and weak points (Kidd and Fisher, 2007; Saarikosi, 2000; Selicato and Maggio, 2011).

These insights imply that CBAs should be used to deliberatively facilitate a process of gradual learning. However, CBA processes are not characterized as such. On a contrary, as mentioned, practitioners perceived CBAs as a final examination, and communication deficits and distrust were experienced among plan owners and evaluators. This suggests that a different approach to CBAs might be necessary, whereby high levels of communication and trust between plan owners and evaluators could form crucial conditions for using the CBA for learning. Through such a learning process, the CBA and the assessed plan would develop in interaction with each other, ideally leading to a better plan (i.e. a better solution, design, or measurement for the formulated problem) and a better assessment (i.e. more accepted by the participants as representing the plan's degree of achievement of the pursued goals). Increasing the levels of communication and trust between participating plan owners and evaluators seem to be crucial conditions for achieving this.

Setup of the Chapter

The aim of this chapter is to reflect on two studies that intended to better understand how to improve communication and trust between plan owners and evaluators in CBA processes in order to facilitate gradual learning. First, in a review of communication planning and organizational learning literature, a theoretical framework was built of mechanisms and interventions that are expected to facilitate comprehensive communication and trust between plan owners and evaluators, see Figure 15.1 (Beukers et al., 2014). This framework, labelled the CBA Dialogue consisted of five communication and trust building interventions in which participants: organize a dialogue space where plan owners and evaluators meet face-to-face; share and discuss the plan and CBA together and simultaneously; are prepared to share and discuss knowledge together; have

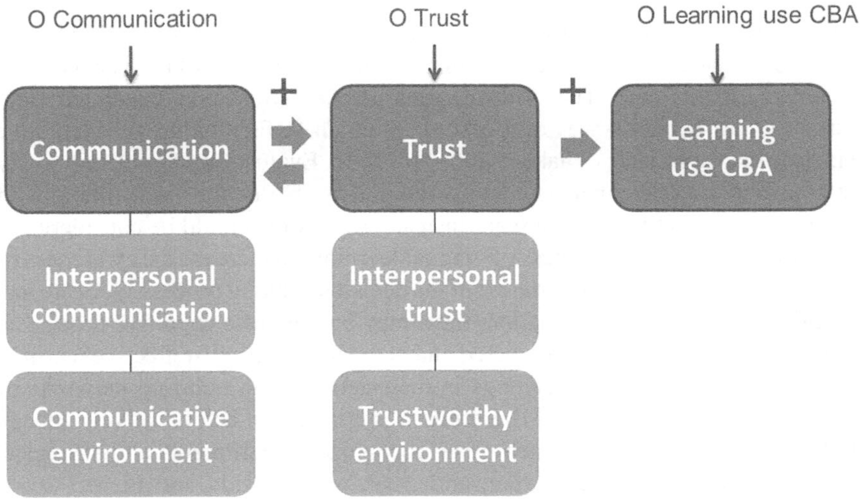

Figure 15.1 Conceptual model of communication and trust conditions for increasing the learning use of CBA

Source: Beukers et al., 2014.

the interaction guided by a moderator; and use dialogue modes (e.g. sketches, simulations, storyboards, role play) presenting the work as 'work in progress'. Second, this framework was applied in two experiential cases to explore if and how these theoretical interventions might work with CBA evaluators and plan owners.

This chapter aims to share the theoretical insights provided by this novel approach to CBA assessments, and how applying this framework in an experiential case study helped to better understand its implications and conditions – narrowing the oft-described gap between theory and practice.

This chapter is organized as follows. The second section elaborates the theoretical background of the framework for the CBA Dialogue, and the five interventions and the communicative and trust mechanisms it claims to trigger. The CBA Dialogue was tested and sharpened in an experiential case study with two sequential cases. The research design of this case study is discussed in the third section, and the outcomes are shared in the fourth section. The chapter closes with reflection and discussion in the fifth section.

The CBA Dialogue: A Theoretical Framework

The CBA Dialogue is a construction of insights from a review of communicative planning and organizational learning literature on how to improve the levels of communication and trust between different types of experts (Beukers et al.,

2014). Whereas many communicative planning and organizational learning studies explain why communication deficits or distrust among, for example, stakeholders form severe issues, fewer studies focus on or are less specific about how communicative and trust issues could be overcome. Nevertheless, this review structured the existing insights and made a distinction between interventions and mechanisms. Thereby, the mechanisms represent how different elements of interpersonal communication and trust function in relation to each other, whereas the interventions inform which concrete actions are needed to trigger these mechanisms.

The insights from these two disciplines were strongly overlapping and, moreover, showed that trust and communication are interrelated variables. 'While trust is necessary for open communication and collaboration, open communication and collaboration are also preconditions of trust' (Laurian, 2009, p. 382).

The review resulted in five interventions that included insights from both communicative planning and organizational learning theory, and that claimed to increase the levels of both trust and communication. The interventions and the related mechanisms they claim to trigger are as follows.

Communication and Trust Interventions and Mechanisms

The first intervention of the CBA Dialogue is to create a dialogue space where planners and evaluators meet face to face (Argyris, 1977; Forester, 1999; Innes, 1998; Nonaka and Konno, 1998; Akgün et al., 2003; Lander et al., 2004; Healey, 2007; Laurian, 2009). This should trigger planners and evaluators to get to know each other and build a relationship. It enables a better estimation of how the other will react in future interactions, thus building up interpersonal trust (Forester, 1987). In so doing, the planners and evaluators may feel safer, share more detailed information, listen more carefully, pay more attention to issues behind immediate agendas, ask and receive critical questions, and collectively change and build assumptions.

The second intervention is to share and discuss the plan and CBA together and simultaneously (Schön, 1983; Healey, 1999; Argyris, 1977; Argyris, 1991; Nonaka et al., 2000; Nonaka et al., 2006; Innes and Booher, 2003; Laurian, 2009). By doing so, the planners and evaluators might better understand each other's standpoints and may see shared interests or opportunities for finding win-win solutions (Forester, 1999). They will know more about each other (and the plan and CBA), which increases the predictability of how they might respond and builds up interpersonal trust. This process continues towards having more open and critical interactions, sharing even more knowledge, and possibly changing one's own perceptions based on the perspectives of others (Healey, 2007; Healey, 2009).

The third intervention involves being prepared for sharing and discussing knowledge together (Forester, 1987; Argyris, 1977; Edelenbos and Klijn, 2007). This might prevent planners and evaluators from becoming unpleasantly surprised or attacked by new arguments. By being prepared, planners and evaluators know how to internalize outside critique without acting defensively and might be stimulated to better articulate their own standpoints and reasoning.

The fourth intervention consists of the interaction being guided by a moderator (Innes and Booher, 2003; Laurian, 2009; Edelenbos and Klijn, 2007). A moderator should stimulate conditions in which: all can speak freely and feel safe; all types of knowledge are considered and included in the discussion; and that the attention of the discussion is on finding shared interest, win-win solutions, and issues behind immediate agendas. When planners and evaluators feel guided and protected by the moderator, they may feel encouraged to share more detailed and diverse knowledge, to be more open towards each other, and to give and receive constructive criticism without responding defensively.

The fifth intervention involves the use of dialogue modes that present the discussed subject as 'work in progress'. The dialogue mode should steer and stimulate the interaction, for example by using sketches, simulation, a story board, or a role play (Forester, 1999). By using dialogue modes, planners and evaluators may feel encouraged to give their standpoints, illustrate their arguments, and make them context specific and explicit. This sharing gives others the opportunity to respond, illustrate, contextualize, and make their points of views explicit (Schön, 1983).

To sum up, these communication and trust building interventions ask for plan owners and evaluators: (1) to meet; (2) to share and discuss the plan and CBA; (3) to be prepared; (4) to have the interaction guided by a moderator; and (5) to use dialogue modes, presenting the work as 'work in progress'. These five interventions together are labelled as the 'CBA Dialogue', because of their interdependency and interrelatedness (Beukers et al., 2015). Thereby, the fifth intervention calls for such interaction to take place when the status of the plan and assessment is 'work in progress', suggesting that it is still open for discussion. Interaction between plan owners and CBA evaluators should thus take place early in the planning process, as an interplay between planning visions and formulating concrete projects, as illustrated in Figure 15.2, instead of as a linear approach with separated planning and assessment processes as prescribed in Dutch planning regulations (Ministry of Transport and Water Management and Ministry of Housing, Spatial Planning and the Environment, 2009).

Testing the CBA Dialogue: An Experiential Case Study

Forming a theoretical framework like the CBA dialogue to improve communication and trust between plan owners and evaluators is one thing. Understanding if and how this approach might work in the practice of CBA assessments of integrated plans is another. In a follow-up study, the CBA Dialogue was therefore applied

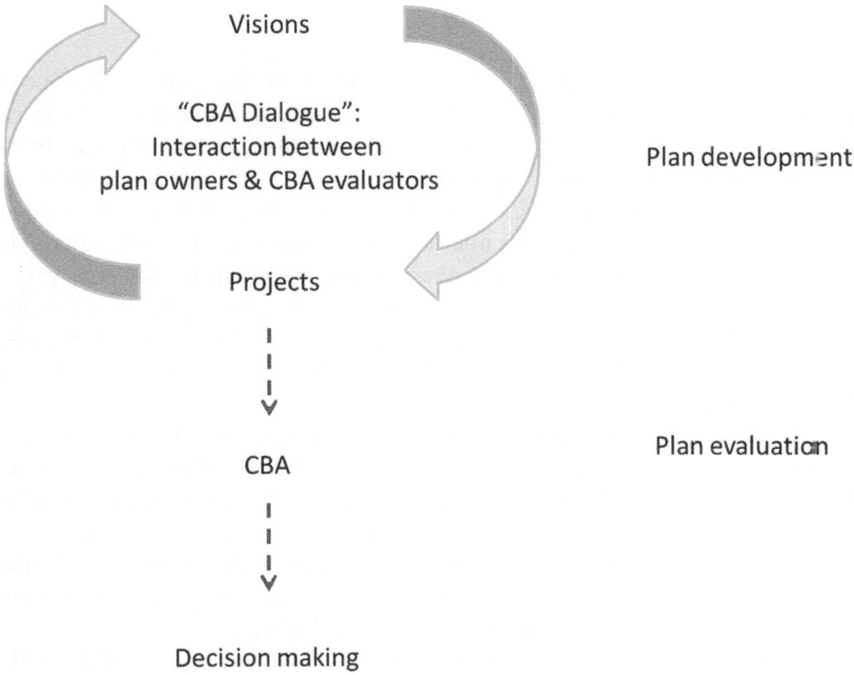

Figure 15.2 Possible use of the CBA Dialogue in planning processes

in two experiential case studies (Beukers et al., 2015). An experiential case study was set up, because it provides a safe environment similar to planning (and CBA) practice in which an alternative approach like the CBA Dialogue could be performed in interaction with its potential users (Straatemeier et al., 2010). The aim of applying and reflecting on the CBA Dialogue in an experiential case study was threefold. First, the experiential case study was conducted to get a better understanding of if and how the communication and trust mechanisms are triggered by the introduced interventions of the CBA Dialogue when confronted with a situation similar to planning (and CBA) practice. Second, participating in an experiential case study allowed the participants to gain experience with the CBA Dialogue. This increased the chance that they might adopt the innovation in their planning practices. Third, by testing the theoretical interventions in a situation similar to practice, its theoretical foundations could be enriched. Thus, the focus of this case study was on whether the CBA Dialogue helps in establishing comprehensive communication and trust between plan owners and evaluators in CBA processes, assuming that these conditions contribute to a use of the CBA for learning.

Two Sequential Cases

The CBA Dialogue was applied in two sequential cases. The context of each case was formed by discussing a CBA setup for an integrated transport plan in the Netherlands that was still in development – early in the planning process. In the first case, the context was the setup of a CBA comparing station area developments at two train stations in Amsterdam. In the second case, the context was the setup of a CBA comparing different routes of a second tramway in the Utrecht region. The participating practitioners in these cases – different participants for each case – were CBA evaluators, plan owners, and a moderator. The participating plan owners were actually involved with the subject at hand while the other participants were not and were invited by the research team. The participants were informed about the academic purpose of the case study.

The cases were monitored through real time observations and analysis of video recordings. After the CBA Dialogue was performed, the participants completed a questionnaire asking them about their experiences in personal communication and trust building during the case, followed by a discussion of their experiences in a group setting. Furthermore, the participants were interviewed for their final reflections a few weeks after the case took place. The case study methodology and setup is more thoroughly discussed in Beukers et al. (2015).

Table 15.1 gives an overview of how the five interventions of the CBA Dialogue were operationalized in both cases, including the lessons learned from the first case that were used to improve the application of the CBA Dialogue in the second case for better communication and trust performances (see section III).

Results of the Case Study

Findings of Case 1

The findings from the first case showed differing results in whether the five interventions helped to improve communication and trust between plan owners and evaluators, and how they may have done so. Most positively influenced were the mechanisms of feeling safe, being open towards each other, listening, asking critical questions, and learning more about the plan and/or CBA. Less positively influenced mechanisms were sharing detailed information, being explicit about the reasoning behind one's standpoints, forming new insights together, and seeing shared interests. Overall, the participants in case 1 didn't seem to really understand each other and the discussion remained somewhat superficial.

Lessons Learned from the First Case for the Second

Based on the group discussion, video analysis, and reflective interviews, the researchers identified four possible reasons for the limited effectiveness of the

Table 15.1 Operationalization of the interventions and lessons learned

Interventions CBA Dialogue	Operationalization case 1	Lesson case 1	Operationalization case 2
I. Meet face-to-face in an open, critical, non-defensive, and sincere way	A dialogue was organized with 6 plan owners (representing the involved departments of the local government), 4 CBA evaluators, and a moderator. The dialogue lasted about one hour.	Devote more time to the discussion	A dialogue was organized with 7 plan owners (representing the involved regional and local governments), 2 evaluators, and a moderator. The dialogue lasted two hours.
II. Interact and discuss the plan and CBA together and simultaneously	The plan (station area development of two Amsterdam train stations) and setup of a CBA were discussed together and simultaneously.	No fundamental changes necessary	The plan (two routes of a second tramway in the Utrecht region) and setup of a CBA were discussed together and simultaneously.
III. Be prepared and trained to interact	The participants were prepared for this interaction through reading a guiding pamphlet together discussing how to approach one another.	Also prepare the content of the discussion in advance	The participants were prepared for this interaction through a guide discussing how to approach one another, and information about the discussed plan was shared in advance.
IV. Have the interaction guided by a moderator	The interaction was guided by an experienced moderator who was prepared for the task by the research team.	Ask the moderator to encourage participants to be explicit and avoid jargon	The interaction was guided by an experienced moderator who was prepared for the task by the research team, and was instructed to pay attention to be explicit and avoid jargon.
V. Use discussion modes presenting the plan and CBA as 'work in progress'	Three questions were used to steer the discussion: What are the expected effects of the plan? How to measure them? And how to include these effects in the CBA?	Use a dialogue mode that stimulates more interaction	The 'Effects Arena' was used to steer the discussion (see Figure 15.3).

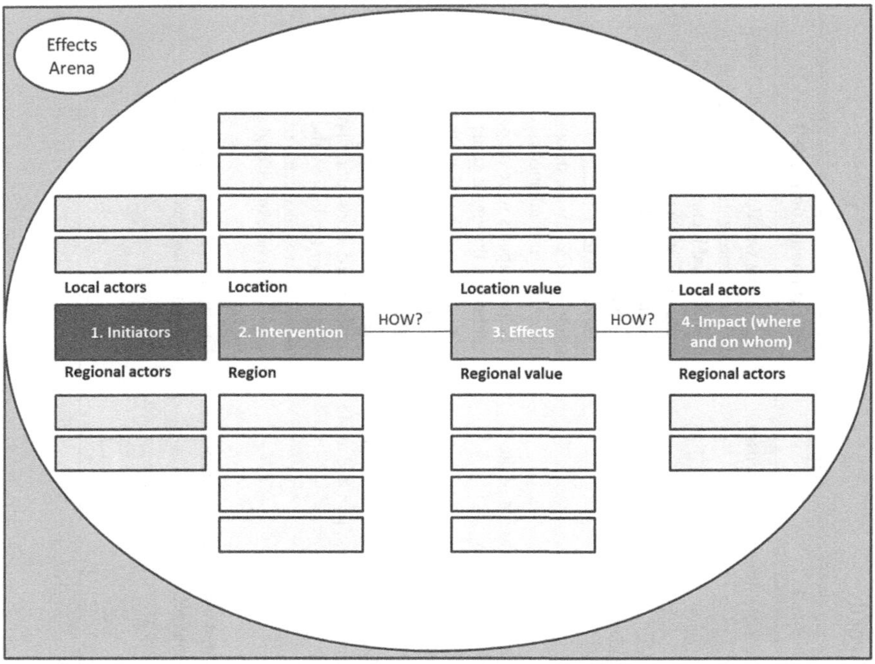

Figure 15.3 The 'Effects Arena' – used in the second case for steering the discussion

Source: Beukers et al., 2013.

interventions in case 1. First, the time dedicated to the dialogue didn't seem to be commensurate with the complexity of the topic discussed. Second, the dialogue seemed to require better preparation of the discussion's content, not just attitudes towards the discussion. It may have been helpful if the plan owners were better aware of the reasoning behind their own plan and its basic ideas and interests, and if the evaluators were more aware of the issues at stake beforehand. Third, the moderator could have paid more attention to urging the participants to be explicit and explain any jargon used. And fourth, the three questions that steered the dialogue in this case study did not appear to be an adequate dialogue mode, as they didn't invite the participants to share their standpoints or react to the views of others. Although other reasons could be given for the limited effectiveness of the interventions in the first case, these four explanations were used to refine interventions 1, 3, 4, and 5 in the second case, whereas no explicit lesson was learned in relation to the second intervention, see also Table 15.1.

Findings of Case 2

Compared to the first case study, the second case study resulted in higher scores for the measured communicative and trust effects of the interventions. Higher levels of communication and trust were measured and related to the performed interventions. The refined interventions seemed to better fit the assignment. The devoted time of two hours was considered enough, the preparation of both the behaviour and the content was seen as sufficient, the moderator was judged to be prepared for her role, and the discussion tool was perceived as suited for its task. Furthermore, the group concluded during the group discussion and reflective interviews that they thought the participants had the appropriate knowledge and experiences to perform the CBA assignment of the integrated transport plan at stake. These factors led to what the participants saw as a fruitful and rich discussion in which diverse knowledge and information was shared and formed, in contrast with their previous experiences with CBAs.

Exchange of Knowledge, Information, and Experiences

The cases showed that the CBA Dialogue helped to facilitate communication and trust building between the participating plan owners and evaluators (Beukers et al., 2015). Thereby, the levels of communication and trust were higher in the second case than in the first, which resulted in a greater and more intense exchange of knowledge, information, and experiences compared to the first. In this exchange, both plan owners and evaluators brought in knowledge that the others did not have and otherwise would not know of.

The evaluators emphasized that the CBA Dialogue helped them to vividly understand the issues behind the plan, through sharing information about conflicting interests rather than relying on a memo. They felt that this could help in working more effectively – if all effects were discussed and explained by both sides and made complete in a dialogue. Moreover, if the participants know the issues better, it is expected that these issues will receive more and more adequate attention in a CBA report. Even if it would not change the final CBA ratio, the analysis could fit better to the understandings of plan owners. Furthermore, the CBA Dialogue gave the evaluators the opportunity to explain the CBA method better and to think of alternative assessment strategies[1] if necessary (for example, paying more attention to qualitative effects). This helped the plan owners to understand the CBA reasoning, which prepared them for the further assessment process. Moreover, it made the CBA less of an insider's tool. It is expected that this

1 Spatial quality was an important goal in the second case. The evaluators explained during the dialogue how effects on spatial quality would be assessed in the CBA, which would be done minimally since these effects are difficult to monetize. Therefore, the group discussed that these effects should be assessed through another, more qualitative method to assure receiving the right attention.

could increase the overall support and acceptance of the assessment. Finally, the CBA Dialogue helped to create a critical view of the plan and sharpen the problem analysis; a CBA evaluator felt that his input made the dialogue more 'rational' and pinpointed the necessity of having a problem analysis in the first place. However, it was noticed that having all stakeholders at the table would be necessary to get the complete picture, which was not the situation in the case study.

Preconditions of the CBA Dialogue

The case study also pointed at some crucial preconditions for the CBA Dialogue to be effective. Dedicating enough time to the discussion turned out to be necessary for discussing such complex matters as integrated transport plans and the CBA. Furthermore, participants need to be prepared for how to behave in such an interaction (being open and critical, but not defensive) and for the content of the discussion. The latter is necessary to be explicit about one's own reasoning, standpoints, and role in the discussion. Also, the research showed the importance of having the discussion guided by a highly skilled moderator, one who is able to create the right atmosphere of openness and who stimulates the asking of questions and explicitness in the discussion. Likewise, the chosen discussion mode can make or break the discussion. The research showed that the discussion mode should, on the one hand, invite different views and stimulate others to react on it, and, on the other hand, structure the discussion. The 'Effects Arena' that was used in the second case study had all of these qualities and was therefore of great importance to the discussion.

Conclusion and Reflection

This chapter started with discussing the problematic assessment of integrated transport plans in a CBA, resulting in many process-related issues: the use of the CBA as a final assessment and communication deficits and distrust between plan owners and evaluators. Literature, though, pointed at the necessity of organizing a process of gradual learning that includes different perspectives and stakeholders when assessing integrated plans. Moreover, high levels of interpersonal communication and trust were described as crucial conditions for this process. This chapter discussed two research studies that aimed to tackle this issue. First, a CBA Dialogue was constructed based on insights from communicative planning and organizational learning literature, with five interventions that might trigger the necessary mechanisms for comprehensive communication and trust between plan owners and evaluators, assuming that this would support a learning use of the CBA. Second, this CBA Dialogue was applied in an experiential case study in two sequential cases to better understand if and how it might improve communication and trust between plan owners and evaluators. The chapter illustrated that an application of theoretical innovations in practice does not work without some fine-tuning. The experiential case study, with two sequential cases, was necessary for this transition and narrowed the gap between theory and practice, improving the application of theoretical interventions in practice by making them more context-specific.

Aside from its advantages, the setup of the experiential case study did limit the validity and reliability of the research results: a small sample was used (two cases with in total 19 participants), the participants selected themselves (they were willing to cooperate), and they knew they were participating in an academic experiment. Therefore, the possible impact of their personalities was not taken into account. Furthermore, the case study was largely an academic setting. How the communication- and trust-improving interventions could function in real planning contexts, and if they support the use of CBA as a learning tool, warrants more investigation. Moreover, the interventions were performed only once in both case studies. Participants suggested that they should be performed several times during a CBA process and possibly with various participants, such as citizens or politicians.

Nevertheless, the case study showed that the CBA Dialogue indeed facilitated communication and trust between the participants, whereby the second case was more effective in this than the first. Through the high levels of communication and trust, an exchange of knowledge, information, and experiences was triggered, which helped the participants to get a more complete picture of the issues at stake, work more effectively, and gain a more critical view of the plan and analysis. The case study therefore indicated that organizing a deliberative (CBA) dialogue early in the planning process, when the plan and assessment are still 'works in progress' and open for discussion, is worth the effort. In this phase, an exchange of knowledge, information, and experiences between the participating evaluators and plan owners could improve both the plan and assessment and trigger a process of gradual learning. However, the discussed case study did not investigate this. A subject for further research could be if and how an early interaction between the planning and assessment processes leads to gradual learning, and influences the development of integrated land use transport plans and (CBA) assessments.

Furthermore, the planning context appeared to be an important factor in the case study influencing the exact form and application of the CBA Dialogue: how much time is needed, which questions need to be raised, what instructions are necessary for the participant roles, what should be the level of preparation, who needs to participate, and what discussion tool needs to be used. This, too, could be subject to further research, to better understand what planning context determines what CBA Dialogue characteristics. Further investigation is thus needed to understand how a CBA Dialogue might influence the communication and trust conditions throughout the whole CBA process, to what extent a CBA Dialogue might influence a learning use of the CBA when other factors like power imbalances between participants are at play, and how the planning context influences its effectiveness.

References

Allmendinger, P. 2002. Towards a post-positivist typology of planning theory. *Planning Theory*, 1, pp. 77–99.

Annema, J.A., Koopmans, C. and van Wee, B. 2007. Evaluating Transport Infrastructure Investments: The Dutch Experience with a Standardized Approach. *Transport Reviews*, 27, pp. 125–50.

Akgün, A.E., Lynn, G.S. and Byrne, J.C. 2003. Organizational learning: A socio-cognitive framework. *Human Relations*, 56, pp. 839–68.

Ackerman, F. and Heinzerling, L. 2002. Pricing the priceless. Cost-Benefit Analysis of environmental protection. *University of Pennsylvania Law Review*, 150(5), pp. 1553–84.

Argyris, C. 1977. Double loop learning in organizations. *Harvard Business Review*, September-October, pp. 115–25.

Argyris, C. 1991. Teaching Smart People How to Learn. *Harvard Business Review*, 4.

Bachmann, R. and Inkpen, A.C. 2011. Understanding Institutional-based Trust Building Processes in Inter-organizational Relationships. *Organization Studies*, 32, pp. 281–301.

Banister, D. 2008. The sustainable Mobility Paradigm. *Transport Policy*, pp. 73–80.

Bertolini, L., Le Clercq, F. and Kapoen, L. 2005. Sustainable accessibility: A conceptual framework to integrate transport and land use plan-making. Two test-applications in the Netherlands and a reflection on the way forward. *Transport Policy*, 12, pp. 207–20.

Beukers, E., Bertolini, L. and Te Brömmelstroet, M. 2012. Why Cost Benefit Analysis is perceived as a problematic tool for assessment of transport plans: A process perspective. *Transportation Research Part A: Policy and Practice*, 46, pp. 68–78.

Beukers, E., Bertolini, L. and Te Brömmelstroet, M. 2014. Using cost benefit analysis as a learning process: Identifying interventions for improving communication and trust. *Transport Policy*, 31, pp. 61–72.

Beukers, E., Bertolini, L. and Te Brömmelstroet, M. 2015. An assessment of interventions for improving communication and trust in Cost Benefit Analysis processes. *Impact Assessment and Project Appraisal*, 33(1), pp. 28–42.

Curtis, C. 2008. Planning for sustainable accessibility: The implementation challenge. *Transport Policy*, 15, pp. 104–12.

De Jong, M. and Geerlings, H. 2003. Exposing weaknesses in interactive planning: The remarkable return of comprehensive policy analysis in the Netherlands. *Impact Assessment and Project Appraisal*, 21, pp. 281–91.

Edelenbos, J. and Klijn, E.H. 2007. Trust in Complex Decision-Making Networks. A Theoretical and Empirical Exploration. *Administration and Society*, 39, pp. 25–50.

Eliasson, J. and Lundberg, M. 2010. Do cost-benefit analyses influence transport investment decisions? Experiences from the Swedish transport investment plan, 2010–2021. 12th World Conference on Transport Research, Lisbon.

European Conference of Ministers of Transport, 2004. *Assessment and Decision Making for Sustainable Transport*. Brussels: European Conference of Ministers of Transport.

Ferreira, A., Beukers, E. and Te Brömmelstroet, M. 2012. Accessibility is gold, mobility is not: A proposal for the improvement of Dutch transport-related cost-benefit analysis. *Environment and Planning B: Planning and Design*, 39(4), pp. 683–97.

Forester, J. 1987. Planning in the Face of Conflict. *Journal of the American Planning Association*, 53(3), pp. 303–14.

Forester, J. 1999. *The Deliberative Practitioner*. Encouraging Participatory Planning Processes, Cambridge: The MIT Press.

Handy, S.L. 2008. Regional transportation planning in the US: An examination of changes in technical aspects of the planning process in response to changing goals. *Transport Policy*, 15, pp. 113–26.

Healey, P. 1999. Institutionalist Analysis, Communicative Planning, and Shaping Places. *Journal of Planning Education and Research*, 19(2), pp. 111–21.

Healey, P. 2007. *Urban Complexity and Spatial Strategies: Towards a Relational Planning for Our Times*. London: Routledge.

Healey, P. 2009. In Search of the 'Strategic' in Spatial Strategy Making. *Planning Theory and Practice*, 10, pp. 439–57.

Heeres, N., Tillema, T. and Arts, J. 2012. Integration in Dutch planning of motorways: From 'line' towards 'area-oriented' approaches. *Transport Policy*, 24, pp. 148–58.

Hull, A., Alexander, E.R., Khakee, A. and Woltjer, J. (eds) 2011. *Evaluation for Participation and Sustainability in Planning*. London: Routledge.

Innes, J.E. 1998. Information in Communicative Planning. *Journal of the American Planning Association*, 64, pp. 52–63.

Innes, J.E. and Booher, D.E. 2003. 'Collaborative Policymaking: Governance through Dialogue'. In Hajer, M.A. and Wagenaar, H. (eds) 2003. *Deliberative Policy Analysis. Understanding Governance in the Network Society*. Cambridge: Cambridge University Press.

Khakee, A. 2003. The Emerging Gap between Evaluation Research and Practice. *Evaluation*, 9(3), pp. 340–52.

Kidd, S. and Fischer, T.B. 2007. Towards sustainability: Is integrated appraisal a step in the right direction? *Environment and Planning C: Government and Policy*, 25, pp. 233–49.

Lander, M.C., Purvis, R.L., McCray, G.E. and Leigh, W. 2004. Trust-building mechanisms utilized in outsourced IS development projects: A case study. *Information and Management*, 41, pp. 509–28.

Laurian, L. 2009. Trust in Planning: Theoretical and Practical Considerations for Participatory and Deliberative Planning. *Planning Theory and Practice*, 10, pp. 369–91.

Mackie, P. 2010. *Cost-Benefit Analysis in Transport: A UK Perspective*. International Transport Forum, Mexico.

Mackie, P. and Preston, J. 1998. Twenty-one sources of error and bias in transport appraisal. *Transport Policy*, 5, pp. 1–7.

Martinsen, J.A., Odeck, J. and Kjerkreit, A. 2010. Why Benefit-cost analyses matter less and how it can be improved for decision making in the transport sector – Experiences from the Norwegian National Transport Plan 2010–2019. Association for European Transport and contributors.

Ministry of Transport and Water Management and Ministry of Housing, Spatial Planning and the Environment. 2009. *Spelregels van het Meerjarenprogramma Infrastructuur, Ruimte en Transport*. The Hague: Ministry of Transport and Water Management and Ministry of Housing, Spatial Planning and the Environment.

Mouter, N., Annema, J.A. and Van Wee, B. 2013, Ranking the substantive problems in the Dutch Cost-Benefit Analysis practice. *Transportation Research Part A*, 49, pp. 241–55.

Naess, P. 2006. Cost-Benefit Analyses of Transportation Investments. Neither Critical nor Realistic. *Journal of Critical Realism*, 5, pp. 32–60.

Nonaka, I. and Konno, N. 1998. The Concept of 'Ba': Building a Foundation for Knowledge Creation. *Californian Management Review*, 40, pp. 40–54.

Nonaka, I., Toyama, R. and Konno, N. 2000. SECI, ba and leadership: A Unified Model of Dynamic Knowledge Creation. *Long Range Planning*, 33, pp. 5–34.

Nonaka, I., Von Krogh, G. and Voelpel, S. 2006. Organizational knowledge Creation Theory: Evolutionary Paths and Future Advances. *Organizational Studies*, 27, pp. 1179–208.

Page, M., Kelly, C., May, A., et al. 2009. Enhancing Appraisal Methods to Support Sustainable Transport and Land Use Policies. *European Journal of Transport and Infrastructure Research*, 9, pp. 296–313.

Saarikoski, H. 2000. Environmental impact assessment (EIA) as collaborative learning process. *Environmental Impact Assessment Review*, pp. 681–700.

Sager, T. and Ravlum, I.A. 2005. The political relevance of planners' analysis: The case of a parliamentary standing committee. *Planning Theory*, 4, pp. 33–65.

Salling, K.B. and Banister, D. 2009. Assessment of large transport infrastructure projects: The CBA-DK model. *Transportation Research Part A*, 43, pp. 800–813.

Schön, D.A. 1983. *The Reflective Practitioner. How Professionals Think in Action*. Aldershot: Ashgate.

Selicato, F. and Maggio, G. 2011. 'The Evaluation Process for a New Planning Culture: Regulatory Compliance and Learning Opportunities'. In A. Hull, E.R., Alexander, A. Khakee and J. Woltjer, J. (eds) 2011. *Evaluation for Participation and Sustainability in Planning*. London: Routledge.

Ševcíková, H., Raftery, A.E. and Waddell, P.A. 2011. Uncertain benefits: Application of Bayesian melding to the Alaskan way viaduct in Seattle. *Transportation Research Part A*, 45, pp. 540–53.

Straatemeier, T. and Bertolini, L. 2008. Joint Accessibility Design: Framework Developed with Practitioners to Integrate Land Use and Transport Planning in the Netherlands. *Transportation Research Record*, 2077, pp. 1–8.

Straatemeier, T., Bertolini, L., Te Brömmelstroet, M. and Hoetjes, P. 2010. An experiential approach to research in planning. *Environment and Planning B: Planning and Design*, 37, pp. 578–591.

Willson, R. 2001. Assessing communicative rationality as a transportation planning paradigm. *Transportation*, 28, pp. 1–31.

PART IV:
Evaluating Planning Intervention, Institutions and Spatial Change

Chapter 16

Evaluating Behaviour Change in Workplace Communities

Angela Hull

Introduction

Local communities have rarely figured in the triumvirate State-Private Sector-Civil Society governance nexus over the last century. Strong state governments with corporatist underpinnings have created weak local communities rarely able to lobby for changed distribution of public sector resources (Gaventa, 2004). More recently, neoliberal ideologues have successfully promoted less government intervention, individual freedoms, and a stronger role for the private sector in the provision of services (Hajer, 2003; Swyngedouw, 2005). As a result, public service provision has become more flexible and discretionary creating uncertainties concerning the role of the state and public provision for welfare and societal well-being (Hajer, 2003). This laissez-faire model of 'governance beyond the state' creates a challenge for the public sector in changing individual and organizational consumption practices and choices in response to external crises such as climate change (Swyngedouw, 2005, p. 1993).

In the past when the governmental gaze was cast on local communities it was to promote acceptable behaviour through giving information and reinforcing social norms of behaviour. Substantial state funding in the twentieth century has been expended on low income communities, ignored by private investors, in order to remove the barriers to investment and to encourage higher levels of employability. In spite of an aspired holistic approach to regeneration, the many facets of community empowerment were difficult to get right (Hull, 2006). More success has come from single-issue campaigns such as smoking, road safety, waste re-cycling, and school cycling initiatives (Scottish Government Social Research, 2011). There are many best practice examples of how public organizations have engaged with citizens. The focus of these narratives is on evaluating the process of engagement, rather than the substantive outcomes, emphasizing the dissemination of advice, raising environmental awareness, and hand holding citizens through bureaucratic procedures (Morris, 2010; Peters et al., 2010; Davies, 2012).

More recently, government funding has been made available to households to reduce the dependency on carbon solutions covering initiatives such as improved loft installation, the subsidisation of renewable technologies and the encouragement of walking and cycling transport modes. Government exhortations have emphasized the responsibility of every citizen and organization to mitigate and adapt to climate change. The urgency of carbon reduction outcomes has been reinforced by government climate change legislation, carbon reduction targets and strategies to achieve them.

The evaluation of these government financed behaviour change initiatives has identified several contextual barriers to internalizing new behaviours in participants. Firstly, behaviour change policies in 'developed' countries tend to compete with 'embedded' economic incentives to increase societal material production and consumption (Seyfang and Smith, 2007; Bailey and Wilson, 2009; Webb, 2012). Often the justification of policies to reduce CO_2 emissions are presented in the terminology economists can understand, such as carbon commodification strategies to address the market failure of climate change. This raises the question of how alternative strategies and scenarios to consume less or more responsibly can influence consumerism as a way of life? (Webb, 2012). Secondly, the gradual retreat of the government sector from social responsibility for civil society outcomes in the neo-liberal model of governance produces fragmented and non-continuous approaches to social welfare (Bailey and Wilson, 2009). If governments rely on a voluntaristic approach to behaviour change, how can climate change adaptation policies reach beyond the committed members of the public to build up community capacity for change? (Middlemass and Parrish, 2010; Peters et al. 2010). Thirdly, since a number of consumer behaviours need to change – such as food, travel, and energy use – how can understanding of the connections between consumerism and environmental degradation be dissipated through civil society? Public surveys suggest that understanding of how individual choices impact on CO_2 emissions and resource distribution is rather shallow (von Borgstede et al., 2013).

The UK Governments are tapping into a mainly middle class activism to protect local service quality through dialogue with service providers and the establishment of neighbourhood community councils tasked with representing the concerns and values of their community (Matthews and Hastings, 2013). This is just one manifestation of social movements, such as the transition movement, that have spawned interest in quality of life and environmental issues (Transition Towns Network, 2008). Relying on Green Party members for overall political control, the Scottish Government set up the Climate Challenge Fund (CCF) making nearly £66m of grants available to community groups between 2008–2015 to actively encourage people to adopt new low carbon behaviours and measures covering the domains of energy, food growing and purchasing, transport and waste (Scottish Government, 2011). Over 600 community groups have received funding for 1–3 year programmes in a bottom up, hands-on experiment in promoting greener lifestyles. What characterizes this experiment as unique, as far as the author is aware, is the informality of the government approach in having faith that environmentally committed members of the public can influence the hearts and minds of their neighbours where other government mechanisms have failed.

The author has been a community activist in several of these Scottish government funded community projects. The questions this chapter asks are: (i) Can a community of interest in a workplace setting influence the values and operating procedures of that institution? and (ii) How would you evaluate the success of such an exercise?

This chapter takes an ethnographic action research approach to evaluating one of the CCF projects funded at Heriot-Watt University in Edinburgh, Scotland – the author's place of work. The critical questions here for institutional analysis are how the problems were defined and what are their implications; how are goals, objectives and targets advanced; how has the community group expanded and gained wider commitment and has this had the effect of reducing CO_2 emissions from the community at Transition Heriot-Watt? The key issue for evaluation is what type of data should be collected that would reasonably demonstrate that the project had achieved its goals? This latter issue has bedevilled the claims for the success of the first few tranches of CCF funding (Scottish Government, 2011).

The next section of this chapter focuses on developing an appropriate evaluation framework, firstly by reviewing the social psychology and behavioural economics literature to understand individual theories of behaviour choice. Then empirical research on social learning and organizational behaviour are reviewed to identify a research methodology that marries our understanding of individual choice sets with a realist perspective on how external contextual factors may influence behaviour choices. The evaluation findings focus on how the organizational issues of capacity and commitment affect the progress of the project. Finally, the findings and the research design are discussed.

Understanding Behaviour Change

The theoretical approaches to behaviour change can be characterized as primarily focused on either individual or organizational decision formation. The former has been addressed in social psychology and behavioural economics theories (Ajzen, 1985; Bamberg et al., 2011; Gadenne et al., 2011; Davies, 2012) whilst the latter domain has attracted the attention of organizational and policy analysis theorists (Hull, 2006, 2010; Middlemass and Parrish, 2010; Shove, 2010). The starting point for many of these theories is that individuals make rational decisions following the evaluation of information on the choice options available to them and the predicted costs. Psychologists have pointed to the importance of personal norms and attitudes towards the decision object or action, the construction of personal identity, and the social norms of the group(s) the individual aspires to belong to. Individuals are characterized as constructing an identity which they then seek to maintain in the face of external threats or challenges (Thøgersen, 1999; Breaknell, 2010; May, 2011; Middlemass, 2012; Murtagh et al., 2012). Sociologists have focused more on the dynamic nature of the interaction between personal norms, the expectations or norms of the social groups or social worlds individuals 'inhabit', the motivation to act, and the decision choice. Research has shown how individuals participate in several different social worlds, internalizing the expectations and norms of these groups, often revealing a different side to their personality or identity in each one (Stryker and Burke, 2000 in Murtagh et al., 2012; Bamberg et al., 2011).

Behaviour choices are defined as mainly routine, habitual and, thus, largely sub-conscious. Individual routines can, however, be disturbed by changes in one's personal environment as a result of new costs or regulations, and/ or lifecycle and employment changes, for example. These may be perceived as opportunities or threats by the individual leading to a reassessment of their self-identity and self-efficacy (Murtagh et al., 2012) or their routine behaviour. Behaviour change theorists identify these 'conditions of uncertainty' as defining moments for the individual since the adjustment to their changing personal environment increases their responsiveness to new information about how others in their social group or reference group cope with the situation (Griskevicius et al., 2008, Department of Transport, 2011; Metcalfe and Dolan, 2012).

Pro-environmental behaviours may result from this conscious reassessment and deliberation about how to readjust ('accommodation') or how to incorporate new practices or components ('assimilation') into their identity structure (Breaknell, 2010). Breaknell notes that the adoption of change may be undermined by the transaction costs (e.g. physical, mental, and financial) and information barriers that prohibit people from behaving in a pro-environmental way. Research has also shown that there is a 'value action gap' or 'cognitive dissonance' between an individual's values and their behaviour (Thøgersen, 1999).

The social psychology literature, thus, foregrounds the heuristic value of researching how attitudes are formed and what motivates individuals to behave the way they do. This is the approach also taken by social marketing companies to influence consumer behaviour by targeting consumers (segmentation) based on their attitudes and historic consumption patterns (Bamberg et al., 2011; Davies, 2012). When the promotion and the messenger are salient to the consumer, with a clear message showing how the new product adds to their lifestyle, and does not detract, using cues (sights, words and sensations) behaviour can be influenced (Department of Transport, 2011).

We can surmise that individuals who interact within an organization follow certain routines in responding to the demands of their work environment (Middlemass and Parrish, 2010; Shove, 2010; May, 2011; Middlemass, 2012). May (2011) and Middlemass (2012) theorize that some interactions become fixed or embedded in everyday life and are routinely repeated based on the social norms of the social group. Shove (2010) and Hull and Pereira (2010) highlight the influence of the wider external environment, the infrastructure, the incentives and the convenience of change.

Theoretical enhancement has sought to expand our understanding of the process of individual behaviour change suggesting that there are stages in the transition to new behaviours (Prochaska and Diclemente, 1983; Bamberg et al., 2011; Scottish Government, 2011; Albert et al., 2012; Davies, 2012; Middlemass, 2012). Bamberg et al. (2011, p. 231) suggest that 'behavioural change is a transition through a time-ordered sequence of stages reflecting the cognitive and motivational difficulties people encounter in implementing a general behavioural change goal into concrete actions'. Each of the models in Table 16.1 have trigger points or stages when the individual or organization is considering the desirability of behaviour change, the planning of new behaviours and who to involve, and the realization/evaluation of the change.

Table 16.1 Sequence of transition stages in behaviour change

Prochaska and Diclemente (1983)	Bamberg et al. (2011)	Scottish Government (2011)	Albert et al. (2012)
Pre-contemplation	Pre-contemplation	Acceleration	Framing planning and Representation
Contemplation	Contemplation	Facilitation	Process and Evaluation
Preparation	Preparation	Activation	Change and Impact
Action	Maintenance	Consolidation OR	Elaboration
Maintenance		Conversion	Decision

Source: Author's summary.

Researchers have employed theories on social learning to gain a deeper understanding of the influence of social environment factors on behaviour change (Albert et al., 2012; Davies, 2012; Middlemass, 2012). These models start to address the contextual or institutional factors embedded in institutional practices which might create 'lock in' or resistance to change (Webb, 2012).

Middlemass (2012) examines how an individual experiences a community based project. Her research shows that the lifestyle history of the individual before joining a project is a key influence on the extent of behaviour change in that previously unengaged individuals with sustainable values are more likely to experience more behaviour change than an individual recently/ historically engaged in such projects. Empirical research by Albert et al. (2012) suggests that the power, knowledge and creativity of the participants may influence the social learning process, whilst Davies (2012) highlights the potential role of participatory evaluation and dissemination in generating social learning outcomes for participants.

Middlemass and Parrish (2010) apply the concept of social learning to understand the capacity of a community organization to change behaviour. They employ a Giddensian framework on structure and agency which theorizes that the structure or context is '*the tool to provide transformative action*' (ibid., p. 7561).

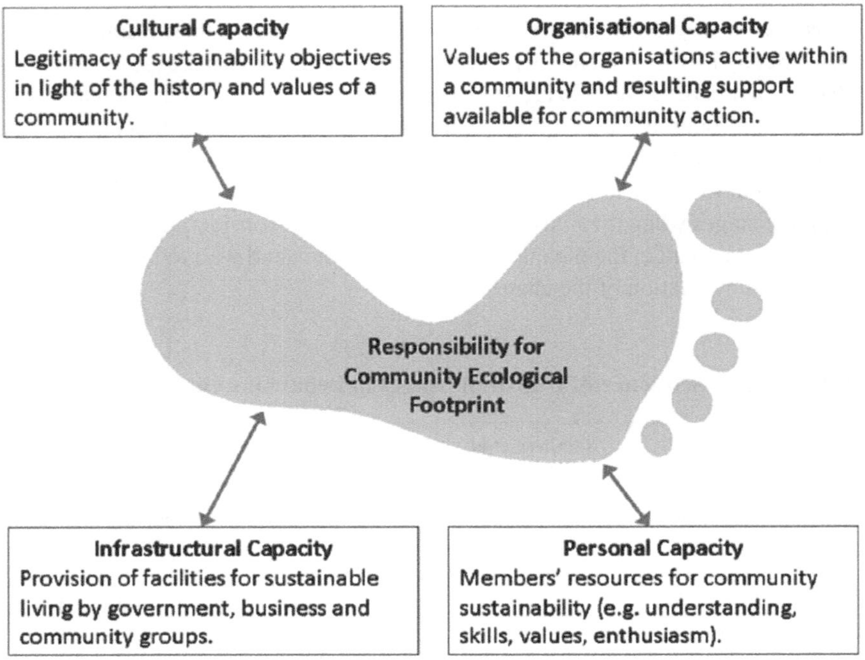

Figure 16.1 Theoretical framework for understanding community capacity
Source: Reprinted from *Energy Policy*, 38, Middlemass and Parrish, 'Building capacity for low-carbon communities: The role of grassroots initiatives', p. 7561, © 2010, with permission from Elsevier.

They theorize that the effectiveness of a community organization in achieving behaviour change is influenced by four dimensions as shown in Figure 16.1 below: cultural capacity, organisational capacity, infrastructural capacity, and personal capacity. They posit that 'if any of the capacities are weak, the community's ability to fulfil responsibility for the ecological footprint diminishes' and vice versa (ibid., p. 7561).

Evaluation Methodology

This chapter hypotheses that Heriot-Watt university's organizational values, the organizational capacity and support provided by the university to Transition Heriot-Watt (THW) will be the key factor determining the implementation of sustainable values and behaviours amongst the community of staff and students at the University of Heriot-Watt's Edinburgh campus. Middlemass and Parrish (2010, p. 7562) theorize that the 'values held by formal organisations that are active in the community, how far these are aligned with attempts to encourage sustainability, and the sources and support available through such organisations to stimulate change' are a key dimension of a community group's ability to effect change. Middlemass (2012, p. 279) suggests that more research 'on the means by which cohesive community-based organisations are created' would help our understanding of the relative importance of organizational capacity in the success of community-based approaches. Middlemass and Parrish (2010, p. 7562) identify that one of ways in which 'grassroots action' can influence the 'community' is through 'people seeking to change the social structures that they inhabit'. They envisage that this could potentially influence all the four capacities in Figure 16.1 above.

THW has received four years of funding (2011–2015) from the Scottish Government's Climate Challenge Fund (SGCCF) totalling £435k. A review of (3 per cent) of completed community projects funded through the SGCCF was carried out in 2011 to assess the range of project outcomes (Scottish Government, 2011). The report included interesting findings about process, such as the importance of 'hand holding' and 'intensive personal support' necessary to elicit change from those moderately interested in climate change. A key finding on how individuals can move through the transition stages is that:

> Projects had most success with participants open to change. They overcame inertia (accelerated change), opened up new possibilities that people might not have considered (activated) and supported participants in working through change processes and barriers which they may have found daunting without projects' help (facilitated). Where participants had already made changes, the projects' influence reinforced their behaviour (consolidated). However, projects rarely succeeded in convincing those to change who saw no merit in it (converted). (Scottish Government, 2011, p. 4)

Since it was politically important to justify the substantial government spend on these community-based approaches to sustainable outcomes, the evaluation report also identifies the '*unique contributions*' of this approach (Scottish Government, 2011, p. 95):

> The local scale appears to be a meaningful one at which to take action on climate change: at one large enough to make its impact seem worthwhile and small enough to create a sense of individual responsibility. (Ibid., p. 95)

Two limitations were noted, but not explained in the report: community-based projects can be of small scale and that 'there are external barriers to pro-environmental behaviour change' (ibid., p. 95). Three organizational issues were raised during the discussion of the research findings. Firstly, it was noted that successful projects had embedded an 'organisational learning culture' in their project teams which led to the application of this new knowledge through '*adapt*[ing project activities] *in line with new insights*' (ibid., p. 73). Secondly, 'environmental projects led by existing organisations with no explicit environmental remit may be open to the risk that the organisation's central aims (e.g. training and employability) are prioritised over the environmental aims of the project' (ibid., p. 73). Thirdly, the importance of leaving 'a strong legacy for the community' is mentioned so that the achievements and momentum can be 'maintained into the future' (ibid., p. 92).

The evaluation is subjective in nature since the author uses ethnographic research techniques to provide an insider account of the THW processes of change. The bias in this approach is acknowledged and mitigated to some extent through the use of multiple methods and the careful scrutiny of each data source and the triangulation of all types of recorded data (Table 16.2). Rich data has been collected from several sources since the establishment of THW in February 2011. Travel survey data has been collected since October 2009. Some of this data provides contextual information only for the main evaluation of organizational processes in this chapter. The analysis has included summarizing and categorizing data and the author's interpretation of verbatim quotes. The aim is to build up as complete a picture as possible given the resources available to understand the overall context and the component parts.

By focusing on the organizational context in which THW operates, the author is theorizing that context, habitual practices and incentive systems create barriers to change in line with Ostrom (1986; 1990); Hull (2006); Middlemass and Parrish (2010) amongst others. The author is aware that multiple competing accounts of the organizational context at Heriot-Watt are possible. Future research will aim to explore these differing accounts.

Table 16.2 Data sources used in Transition Heriot-Watt evaluation

Methods	Data types	Amount of data
Surveys pre- and post-planning process on travel behaviour; incentives to change travel behaviour; willingness to engage in THW activities; and willingness to change behaviour	Returned questionnaires and pledge cards	3 web-travel surveys plus facilities surveys 1 community garden survey 1 survey of behaviour change
Focus group	Transcript notes	2 focus groups ca. 3 hours External facilitator 2 × 10 participants
Semi-structured interviews	Transcript notes	7 interviews ca. 60 min External facilitator
Surveys on stated levels of support for relevant decisions	Responses recorded in meetings and from emails	10 participants
Documentary review	THW reports, minutes and website	Collected since April 2011
Observations	Meeting notes	Author participation

Source: Authors own.

Case Study Description

As explained above, THW is funded through the SGCCF which provides revenue funding (personnel and incidental expenses) but not the costs of infrastructure to community groups in Scotland. The first informal meeting of THW in December 2010 was convened by two researchers in the School of the Built Environment who had been working with the current THW Chair to start to initiate deeper thinking and commitment by the university on sustainable transport issues. These three were promoting car parking restrictions, and access for the city car club, on the campus. The first bid for funding, thus, focussed on car sharing, cycling training and maintenance, and environmental education. THW was formally established as a community group within the university with 14 officers in February 2011 (Table 16.3 overleaf).

This section seeks to describe the case study through the events, processes, actors' roles/ interactions and consequences and then evaluates how the university's organizational values, the organizational capacity and support given to THW has hindered or galvanized effective action to release implementation of more sustainable values at the Edinburgh Riccarton campus.

The research since early 2011 has collected data on the organizational capacity and commitment of Heriot-Watt University to behaviour change, the alignment of values between THW and the university, and finally the influence of THW on university practices. These three issues are described below in turn. To put the discussion below in the context of the main operational events and input of external resources Table 16.3 provides a summarized THW timeline from the first instigation of action in December 2010 until the appointment of the second project manager in July 2013.

Organizational Capacity and Commitment

THW has operated closely with the managers of the campus services (Estates, Hospitality, Sport and Exercise, Students' Union, Procurement) and various academic departments aiming to influence university practices and knowledge dissemination to students. Agreeing the norms and rules of engagement for THW with the campus services as a new 'community' project has been problematic throughout the project. Support for the project at the initial bid stage in January 2011 was not forthcoming from Estate Services and later, when a second bid for funding, was conditional on the project not affecting their operations. Hiding within the teaching side of the university in the first year of operation, it could be tolerated as a student project that promoted awareness-raising of climate change and resource efficiency through film nights, Bike Week events, and support for a student-led Food Society.

When new funding was obtained for projects (community garden, bike storage, energy initiatives) in March 2012 that required space on the campus this ruffled the feathers of the second tier operational managers. The students in the

Table 16.3 Key operational stages in Transition Heriot-Watt's history

Date	Event	Actors	Action decision	Topic area
Dec 2010	Collaborative meeting	2 researchers; 1 academic	Bid for external (CCF) funding	Sustainable transport
Feb 2011	THW established	6 academics; 5 researchers; 1 student activist; 1 support staff	Establish THW constitution and Steering Group (SG)	Sustainable behaviour at HWU
Apr 2011	£48k grant from CCF	SG composed of: 4 academics; 2 researchers; 2 students	Appoint PT co-ordinator plus student interns. Establish communications working group (WG) and ?	Sustainable transport
May 2011–Feb 2012	PT staff appointed in May–June (1.3 FT)	THW SG and staff; Student volunteers; THW Estates staff	Programme of networking and awareness-raising of sustainable behaviour; travel data collection	Sustainability; sustainable travel
Nov 2011	SG meeting	THW SG and staff	Bid for external (CCF) funding	Recycling/ re-use of discarded items at Student Halls; Food and energy awareness; sustainable travel
Jan 2012	CCF bid submitted	University Sec; 3 departmental heads (Research; Estates; Built Environment); THW Chair and Treasurer	University to distance itself from THW	Perceived risk of THW project causing negative reputation and raising student expectations
Mar 2012	£380k grant from CCF	Project co-ordinator and THW Chair	Negotiate project home with Head of Professional Services	Embed THW in university structures
May 2012	New Steering Group appointed	1 senior Manager (Professional Services); 4 dept heads (Estates, Centre for Sport, Procurement, Hospitality); 2 academics; 1 researcher; 2 Support staff (Estates; Student's Union); 3 THW staff; 3 student interns	Provide strategic management for THW and ensure professional services are involved	Embed THW in university structures
May 2012	2 Development officers appointed (THW paid staff = 4FT)	THW SG	Develop cycling and food programme of events	Cycling and Food
Aug 2012	Development officers claim there is a lack of management and direction	THW Chair; Head of Centre for Sport (HCfS); Project co-ordinator; Development officers	Review line management; THW Chair/HCfS to co-manage with project co-ordinator	Strategic Management

Oct–Dec 2012	External consultant engaged	THW Chair/ HCfS; THW SG	Consultant to review THW operations and enhance strategic management	Strategic Management
Jan 2013	Steering Group meeting	THW SG	HWU Environmental Manager agrees to manage halls reuse project using University resources and to coordinate energy awareness programme	Legacy
Jan 2013	Project Coordinator resigns	THW Chair/ HCfS; Project co-ordinator	HCfS decides to seek permission to re-appoint urgently	Strategic Management
Jan 2013	Development officers find it difficult to contact student interns	THW Chair/ HCfS;	Interns' contracts changed so that they have specific attendance times	Strategic Management
Mar 2013	Meeting with external consultant; engaged by development officer	THW Chair; 2 x development officers; External consultant	Consultant offers to help development officer to prioritize SG potential projects based on MacKenzie-Mohr Rating scheme	Strategic Management
May 2013	Reverse Vending Machines installed	Hospitality Manager	3 machines installed	Waste recycling
May 2013	Presentation of Project priority matrix to Steering Group meeting	THW SG; External consultant	Principal's litter pick omitted. SG agree to run this and that project scoring system be shared with SG	Strategic Management
June 2013	Steering Group meeting discussion of terms of Reference	THW SG	SG Terms of Reference not accepted by Steering group. Students' Union CE agrees to rewrite constitution and TOR	
July 2013		THW Chair/ HCfS; HWU Hd of Human Resources	New project coordinator appointed	

Source: Author's own.

Food Society had suggested converting the historic walled garden, now used as a car park for the conference centre, back to its original use. This was included in the second bid as the community garden project and support from the university principal, the HWU Site Conservation committee, and 99 students was secured. The 12 months following the award of funding from the CCF was marked by filibustering by Estates with their THW steering group members claiming that 'there were reputational issues to keep in mind' (respondent 1) and that in addition to a feasibility study the business case would need to be appraised with three costed options. During this period, with continuing uncertainty over whether HWU would donate the space, the principal offered the walled garden attached to his university residence as the THW community garden. A feasibility study for this new space was carried out with the help of Estates but it failed to be presented to the relevant university committees due to staff illness. Unbeknown to staff members, two events were rolling out behind the scenes. Firstly, Estates blocked the entrance to the principal's walled garden with a 2 metre high security metal fence and secondly, the Centre for Sport and Exercise included a community garden in the historic walled garden in their winning bid to the Scottish Government to host the National Performance Centre for Sport.

Whilst the university has allowed senior managers to work with THW, their continuing role on the Steering Group is seen as necessary to limit any damage to the university's reputation and to provide a watching brief to make sure that THW does 'not mess up' (External consultant, 2012). Steering group members themselves have openly voiced that they are on the steering group to ensure that THW does not damage the reputation of the community (respondents 2+3) and concern has been expressed that everyone on the steering group has a bias- either professional or job related (respondent 2). One member suggested that it is a pseudo-environmental committee for the university without having the buy-in (respondent 1).

Table 16.3 shows the changing composition of the Steering Group in response to negative feedback from key university managers (Human Resources, Estates, Research Services, School of the Built Environment) and the re-positioning of THW from the School of the Built Environment to a more central position in the Centre for Sport and Exercise within the Professional Services department. Only four people from the initial steering group still have a place on the current steering group. Six members have left the university; one has died; and one did not wish to be involved anymore.

An external consultant was appointed in late 2012 to identify how the Steering Group and THW staff could work more effectively together. Following interviews with staff and some of the Steering Group she concluded that there was low ownership by the Steering Group of THW aims and objectives, and that university stakeholders were not brought in to the degree needed to change behaviour. As a result of this process of introspection, staff members argued for a focus on those projects, identified by the Steering Group, which scored highly on the

McKenzie-Mohr scoring matrix that had both high potential to reduce carbon impact and could demonstrate organizational support (McKenzie-Mohr, 2000a, 2000b). At the 15 May 2013 Steering Group, five projects (food waste; business travel; procurement; building energy; promotion of university good practices) were promoted for inclusion in the next two years' programme. These five projects were amongst 23 suggested by the Steering Group. The Steering Group were reluctant to endorse the five selected projects since they were not party to the scoring matrix used. Two members stated that they could not understand what the added value to the university would be in THW's involvement in these projects and, therefore, could not invest in something they did not understand the value of (respondent 1+3); one of whom could not see where 'THW fits in since except for one project they were all being done by the university' (respondent 3). During a heated discussion of the added value of THW, the Chair suggested that the success of THW halls reuse project in summer 2012 had mobilized Estates to run this initiative themselves in 2013. Both Estates members of the Steering Group disagreed with this assessment. Steering Group members challenged the staff member present and external consultant, who was in attendance, on their knowledge of behaviour change with one member saying 'you say you know what is interesting but so do we' (respondent 1). The same person said that any energy initiatives would be coordinated through him and if he needed help he would come to THW.

Despite the mediation skills of the external consultant and the Chair, the opposition of the staff member to sharing the scoring matrix used to select the preferred main projects, brought to the fore that the governance and existence of THW was still not endorsed by university operational senior managers. The external consultant had initially been brought in by the Chair and the Head of Centre for Sport and Exercise (HCfSE) to help the Steering Group to agree a new THW Vision, Purpose and Objectives and to play a more strategic proactive role in THW. These two Steering Group members also decided to provide a shadow management support role for the project co-ordinator, who had found it challenging to manage the new staff appointed following the success of the second bid. The two new development officers who were appointed in the summer of 2012 complained after three months that there was no effective line management and that THW lacked clarity of purpose. The project coordinator left the project in January 2013.

Alignment in Values

HWU's approach to sustainability and resource efficiency appears dominated by economic and operational considerations. The universities Environmental Manager produced a sustainability strategy (set of principles) in 2008 for the university which remained in draft until autumn 2012 when it was placed on the Estates website. One senior manager interviewed claimed that the strategy was not owned by any other department and that it was not on the university's agenda for action: 'university sustainability policy is seen as the Environmental manager's policies – no one checks up on this' (respondent 4).

It was with this backdrop of university scepticism towards the aims and goals of THW, that the external consultant was employed to run a series of half day workshops to help align THW more closely with university values. Ten of the 16 Steering Group members participated in a half day workshop/focus group run by the external facilitator in December 2012. The external consultants' view of the THW Steering Group was that they were working to too many different agendas and not proactive enough in support of THW staff. The aim of the workshop was to get the steering group to agree what the overarching purpose of THW was and to agree clear terms of reference for THW.

The facilitator's two-page summary of the THW Vision and Purpose agreed by 10 steering group members was unfortunately not accepted by those who attended the next Steering Group meeting in January 2013. One operational manager claimed that the workshop vision did not fit in with university policies and procedures.

Only the following lines were accepted by the Steering Group:

> Transition Heriot-Watt Edinburgh VISION STATEMENT
>
> By 2023 Transition Heriot-Watt has assisted the process of Heriot-Watt University and its community to become a world leading exemplar of a sustainable learning community.
>
> Transition Heriot-Watt Edinburgh PURPOSE to help bring about this vision
>
> Transition Heriot-Watt Edinburgh recognises that its overall purpose is to act as a catalyst to bring about ownership of holistic sustainability amongst the community at Heriot-Watt University. (Steering Group minutes 21/01/13)

Against this strategic approach to improving the organizational capacity of THW, it is illuminating to understand how the values of different members of the Steering Group have been expressed in meetings and emails. The value statements have been collected over a 20 month period from January 2012. Table 16.4 below summarizes the value statements of Steering Group members active in meetings and between meetings through emails. The categorization in Table 16.4 is organized according to how these are aligned with THW attempts to encourage sustainability. Stakeholder values that align with a community organization's stated purpose are a key dimension of a community group's ability to effect change in Middlemass and Parrish's model in Figure 16.1 above.

Table 16.4 Summary of stated levels of support for relevant decisions (January 2012–August 2013)

Type of value held	Example	Number of:	
		Participants	Cases
Introduction of university procedures or purpose as a barrier	The community garden proposal 'requires costed option appraisal – 3 options … There's always reputational issues and the ongoing legacy costs'.	3	9
Criticism of THW (staff or purpose)	'The THW Vision statement does not fit in with university policies and procedures'.	7	38
University support to THW	'Barriers can easily be moved- pushed sideways with the relevant risk assessment. We can leap over these barriers'.	8	18
Opportunity suggestion	THW can 'give advice on which appliance to buy – a dummy's guide to schemes people can qualify for'.	3	11
Positive suggestion to overcome perceived barrier	'Go to x and y and say: "How as a food officer can I help you?" and come out with ideas'.	4	18

Source: Author's own.

Most of the recorded text in Steering Group meetings is the presentation of facts about internal and external events. Table 16.4 only records the value statements of Steering Group members. These are presented as a continuum with the most negative values towards THW presented first with increasing positive value statements following. One member made 25 value statements in Steering Group meetings over the course of 12 months; this represents 48 per cent of the value statements in Steering Group meetings. This member made statements in each of the categories below, even contradicting himself in meetings.

THW Influence

It is difficult to be certain how THW has influenced the working of the university and what additional value it has had. An initiative of THW is always mediated by the established power coalitions within the university which work to stamp out anything that they cannot control. There are a few activities started by THW which have been taken over by the university. One of these is the Halls Reuse programme which the Estates department now carries out with the help of a national charity. There are other examples of the university partnering THW in environmental awareness-raising events (Bike Week, Green Energy Week) and publicising environmental aspirations previously hidden in departments (Sustainability Policy). Nearly two years into the operational timeline for THW the University Estates Services published HWU's carbon management plan first devised in 2008. This states:

> Our low carbon vision
>
> Our vision for the next decade is to become a world-leading university which will produce the next generation of global leaders in business and technology with a sound understanding of the principles of sustainable development and in particular the need to reduce our Carbon Dioxide emissions and those of other greenhouse gases …
>
> Creating a learning and working environment where carbon and its associated costs are fully appreciated and actively managed will help to instil this. To deliver this ambition, we are committed to growth and investment, both in staff and infrastructure whilst at the same time reducing our relative carbon footprint by 22.5% in the next 5 years. This Carbon Management Plan and the on-going support of the Carbon Trust will provide the framework to allow us to achieve our carbon management ambitions. (HWU website, 2012)

The change in the university's sustainability aspirations comes out clearly in the analysis of replies to staff and student questions to the University Principal ('Points for the Principal') that are answered by the University Secretary.

It was noted in 2010 that 'it is simply not possible to make a robust business case sufficient to support [electricity through wind turbines or solar panels] being taken' because of the 'high initial capital costs … [and] lengthy payback periods involved'. In the same year the provision of allotments was not

> considered viable as they would require a change to the University's Planning consent; a capital cost to screen and fence off an appropriate area and ongoing resources to provide lease type agreements with the appropriate policing […] they could prove to be a lot of extra work for our grounds team and pest controllers; be unsightly and have a negative impact on our biodiversity while also increasing the risk of pollution.

In 2011, the reply to a question asking why staff are not encouraged to switch off their monitors/printers/PCs when not in use was that individuals should be encouraged to switch off IT equipment in general provided there is no detrimental effect but that computer base units require more care and attention and that this decision should be left to the university IT controller. Another question on why the university does not do more recycling of plastic bottles elicited the statement that plastics prices are very volatile and that 'the high volume, low weight nature of plastic does not easily lend itself to economic handling, transportation and recycling'. The questioners were assured that Estates Services would be carrying out an Energy Awareness Programme in the near future and that more plastic recycling facilities would be sought.

In 2013 the Principal was asked why there were no recycling facilities in the new halls of residence. Recycling rooms were either adjacent or outside the new buildings. 'Space constraints meant it would be unpractical to have recycling facilities in every kitchen'. Similarly grey water recycling systems have not been installed because 'there are issues around embedded carbon in respect of grey water system installations that detract from the perceived benefits of reducing the use of mains water'. In the new halls of residence these systems were 'not taken forwards for carbon, economic and operational considerations'.

Another question to the Principal in 2013 was what efforts are being made to improve energy efficiency at the Edinburgh campus? The reply was that the university had made significant improvements (8.3 per cent absolute carbon emissions reduction) in energy efficiency in the 2011/12 year being placed the highest ranking Scottish University for carbon dioxide emissions reduction that year coming 367th out of 2100 participating agencies in the UK by the assessors of the Carbon Reduction Commitment's Performance League Table. This followed 'significant investments' in replacing old boilers and heat distribution plant. THW was also mentioned as helping to improve energy efficiency:

> Transition Heriot-Watt, a community group assisting with the promotion of sustainability on campus, has helped organise many campaigns, including a Carbon Diet Challenge and the Student Switch Off Campaign in collaboration with the Student Union and Estates Services. Transition Heriot-Watt also runs sustainable commuting initiatives such as bringing Dr Bike on campus, organising social cycle rides, offering free eco-driving lessons and cycling training sessions. The group have helped to expand end of term re-use schemes to assist the activities of Hospitality Services, Estates and the Chaplaincy, and are looking at developing a community garden on campus, as well as arranging low carbon cooking lessons.
>
> Transition Heriot-Watt is looking for more volunteers to carry out energy audits and to plan a programme to challenge Schools to reduce their building energy consumption.

There were many small changes being contemplated and implemented by the University coming to light in 2013. The university had agreed to produce an environmental awareness strategy and a sustainable procurement strategy. The university were invited to bid for funds to trial 3 Reward and Recycle plastic bottle installations on the campus by the Scottish Government's Zero Waste Scotland Office. In addition, they were looking at costing a small composting installation and at bidding into a project to install 250Kv of photovoltaic panels on the flat roofs of several campus buildings.

Evaluation and Findings

This final section draws together the findings from the case study and the influence THW appears to have had on outcomes. The data presented above on changes to organizational capacity and commitment, and the alignment of values is assessed against Middlemass and Parrish's (2010) theoretical framework for understanding community capacity (Figure 16.1 above).

The focus of this chapter has been on understanding the powerful effects of context, habitual practices, lock in and incentives embedded in organizational systems. The chapter first reviewed the literature on models of individual behaviour change and found these aspects lacking from the theories. Theories of social learning and institutional analysis were found to have more heuristic value for understanding the mechanisms fundamental for achieving behaviour change. In this respect, Middlemass and Parrish's (2010) conceptualization of the importance of the organizational capacity of a community group and Prochaska and Diclemente's (1983) conceptualization of the stages, or trigger points, of behaviour change provide a helpful lenses in which to analyse the case study findings.

THW has spent substantial time wallowing in the first two transition stages of pre-contemplation and contemplation and are only just formulating or preparing a well-justified action plan (see tables 16.1 and 16.3). This formulation of stages of transition suggest that gaining the support of other organizations active within the community are critical for reaching the stages of preparation for action and activation. It is important that the initial goals defined by a community organization such as THW are subjected to critical scrutiny and in the process of problem abstraction agreement is reached on how the goals should be reinvented or refined so that broad consensus on how to frame action can be reached.

In this case study, THW were initially kept away from any university operational issues and was seen more as a student society. The appointment of an external consultant was key for THW to understand the mismatch of values within the Steering Group, heavily staffed with departmental heads fearful of what THW wished to change. This second tier of university management appeared intent on containing this experiment in horizontal participatory governance on their campus which some on the Steering Group thought would lead to a 'bad reputation' for the university and raise student expectations. By raising this as an issue, the external consultant brought the mismatch in values out into the open for the Steering Group to address. The change in organizational practices that resulted (Mackenzie-Mohr strategic focus; new Steering Group members) helped to align the values of the Steering Group and create support for THW actions.

Ostrom's (1986; 1990) and Bailey and Wilson's (2009) theorization on the importance of the power relationships between stakeholders also finds much resonance in this empirical work. Table 16.4 shows how the power game was played out between Steering Group members. When time was found to understand the different perspectives Steering Group members had this led to a focus on facilitating the sustainable visions of university departmental heads that wanted to work with THW, and the use of the group's collective knowledge on how to engage participants and elaborate the risks of current behaviour and impacts. There has been a social learning process that has led to 'enhanced awareness, altered agendas, and better social relations' (Davies, 2012, p. 23).

THW is starting to work effectively with sections of the university administration. Existing research on transition theory highlights that niches of innovation in an organization often stay as niches and never become part of the regime of actors or stakeholders (Seyfang and Smith, 2007). The research findings have highlighted the problems contingent to a divided steering group and suggest that resistance to change in an organization should never be underestimated. THW has put substantial effort into consensus-building with the host organization, new staff have been appointed with appropriate skills, a programme of next steps has been elaborated, the implementation of which looks more optimistic.

References

Aizen, I. 1985. 'From intention to action: A theory of planned behaviour'. In J. Kuhl and J. Beckmann (eds) 1985. *Action Control: from Cognition to Behaviour*. Heidelberg: Springer, pp. 1–39.

Bailey, I. and Wilson, G.A. 2009. Theorising transitional pathways in response to climate change: Technocentrism, ecocentrism and the carbon economy. *Environment and Planning A*, 41, 2324–41.

Bamberg, S., Fujii, S., Friman, M. and Gärling, T. 2011. Behaviour theory and soft transport policy measures. *Transport Policy*, 18, 228–35.

Breakwell, G.M. 2010. Resisting representations and identity processes. *Papers on Social Representations*, 19, 6–1.

Davies, N. 2012. What are the ingredients of successful travel behaviour change campaigns? *Transport Policy*, 24, 19–29.

Department for Transport, 2011. *Behavioural Insights Toolkit*. London: Department for Transport.

Gaventa, J. 2004. *Representation, Community Leadership and Participation: Citizen Involvement in Neighbourhood Renewal and Local Governance*. Neighbourhood Renewal Unit. London: Office of Deputy Prime Minister.

Griskevicius, V., Cialdini, R.B. and Goldstein, N.J. 2008. Social Norms: An underestimated and underemployed lever for managing climate change. *International Journal for Sustainability Communication*, 3, 5–13.

Hajer, M. 2003. Policy without polity? Policy analysis and the institutional void. *Policy Sciences*, 36, 175–95.

Heriot-Watt University (n.d.). *Carbon Management Plan* [online]. http://www.hw.ac.uk/estate/carbon-management.htm [accessed January 2013].

Heriot-Watt University. 2013. *Secretary's Board Bulletin*, Issue 15, July 2013 [online]. http://www.hw.ac.uk/committees/secretarys-board/minutes.htm [accessed 12 July 2013].

Heriot-Watt University. 2013. *Points for the Principal* [online]. http://www.hw.ac.uk/internal/principal/faqs.htm [accessed 3 August 2013].

Hull, A.D. and Perreira, M. 2010. *Encouraging Development in an Economic Downturn*. Edinburgh: Scottish Government, Housing and Analytical Services Division.

Hull, A.D. 2006. Structures for communication and interpretation in Neighbourhood Regeneration: Mainstreaming the Housing Action Trust philosophy. *Urban Studies*, 43(12), 2317–50.

Jackson, T. 2005. *Motivating Sustainable Consumption – A Review of Models of Consumer Behaviour and Behaviour Change*. Report to the Sustainable Research Network. London: Policy Studies Institute.

Matthews, P. and Hastings, A. 2013. Middle-Class Political Activism and Middle-Class Advantage in Relation to Public Services: A Realist Synthesis of the Evidence Base. *Social Policy & Administration*, 47(1), 72–92.

May, V. 2011. Self, belonging and social change. *Sociology*, 45(3), 363–78.

McKenzie-Mohr, D. 2000a. New ways to promote proenvironmental behavior: Promoting sustainable behavior: An introduction to community-based social *marketing. Journal of Social Issues*, 56(3), 543–54.

McKenzie-Mohr, D. 2000b. Fostering sustainable behavior through community-based social marketing. *American Psychologist*, 55(5), 531–7.

Metcalfe, R. and Dolan, P. 2012. Behavioural economics and its implications for transport. *Journal of Transport Geography*, 24, 513–11.

Middlemass, L. 2012. The effects of community-based action for sustainability on participants' lifestyles. *Local Environment: The International Journal of Justice and Sustainability*, 16(3), 265–80.

Middlemass, L. and Parrish, B.D. 2010. Building capacity for low carbon communities: The role of grassroots initiatives. *Energy Policy*, 38(12), 7559–66.

Morris, C. 2010. *Area-based Schemes: Lessons Learnt from Fife Solar*. Edinburgh: Energy Savings Trust.

Murtagh, N., Gatersleben, B. and Uzzell, D. 2012. Multiple identities and travel mode choice for regular journeys. *Transportation Research Part F: Traffic Psychology and Behaviour*, 15, 514–24.

Ostrom, E. 1986. 'A method of institutional analysis'. In F.X. Kaufmann, G. Majone and E. Ostrom (eds) 1986. *Guidance, Control, and Evaluation in the Public Sector.* Berlin: Walter de Gruyter, pp. 459–75.

Ostrom, E. 1990. *Governing the Commons: The Evolution of Institutions for Collective Action.* Cambridge, MA: Cambridge University Press.

Peters, M., Fudge, S. and Sinclair, P. 2010. Mobilising community action towards a low-carbon future: Opportunities and challenges for local government in the UK. *Energy Policy*, 38(12), 7596–603.

Scottish Government Social Research. 2011. *International Review of Behaviour Change Initiatives: Climate Change Behaviours Research Programme*. Edinburgh, Scottish Government.

Seyfang, G. and Smith, A. 2007. Grassroots innovations for sustainable development: Towards a new research and policy agenda. *Environmental Politics*, 16(4), 584–603.

Shove, E. 2010. Beyond the ABC: Climate change policy and theories of social change. *Environment and Planning A*, 42(6), 1273–85.

Stern, P.C. 2000. New Environmental Theories: Toward a Coherent Theory of Environmentally Significant Behavior. *Journal of Social Issues*, 56, 407–24.

Stern, P.C., Dietz, T., Guagnano, G.A. and Kalof, L. 1999. A Value-Belief-Norm Theory of Support for Social Movements: The Case of Environmentalism. *Human Ecology Review*, 6, 81–98.

Swyngedouw, E. 2005. Governance Innovation and the citizen. The Janus face of Governance-beyond-the-State. *Urban Studies*, 42(11), 1991–2006.

The Scottish Government, 2011. *Review of the Climate Challenge Fund – Research Findings*, No. 8/2011 [online]. http://www.scotland.gov.uk/Publications/2011/06/13142135/1 [accessed 4 July 2013].

Thogersen, J. 1999. The Ethical Consumer. Moral Norms and Packaging Choice. *Journal of Consumer Policy*, 22, 439–60.

Transition Towns Network. 2008. Transition Towns Wiki [online]. http://www.transitiontowns.org [accessed 31 March 2009].

von Borgstede, C., Andersson, M. and Johnsson, F. 2013. Public attitudes to Climate Change and Carbon Mitigation. *Energy Policy*, 57, 182–93.

Webb, J. 2012. Climate Change and Society: The Chimera of Behaviour Change Technologies. *Sociology*, 46(1), 109–25.

Chapter 17
Evaluation of Environmental Policy Integration in Swedish Structure Plans

Ann Åkerskog, Sylvia Dovlén and Abdul Khakee

Introduction

The traditional concern in planning evaluation has been on the potential substantive impact of plans – costs and benefits of the plans, of these instrumental values together with distributional impact, e.g. in planning balance sheet; and these two impacts together with environmental consequences, e.g. in community impact analysis (Lichfield, 1996). The integration of sustainability aspects, however, has met with a large number of problems. With increasing attention on the discourse-based logic in planning evaluation interest has also been directed towards how plans succeed in integrating different sustainability factors. The point of departure for this frame of evaluation is that plans narrate a set of knowledge practices.

Integration of environmental objectives in planning poses several challenges because environmental issues, as opposed to other policy fields, are not bound by geographical, administrative or teleological boundaries. Environmental issues are multi-level, multi-facetted, controversial and complex and environmental policy evaluation provokes controversies (Crabbé and Leroy, 2008). In reading plans in order to assess the level and significance of the integration of environmental objectives with other development objectives, we have to make several simplifying assumptions. We assume that plans represent a set of knowledge practices that physically construct a material reality (Tett and Wolfe, 1991). We also assume that plans enable us to reduce and understand the complexity of environmental problems to a few key variables or themes. Evaluating the integration of environmental policy with other development policies is a comprehensive exercise that includes rationalistic as well as interactive perspectives (Alexander, 2011) and therefore reading plans in order to assess the integration of environmental objectives in urban development planning requires multiples approaches.

In this chapter we present two case studies; in the first case study we examine plans to find out if they have fulfilled the Strategic Environmental Assessment (SEA) directives issued by the European Union (EU) and if so how does this promote the integration of environmental issues in Swedish urban planning. In the second study we read plans in order to assess the actual impact of integrating national climate change and energy efficiency goals to municipal comprehensive plans.

In Sweden the main responsibility for spatial planning is entrusted to the 290 municipalities. These local authorities have a high degree of autonomy and a monopoly in planning, which means that the central government can only influence spatial planning through legislation and guidance and the monitoring that is carried out by the 21 County Administrative Boards (CABs). This strong position held by the local authorities represents the Scandinavian model in public administration as well as politics (Khakee and Elander, 2001).

Spatial planning in Sweden is regulated by the Planning and Building Act (PBA) (SFS, 1987:10) and the Environmental Code (SFS, 1998:808). The municipal comprehensive plans (MCP) cover the entire area of a municipality and provide recommendations for the long-term use of land and water and the management of the built environment, but are not legally binding documents. The MCPs can be equalled to structure plans in the UK or strategic spatial plans in many western European countries, in that they consist of legal and administrative rules and institutional arrangements for guidance in locating investments, regulating building development and the use of land paying specific attention to nature and cultural conservation (Khakee, 2000, p. 136). More recently, MCPs are required to integrate environmental objectives as stipulated in the European Union directive on Strategic Environmental Assessment (EU SEA Directive, 2001/42/EC) and the central governments Environmental Quality Objectives that are in turn based on various UN conventions and their interpretations at the European Union level (Dovlén and Khakee, 2011).

The major aim of this chapter is to study the integration of environmental objectives in the MCPs. More specifically, with the help of two case studies, we examine how the EU SEA Directive affected the contents of a selected number of MCPs and how one specific national Environmental Quality Objective namely with respect to energy efficiency was integrated in the MCPs.

Besides this introduction the chapter consists of four sections. The following section contains brief reflections on the theoretical issues that have been discussed in planning evaluation research on the integration of environmental objectives in planning. The two case studies are presented in the third and the fourth section respectively. Finally we present the major conclusions of our study, discuss broader issues regarding the need for comparative evaluation.

Interpretation and Integration of Environmental Objectives in Planning

The recognition that a plan is more than just a policy document consisting of goals, policies and programs has implied that planning evaluation is no longer seen as an exercise in finding out the most favourable relationship between goal achievement and the use of resources. This shift in emphasis in planning evaluation is clearly evident from Lichfield (1998) where he acknowledged that if planning evaluation has to remain robust it would need to be flexible and take into consideration interactive perspectives (Lichfield, 1998). Similarly Voogd (1998) admitted that if planning evaluation methods were to be useful as the society moves from representative to participatory democracy they shall have to pay more attention to interactive perspectives.

Even among planning researchers there has been an increasing concern about the changing implications of planning instruments. Hoch (2002) perceives planning as process where planners compose narratives, put forward arguments, and envision relationships that speak to various audiences, such as decision makers, critics, colleagues, citizens, officials and themselves. Similarly Mandelbaum (1990) suggests that the authors of plans are affected by the knowledge that their work will be read and interpreted by a varied audience. Gunder and Hillier (2004) describe plan making as crucial in order to safeguard that the facts are made to speak by a network of discursive devices. Hajer (2009) describes a plan as an ensemble of notions, ideas, concepts, and categorizations through which meaning is ascribed to a phenomenon.

Healey (2007) makes use the concept 'frame' as a system of meanings that organizes what we know and helps us to transform fragmentary information into a structured and meaningful whole. It further provides conceptual coherence, a direction for action and a basis for persuasion.

Against the backdrop of these views the issue of the performance of plans and planning has received increasing attention. One interesting methodological approach that has been found useful in evaluating strategic or structure plans makes use of the concept conformance vs. performance (Faludi, 2012; Alexander, 2011; Faludi, 2010; Mastop and Faludi, 1997; Talen, 1997; Mastop and Needham, 1997; Faludi and Altes, 1997; Alexander and Faludi, 1989). There is a general consensus that the conformance- criteria are relevant in determining whether 'the material outcome conform to intentions' (Faludi, 2012, p. 138). Conformance-criteria are 'less tenable' when planning activity is extended to 'decision-centred planning' or 'strategic planning' or equated to planning process or coincides with any of the views stated in the previous paragraph. Therefore most of the attention in ex ante evaluation has focussed on the performance-criteria in order to determine the effectiveness of plans. The performance-criteria that have been used are quite flexible and allow a variety of combinations to examine effectiveness of plans. Whereas Faludi and Mastop (Faludi, 2012, p. 140) mention three main conditions ('thorough understanding of planning object', 'continuing relevance to the evolving situation' and helping to define 'operational decision situations'); Mastop and Needham (1997, pp. 885–7) mention several more conditions e.g. judging the importance of planning doctrine, reasons why strategic plans are made, skill and awareness among plan-makers in organizing the planning process, the extent to which operational decisions rely on broader policy context, tangible or significant effect of plans on subsequent decisions and measuring the input to achieve performance.

In this chapter we are not evaluating strategic plans per se; we are evaluating the consequences of SEA requirements expressed in the Swedish legislation and the attained results as found in the MCPs. In the first study we make use of a number of system and process criteria (see e.g. Fischer, 1995). Systems criteria refer to questions in the checklist on legal basis, integration, guidance and coverage. Process criteria on the other hand are related to questions put on alternatives, screening, scoping, report preparation, review, mitigation and public consultation.

In short we are trying to find out how well did a selected number of municipal plans perform in relation to national prescription with regard to the SEA. We also examine if plans espouse the goals in the SEA legislation.

In the second case study, we go further in our evaluation to answer how well do MCPs perform with regard to future decisions. For this purpose we use the discursive approach (see e.g. Runhaar, et al. 2013). The phenomenon investigated in the second case is sustainable energy use i.e. measures that reduce greenhouse gas emissions and/or contribute to an efficient energy systems. The evaluation focuses on how climate change and energy efficiency are framed in visions and strategies and if linkages are made between formulations in vision/strategies/ objectives and implementation measures including land-use development. In other words we examine how well did the plans perform in defining operational measures in the form of programmes and projects (see Alexander, 2011).

Implementation of SEA in Spatial Planning in Sweden

Background

The European Union (EU) Directive 2001/42/EC on the assessment of the effects of plans and programmes on the environment (the EU SEA-directive) that came into effect on 21 July in 2001 has started a policy implementation process in the 27 member states. The EU SEA Directive is being implemented through national legislation, guidelines and education. The main area for the implementation of the SEA Directive in Sweden has been spatial planning in which the Swedish municipalities and their comprehensive planning activities play a major role.

The policy, outlining the reformation of the planning legislation, of which SEA policy is one part, proposes the integration of environmental issues and sustainable development into MCP. The implementation of SEA Directive has introduced detailed legislative requirements for the procedure as well as the contents of the Environmental Report (ER) that is included in the MCP. These requirements include:

1. Scoping consultations with the County Administrative Boards;
2. Description of relevant environmental quality goals and how they have been taken into consideration;
3. Presentation of reasonable alternatives;
4. Explanation of how environmental aspects have been integrated in the plan and the motives as to why a specific plan alternative has been selected;

5. Preparation of a special document describing how SEA and the comments from public consultations have been taken into consideration; and
6. Specification of how municipalities acquire and use knowledge concerning environmental impacts from the plans.

For our case study, items 4 to 6 are the most relevant points of consideration.

Aim and Approach of the Case Study

Central for this case study are the planning documents, including the ER as they describe the planning process and the outcome of the local practice. The empirical focus of this case study are therefore the contents in 24 MCPs prepared prior to the introduction of the SEA Directive and the first 12 ERs that were included in MCPs following the implementation of the Swedish SEA legislation. We try to answer the following specific questions:

- How much of the demands in the Swedish SEA legislation was implemented in the first 12 plans made making use of goal achievement approach (Hill and Hupe, 2002);
- What similarities and differences can be observed before and after the introduction of the SEA Directive?

The purpose of implementing SEA, as expressed in the preparatory works to the new planning legislation, are formulated as 'increased clarity' and 'improvement' in environmental assessment. To be able to evaluate such objectives, a comparison is needed with former conditions. A comparative evaluation of Swedish MCPs shows how the demands on reporting the consequences of the plan were met before and after SEA legislation.

The 24 pre-SEA MCPs were formally approved between 1996 and 2002. They have been selected from five counties in southern Sweden representing sparsely and densely populated and larger and smaller municipalities respectively. The 12 post-SEA MCPs were the first ones that were ready between 2006 and 2007 (see Table 17.1).

In order to find out whether the anticipated effects of the implementation of SEA in municipal comprehensive planning were achieved, a checklist containing 52 questions was used. The checklist includes questions about:

- the mandatory requirements of SEA;
- the integration of planning and SEA in the planning processes as well as the plan documents;
- actors involved in preparation of MCP and ER.

Most of the questions in the comparative evaluation adhered to the checklist used for evaluating pre-SEA MCPs. However, some new questions were added and some changed because of the interpretation of the SEA-directive.

Table 17.1 List of municipalities and counties included in the comparative evaluation

Before SEA 1996–2002			
Municipality	**County**	**Municipality**	**County**
Gagnef	Dalarna	Stenungsund	
Halmstad	Halland	Kungälv	
Varberg		Kävlinge	Skåne
Skara	Västra Götaland	Eslöv	
Ulricehamn		Höör	
Alingsås		Klippan	
Strömstad		Helsingborg	
Uddevalla		Lomma	
Göteborg		Tomelilla	
Bollebygd		Burlöv	
Lerum		Åstorp	
Tanum		Staffanstorp	
Munkedal		Karlskrona	Blekinge
After SEA 21 July 2006–31 October 2007			
Municipality	**County**	**Municipality**	**County**
Mörbylånga	Kalmar	Hässleholm	Skåne
Vimmerby		Flen	Södermanland
Nybro	Kronoberg	Lindesberg	Örebro
Älmhult		Avesta	Dalarna
Ronneby	Blekinge	Arvika	Värmland
Karlshamn		Örnsköldsvik	Västernorrland

Main Findings

The following issues were considered central in evaluating the integration of environmental aspects in the MCPs:

- Description of the planning procedure including consultations;
- Identification, narration and assessment of environmental impacts and presentation of alternatives;
- Delineation of mitigation measures;
- SEA's impact on MCP.

Contents of Impact Assessment in 24 Plans Prior to the Implementation of SEA

In most of the pre-SEA plans, the background description of the environmental conditions in the municipality as well as the environmental objectives, are well described (21 out of 24 plans). Broader considerations of consequences are made in 20 plans. Moreover environmental impacts are described in 11 and assessed in 15 plans, though the notion of significant environmental impact was not in use at that time. Alternative ways to reach the planning goal are presented in 13 plans and in seven of them an assessment of these alternatives is made. Steps taken for monitoring and surveillance of environmental impacts are presented in 16 and mitigation measures are described in 19 of the plans.

Nineteen MCPs take into consideration viewpoints put forward during public consultations. These viewpoints, however, not only reflect environmental but also other impacts. In four plans the impact assessment process is indicated and two plans integrate it with the outline of the planning process. These results indicate that several aspects regarding the integration of environmental perspectives in MCPs were already part and parcel of the municipal planning practice.

Contents of Impact Assessment in 12 Plans Made After SEA Implementation

SEA assumes that significant environmental impacts are identified, described and assessed. The identification and description is done in nine out of 12 ERs included in the plans and seven of the ERs assess these impacts. However, the impacts are seldom singled out as *significant*. The seven plans that did assess environmental impacts do not go so far as to consider cumulative or long-term effects. These results should be judged with regard to *significant environmental impacts* which constitute the most important factor for screening as well as for scoping the environmental report.

Alternatives are identified in six environmental reports, and in five of these the alternatives are also described and assessed. These alternatives relate to design, location or scale of planned development. In none of the 12 MCPs an alternative to the entire plan is discussed. Mitigation measures and monitoring of environmental impacts are described in nine ERs.

Consultations were carried out with the County Administrative Board in eight ERs and with the public in 10 ERs. In eight MCPs ER is presented as a separate report and in four it is integrated in the plan. It is more difficult to see whether the SEA process was integrated with the planning process. However, in two plans it is evident that the processes were actually merged not only time-wise but also through joint consultations. One important result is that the environmental aspects are more fully covered and are not subsumed in socio-economic analysis.

Comparison of Pre- and Post-SEA Results

The concept, *significant environmental impacts*, has put focus on scoping. The latter requires ER to analyse and assess these impacts as opposed to former more general assessment of environmental impacts. Scoping in ER emphasizes environmental effects rather than vision and/or objectives of the plan as was the situation before. The pre-SEA plans had not necessarily an environmental orientation. The change in perspective towards more specific environmental considerations in the post-SEA plans, indicate that environmental issues are integrated to a greater extent in the plans than before.

The SEA-requirements have forced planning practice to adapt new routines. This is particularly evident with regard to how planning is defined by the SEA requirements, the practice of consultation and analysis of prevailing environmental conditions. Requirements such as the summarizing statement describing how environmental aspects have been considered in the planning proposal are still not part of planning practice. The expected increase in transparency of how environmental issues are integrated in the plans has not been achieved. However, SEA requirements are being gradually adapted to the existing practice.

Our comparative assessment also shows that several of the new requirements that were introduced in the Swedish Environmental Code were already applied before the SEA-implementation, especially the systems criteria. The most notable change in practice relates to process criteria. New requirements that were already known to the implementers prior to the SEA-legislation are more likely to be accepted and practiced (see e.g. Mazmanian and Sabatier, 1989). The evaluation of post-SEA plans indicates that prerequisites for the integration of environmental aspects in municipal planning have been strengthened. However, the Swedish municipal planning has a long way to go before environmental criteria get the same attention as socioeconomic consideration in urban development in reality.

Climate Change and Energy Efficiency Measures in MCPs

Background

According to the central government, the Swedish national energy policy shall follow the principles outlined by the European Union. Energy efficiency is claimed to be one key area for transforming Sweden into a sustainable society. Energy is thus a key parameter in the national environmental objective for reducing the climate impact.[1] In Sweden as well as in many other countries the major cause of

1 'Concentrations of greenhouse gases in the atmosphere must be stabilized at a level that will prevent dangerous anthropogenic interference in the climate system. Sweden, together with other countries, must assume responsibility for achieving this global objective' (Swedish EPA, 2013).

the greenhouse effect is the use of fossil fuels for generating heat, electricity and in transport (Swedish EPA, 2013). It is therefore crucial to strengthen the regional and local efforts for more efficient use of energy (Energimyndigheten, 2010).

In 2002, the Swedish Energy Agency (SwEA) initiated a program called 'Sustainable Municipalities' (SusMun) which is based on the national energy policy goals. According to SwEA, SusMun represents a unique cooperation between the SwEA and a selected number of municipalities. By participating in the program, these municipalities have access to expert support in the form of knowledge exchange, education, policy networks and implementation tools in the energy field. The overall aim of SusMum is to improve the capacity of municipalities to apply measures for sustainable energy use and to put energy efficiency on the local political agenda. The program consists of three phases: 2003–2007, 2008–2012 and 2013–2014 respectively. Out of a total of 290 Swedish municipalities 66 joined the program (Energimyndigheten, 2009).

The SusMun-program is built around several themes, one of which is *energy-smart* planning (Energimyndigheten, 2012). The municipalities are offered expert support in such areas as transportation, energy supply and renewable energy in order to improve municipal comprehensive planning (Energimyndigheten, 2009, p. 22). According to SwEA, MCP is the key instrument in order to reduce energy use and carbon footprint (Energimyndigheten, 2012). The municipalities are seen as core actors since they have local planning monopoly.

Besides the MCP, every Swedish municipality is mandated to prepare an energy plan according to the law on energy planning (SFS, 1977:439). The requirement to prepare energy plans dates back to 1977. What is new, however, is the integration of the energy plan with MCP. The SusMun-program aims to make this relationship robust despite the fact that both national and local public administration is fragmented in various policy units or departments; each with its own set of expertise, professional roles, routines and strategies.

Aim of the Case Study

In this chapter we report some findings from the first part of the ongoing research project entitled Transform[2] which is funded by SwEA and is a part of the SusMun program. One aim of the project is to evaluate the outcome of the SusMun programme in relation to how climate change and energy efficiency measures are integrated in MCPs. The case study focuses on the following questions:

- How are climate change and energy efficiency framed in the visions and strategies of MCPs?
- To what extent are climate change and energy efficiency measures integrated in these plans?

2 Ann Åkerskog, Sylvia Dovlén and Tuija Hilding-Rydevik are members of the Transform project team. The project is carried out between 2011–2014.

The Review of Plans

The research study started with examining MCPs prepared by the municipalities that participate in the SusMun programme. MCPs were selected on the criterion that they were formally approved by the municipal council at the end of the first phase of the SusMun programme (2003–2007), 27 such plans met this criterion.

Table 17.2 List of selected municipalities

2007	Avesta	Lindesberg	Mörbylånga	Nybro	Vimmerby	
2008	Alvesta	Falköping	Hudiksvall	Lerum		
2010	Gotland	Helsingborg	Hultsfred	Kil	Linköping/Norrköping	Lund
	Sunne	Torsås	Trosa	Uppsala	Vingåker	Örebro
2011	Storuman					
2012	Nyköping	Värmdö	Örnsköldsvik	Olofström	Lilla Edet	

In the first step of evaluating the MCPs, a review form was constructed that focused on:

- how climate change and energy efficiency measures were approached in visions and strategies;
- if and how the municipalities had prepared a statement of measures for climate change and energy efficiency that were to be implemented.

This first phase of review showed that many of the evaluated plans treated climate change and energy efficiency as an issue isolated from other urban development projects and the planning discourse lacked a holistic view. Moreover, plans with visions and/or comprehensive strategies linked climate and energy issues with implementation measures and land-use proposals. After this initial review, the plans were divided into two groups – those that separated climate and energy from other issues (group 1) and plans that linked energy issues to land-use proposals (group 2).

A closer scrutiny of visions and comprehensive strategies in of plans in group 2 was carried out in the second phase of the evaluation. The visions are expected to serve as an orienting device in order to inspire local actors as well as to prepare a framework within which development projects and regulations could be located (see e.g. Healey, 2007, p. 179) Innes and Booher (2010, p. 108) argue that comprehensive visions and goals are needed to inspire planning participants and make them feel that their participation is worthwhile despite the fact that implementation is mostly carried out through incremental steps. It should be added that the visions in the evaluated plans are different in scope – some are extremely brief in form of slogans, others are more detailed descriptions of a desired future.

Strategies related to visions can be viewed as a frame of reference for action. They offer a sense of direction and provide some parameters within which specific actions can be set as well as ideas about how future opportunities may be grasped and threats avoided (Healey, 2007, p. 183).

Major Findings

Visions and strategies

The first phase of the review showed that 24 of the 27 (89 per cent) reviewed plans had comprehensive visions and of these 18 had a time horizon extending from 2010 to 2035. Six lacked time horizons. There are 17 (71 per cent) visions in the form of general statements about sustainability and sustainable development and growth. Seventeen of the 24 plans with visions include overall strategies but not all of them explicitly refer to climate and energy aspects. Plans that do not have climate change and/or energy efficiency in their visions and/or comprehensive strategies also lack climate and energy efficiency measures linked to land-use development proposals. Only 11 (46 per cent) of the plans have visions and/or comprehensive strategies including climate change and/or energy efficiency. Those 11 plans were more closely reviewed (see Table 17.3 below).

Table 17.3 Plans with visions and strategies on climate change and energy efficiency

Municipality	Time horizon	Visions Ideas and concepts (climate change and energy)	Comprehensive strategies Ideas and concepts (climate change and energy)
Gotland	Vision 2025	'Gotland is a leading island-region in environmental and climate issues' (Gotlands kommun, 2010, p. 6).	No comprehensive strategy concerning climate change and/or energy efficiency.
Helsingborg	Vision 2035	'In 2035 Helsingborg is carbon neutral and energy efficient' (Helsingborgs kommun, 2010, p. 11).	Energy is used in an efficient manner through more effective infrastructure, densification and streamlined energy consumption (ibid., p. 14).
Linköping/ Norrköping	Vision 2030	Lindköping/Norrköping promotes sustainable growth (Linköpings kommun and Norrköpings kommun, 2010, p. 7).	Climate conscious municipal planning. Reduction of greenhouse gas emissions through the use of renewable energy. Adaptation to future climate changes (ibid., p. 17).
Lund	No time horizon	Lund in long-term is economically, socially, environmentally and culturally sustainable society. Ecological aspects form a precondition for all other development (Lunds kommun, 2010, p. 14).	CO_2 emissions reduced by 85 per cent by 2050. Followed by concrete measures e.g. expansion of district heating, in-crease of public transport and cycling routes, increase workplaces in larger urban areas to reduce commuting.

Municipality	Time horizon	Visions Ideas and concepts (climate change and energy)	Comprehensive strategies Ideas and concepts (climate change and energy)
Mörbylånga	Vision 2015	Öland has locally produced energy and make efforts to reduce the use of fossil fuels. The energy is produced in an environmentally sound manner. (Borgholms kommun and Mörbylånga kommun, 2000, p. 2).	Reduce dependency on fossil fuels. Promote wind energy and fossil free heating in urban areas (Mörbylånga kommun, 2007, p. 11).
Nybro	Vision 2030	Nybro is characterized by sustainability in all respects. Climate change is counteracted. (Nybro kommun, 2007, p. 11).	Emissions from fossil fuel eliminated. Consider climate change (ibid., p. 15).
Olofström	No time horizon	Olofström should be a place that our visitors would return to* (Olofströms kommun, 2012, p. 9).	Reduced dependence on fossil fuels. Reduced energy use (ibid., p. 20).
Trosa	Vision 2020	The sustainable development has begun to show results in form of decreased amount of emissions, waste water and use of fossil fuels (Trosa kommun, 2010, p. 5).	No comprehensive strategy concerning climate change and/or energy efficiency.
Uppsala	Vision 2030	Uppsala has responsible resource use to ensure environmental, public health, biodiversity and counteract climate change. Uppsala is moving towards climate neutrality and is ready to meet the effects of climate change (Uppsala kommun, 2010, p. 11).	City area's web of buildings, green areas, transport systems and technical supply is closer to cope with climatic and environmental standards and requirements for city life qualities (ibid., p. 25). Action measures are also included in the plan.
Värmdö	No time horizon	Värmdö use renewable energy, modern and environmentally friendly transport and build eco-friendly. A good access to renewable fuels means that more people can take greater responsibility (Värmdö kommun, 2012, p. 6).	Climate adapted lifestyle. Increased share of public transport and energy efficiency (ibid., p. 25). Energy saving and efficiency should increase 20 per cent by 2020 (ibid., p. 104).
Örnsköldsvik	Vision 2015	Örnsköldsvik is a sustainable, accessible municipality in world class – to live and work in and to visit (Örnsköldsviks kommun, 2012, p. 20).	A climate-smart municipality. Energy is effectively used, sustainably produced and carbon neutral. Action measures are also included in the plan.

Note: * Olofström is the only municipality in Table 17.3 that does not mention sustainable development or environmental concerns in its vision.

Five municipalities, Helsingborg, Mörbylånga, Nybro, Uppsala and Värmdö, had visions and comprehensive strategies that include climate change and/or energy efficiency. Three of them, Helsingborg, Uppsala and Värmdö, present climate change and energy efficiency in visions and strategies in a way that link vision to strategies and implementation measures. These three visions acknowledge, for example, conversion to renewable energy, even if they differ in focus and ambition.

For example, Helsingborg is 'a carbon neutral municipality in 2030' while Uppsala in 2035 'strives to reach climate neutrality'. In these municipalities densification is a development strategy along with investment in renewable energy resources and energy efficiency measures. Residents' role and behaviour are clearly underlined in the strategies. Improved public transport and the development of pedestrian and bicycle paths are also included along with campaigns for energy savings and climate-adapted lifestyle.

The rural municipality, Mörbylånga, focuses mainly on the development of wind power and energy crops in vision as well as strategies. The vision of Nybro declares: 'In 2030 climate change is completely counteracted'. Their strategy to 'consider climate change' acknowledges climate as a concept, but it is not given tangible contents.

Lund and Örnsköldsvik have a general slogan as vision but without concrete formulations about climate change and/or energy efficiency, although the plans have comprehensive strategies. Sustainable development and climate change adaption is the point of departure for Lund. The goal is to reduce greenhouse gases by 85 per cent in 2050. Proposed measures are expansion of district heating, transition to renewable energy, expansion of public transport and cycle paths, and relocation of workplaces in order to reduce commuting trips. Örnsköldsvik has a comprehensive strategy that includes such challenges as 'complete conversion of all energy systems, energy efficiency through the deployment of carbon-free energy and transport systems and use of renewable raw materials in the manufacture of new energy-efficient climate-neutral products' (Örnsköldsviks kommun, 2012, p. 19).

Gotland and Trosa have visions that acknowledge climate change and energy efficiency but their plans lack comprehensive strategies for these issues. According to Gotland's vision 2025, the island-region shall be leading in environmental and climate issues. Trosa aims to decrease substantially the use of fossil fuel in 2020.

The Linköping/Norrköping and Olofström visons lack statements about energy and climate issues but they include them in their strategies. Linköping/Norrköping strategy aims at reducing greenhouse gas emissions. Olofströms strategy is not so extensive but it contains some statements about reducing dependency on fossil fuels.

In short; only five municipalities, Helsingborg, Lund, Uppsala, Värmdö and Örnsköldsvik have visions as well as strategies that frame climate and energy issues in multiple dimensions and organise information into a structured analysis. This in turn provides bases for further concretisation of implementation measures and land use proposals. The remaining six municipalities (Gotland, Linköping/Norrköping, Mörbylånga, Nybro, Olofström and Trosa, see Table 17.3) address climate and energy issues in more general and vague terms.

Implementation measures

Five of the municipalities have integrated climate change and energy efficiency in their MCPs. We will illustrate this with the help of Värmdö's vision, strategies and implementation measures including land-use (see Table 17.4). Värmdö, is located close to Stockholm city, and constitutes the heart of the Stockholm archipelago with about 10,000 of the archipelago's 24,000 islands and skerries.

Table 17.4 MCP of Värmdö: Vision and strategies linked to implementation measures

Vision	Comprehensive strategy	Measures/Land-use
'Here we meet the future by being at the forefront of the necessary adjustment to the long-term sustainability. We use renewable energy, modern and environmentally friendly transport and build eco-friendly'. 'We use renewable energy in an efficient way. We use to a large extent wind, wave and solar power, eco-friendly transport technologies such as track cars. A good access to renewable fuels means that more people can take greater responsibility' (Värmdö kommun, 2012, p. 6).	'Habitations will be accessibly adapted and localized so that you can walk or cycle to a greater extent. The municipal and regional cycle network will be developed'. 'Attractive public transport with buses or in the future tram roads will reduce the need for car commuting' (ibid., p. 26). 'Värmdö must be developed in a new direction with resource and material flows based on reuse, recycling and energy extraction. Use of fossil fuels, a finite resource, will be replaced by energy from the sun, wind and water. The goal to achieve a sustainable Värmdö will be facilitated if growth and economic development can be based on other than material consumption' (ibid. p. 29).	Densification in proposed areas. District heating is expanded in dense areas. Restrictions to counteract 'urban sprawl'. Pedestrian and cycle paths. Park and ride facilities. Bus stops. Energy and resource efficient urban areas. Energy supply is clarified in new local plans Prioritize renewable energy use. Specified areas for wind and wave power.
'Värmdö development must adapt to the impacts of climate change and contribute to climate goals' (ibid., p. 11). 'Challenges for the development: climate-friendly lifestyle; increased share of public transport; more efficient energy use' (ibid., p. 25).	'Värmdö can take responsibility (for global climate change) by the municipality, property owners, businesses and citizens to take action to reduce greenhouse gas emissions. Heating systems must be changed – district heating and new technologies can quickly lead to reduced emissions. New transport infrastructure can lead to radically reduce reliance on fossil fuel' (ibid., p. 32). 'Through planning and exercise of public authority, can individuals be guided towards climate-neutral initiatives. The municipality can, as operator, take measures in several areas that reduce energy dependence' (ibid., p. 32).	The municipality support citizens' efforts to achieve a sustainable lifestyle by focusing on housing, travel, waste and energy use. New working areas with environmental profile. Prioritize environmental business in established working areas. Intensified municipal energy advice to property owners. Prioritize solar, wind and wave power, and heat pump technology, passive technology and environmentally friendly biofuel technology.

This vision depicts a future where climate change and energy efficiency are important development tasks. It points out some challenges for the future where a climate-friendly lifestyle, public transport and efficient energy use are key policy areas for improvements which in turn are linked to a comprehensive strategy for sustainable development. Other land-use measures include new working areas with environmental profile giving priority to environmental businesses and locations for wind and wave power.

The strategy includes guidelines for all the districts with or without detailed development plans. In districts with such plans, objectives and guidelines are stated at two levels. The first level includes general objectives, for example densification 'new housing near advanced public service with opportunity to organize good public transports' and urban areas that are energy- and resource-efficient. Pedestrian and bicycle paths and public transport are prioritized, more park and ride facilities built in the strategic and central locations (Värmdö kommun, 2012, p. 114). In districts without detailed development plans guidelines are so elaborated that even geographical positions for different measures (e.g. pedestrian and bicycle paths, park and ride areas, buss-stop) are specified.

There are about 11,000 holiday homes in Värmdö and most of them are located outside the development plan areas. Proximity to Stockholm has fuelled a development where such homes have been gradually converted into permanent homes resulting in a kind of 'urban sprawl'. In order to counteract the sprawl, the MCP defines a specific number of areas that shall be provided with municipal services including public transport facilities, commuter parking and extension of pedestrian and cycle paths.

Conclusions

In this chapter we have tried to evaluate how well the Swedish municipal plans and planning respond to national legislation with regard to the EU's SEA directive and climate change and energy efficiency and to what extent SEA directive and climate change and energy efficiency goals are integrated in the Swedish municipal comprehensive planning.

Considering the fact that our evaluation deals with the early outcome of the implementation of SEA, we should expect more changes to take place in the future. One of these changes relate to the concept *significant environmental impact*. Although the concept has been treated quite differently by the municipalities and rests upon different bases in the ERs, a change can still be seen how scoping was made before and after the introduction of SEA requirements. Requirements that have not been implemented are the new demands of presenting motives as to why certain alternatives have been produced and selected, and if and how SEA has had an influence on the urban development plan. The lack of the special compilations and summarizing statements indicates this. The absence of this crucial information creates a void rather than an increased transparency in the planning process. It could be a sign of well integrated environmental aspects because these are difficult to separate from other aspects in the planning process or it could also imply that the SEA is not connected to the process at all. ERs seem to have been prepared after MCPs were ready in cases where SEA did not have any effect on the plan and thereby limited possibility for the integration of environmental aspects.

Owing to long time interval in between which municipalities prepare new comprehensive plans and since municipalities have different schedules for preparing these plans, it will take time before the SEA directive is implemented throughout Sweden. Under these circumstances learning from municipalities that have already implemented the SEA directive would be important. Hopefully some kind of 'best-practice' tradition is established in this context. Another reflection from the results of our study is that there is clear evidence of implementers using their own discretion about what to include in the ER. Wherever part of the existing planning practice includes some aspects of the SEA directive, implementers minimize their work by not referring to these aspects. The latter may not be obvious to those who examine or use the planning document.

In short, the early outcomes of implementing SEA directive in Swedish municipal comprehensive planning indicate that many municipalities have initiated a successful process but it shall take time for a more thorough implementation of the directive. It is too early to say as to what extent the SEA directive shall promote sustainable development through the integration of environmental aspects in urban development planning. In short our study not only shows how well municipal planning performs with respect to the implementation of the SEA directive but also how it espouses the SEA legislation.

In the second case study, our evaluation attempts to answer how well urban plans perform in integrating climate change and energy efficiency goals. This is an important aspect in the transformation of Sweden into a sustainable society. Our evaluation of MPCs shows that only five out of 27 investigated MPCs manage to integrate climate and energy issues. The criteria for integration expressing that climate change and energy efficiency should be framed coherently and linkages made between formulations in visions and strategies on the one hand and implementation measures on the other. These five integrative plans contain a coherent story that narrates climate change and energy efficiency linking universal statements and land-use proposals. The other 22 examined plans lack concrete actions for energy efficiency linked to future land use and urban development.

A common property for the five municipalities that integrate climate and energy issues in their MPCs is that these issues have most likely been on the agenda during the entire planning process and have thereby received political legitimacy. Four of the municipalities with growing population and new residential areas focus on densification and improvement of existing building and infrastructures in order to reduce climate impact and energy use.

The other 22 plans treat climate and energy issues in less coherent manner without linking overall goals and land-use proposals. A majority of Swedish plans do not integrate energy efficiency measures in urban development proposals.

Energy issues are often presented as separate chapters or sections in plans and are detached from other policy areas and do not act as planning prerequisites. In short, the number of Swedish plans that are successful in integrating climate change and energy efficiency is relatively small. However, the few that are successful perform well with regard to future decisions in land use development that are *climate-smart* and promote effective use of energy.

Our two case studies show that the performance approach in planning evaluation increases our understanding of the prerequisites and obstacles to integrate environmental legislations in municipal planning. This approach is intricate and demanding since it not only requires how local plans integrate environmental goals but also how planners as well as other stakeholders accept environmental concerns.

References

Alexander, E.R. 2011. 'Evaluating planning: What is successful planning and (how) can we measure it?' In Hull, A., Alexander E.R., Khakee, A. and Woltjer, W. (eds) 2011. *Evaluation for Participation and Sustainability in Planning*. London: Routledge.

Alexander, E.R. and Faludi, A. 1989. Planning and Plan implementation: Notes on evaluation criteria. *Environment and Planning B: Planning & Design*, 16(1), pp. 127–140.

Borgholms kommun and Mörbylånga kommun. 2000. *En vision av Öland år 2015*. http://www.borgholm.se/Bygga%2C+bo+amp+miljo/Samhallsplanering/Oversiktsplanering/Oversiktsplan/se/86/ [accessed 15 January 2013].

Crabbé, A. and Leroy, P. 2008. *The Handbook of Environmental Policy Evaluation*. London: Earthscan.

Dovlén, S. and Khakee, A. 2011. 'Our threatened biodiversity: Interpreting the UN Convention on Biodiversity in Swedish regional development planning with the help of a story line'. In Hull, A., Alexander E.R., Khakee, A. and Woltjer, W. (eds) *Evaluation for Participation and Sustainability in Planning*. London: Routledge.

Energimyndigheten. 2009. *Programbeskrivning för programmet – Forsknings-programmet Uthållig kommun 2009-09-01-2012-12-31*. Eskilstuna: Diarienummer 2009-002109, Energimyndigheten.

Energimyndigheten. 2010. *Energiläget 2010*. ET 2010:45. Eskilstuna: Energimyndigheten.

Energimyndigheten. 2012. *Temaområde energismart planering*. http://www.energimyndigheten.se/sv/Offentlig-sektor/uthallig-kommun/Energismart-planering/ [accessed 13 December 2012].

EU SEA Directive. 2001/42/EC. http://ec.europa.eu/environment/eia/sea-legalcontext.htm [accessed 11 March 2014].

Faludi, A. 2012. Conformance vs. performance: Implications for evaluation. *Impact Assessment*, 7(2–3), pp. 135–51.

Faludi, A. 2010. The performance of spatial planning. *Planning Practice & Research*, 15(4), pp. 299–318.

Faludi, A. and Altes, W.K. 1997. 'Evaluating communicative planning'. In Borri, D., Khakee, A. and Lacirignola, C. *Evaluating Theory-Practice and Urban-Rural Interplay in Planning*. Dordrecht: Kluwer.

Fischer, F. 1995. *Evaluating Public Policy*. Chicago: Nelson-Hall Publishers.

Gotlands kommun. 2010. *Bygg Gotland – översiktsplan för Gotlands kommun 2010–2025*. http://www.gotland.se/imcms/ByggGotland [accessed 15 January 2013].

Gunder, M. and Hillier, J. 2004. Conforming to the expectations of the profession: A Lacanian perspective on planning practice, norms and values. *Planning Theory and Practice* 5(2), pp. 217–35.

Hajer, M. 2009. *Authoritative Governance: Policy-Making in the Age of Mediatisation.* Oxford: Oxford University Press.

Healey, P. 2007. *Urban Complexity and Spatial Strategies. Towards a Relational Planning for Our Times*. Oxford and New York: Routledge.

Helsingborgs kommun. 2010. ÖP 2010 – en strategisk översiktsplan för Helsingborgs utveckling. http://www.helsingborg.se/ImageVaultFiles/id_1866/cf_2/Oversiktplan2010_2010_sbf.pdf [accessed 15 January 2013].

Hill, M. and Hupe, P. 2002. *Implementing Public Policy*. London: Sage.

Hoch, C.J. 2002. Evaluating Plans Pragmatically. *Planning Theory*, 1(1) pp. 53–75.

Innes, J.E. and Booher, D.E. 2010. *Planning with Complexity – An Introduction to Collaborative Rationality for Public Policy*. Oxford: Routledge.

Khakee, A. 2000 *Samhällsplanering. Nya mål, perspektiv och förutsättningar*. Lund: Studentlitteratur.

Khakee, A. and Elander, I. 2001. 'Svensk samhällsplanering i början av ett nytt millennium'. In A Lindbom (ed.) *Den nya bostadspolitiken*. Umeå: Boréa, pp. 219–53.

Lichfield, N. 1996. *Community Impact Evaluation*. London: UCL Press.

Lichfield, N. 1998. 'Trends in planning evaluation: A British perspective'. In N. Lichfield et al. 1998. *Evaluation in Planning. Facing the Challenge of Complexity*. Dordrecht: Kluwer, pp. 1–15.

Linköpings kommun and Norrköpings kommun. 2010. *Gemensam översiktsplan för Linköping och Norrköping*. http://weblisher.textalk.se/linkoping/10gem_op/?page=1&mode=50&noConflict=1 [accessed 25 January 2013].

Lunds kommun. 2010. Översiktsplan för Lunds kommun. http://www.lund.se/Global/Förvaltningar/Stadsbyggnadskontoret/ÖP2010/antagandehandling/ÖP%202010%20till%20webben%20avdelning%201.pdf?epslanguage=sv [accessed 15 January 2013].

Mandelbaum, S.J. 1990. Reading plans. *APA Journal*, 56(2), pp. 350–56.

Mastop, H. and Faludi, A. 1997. Evaluation of strategic plans: The performance principle. *Environment and Planning B*, 24, pp. 815–32.

Mastop, H. and Needham, B. 1997. Performance studies in spatial planning: The state of the art. *Environment and Planning B: Planning and Design*, 24(69), pp. 881–8.

Mazmanian, D.A. and Sabatier, P.A. 1989. *Implementation and Public Policy*. Lanham, MD: University Press of America, Inc.

Mörbylånga kommun. 2007. *Mörbylånga kommun – översiktsplan 2007*. http://www.morbylanga.se/Miljo-Boende/Bygga-Bo/Oversiktsplan/ [accessed 15 January 2013].

Nybro kommun. 2007. Översiktsplan för Nybro kommun. http://www.nybro.se/templates/Page____1702.aspx [accessed 15 January 2013].

Olofströms kommun. 2012. *Nära till allt! Översiktsplan Olofströms kommun*. http://www.olofstrom.se/System/FileArchive/97/File_27711.pdf [accessed 15 January 2013].

Planning and Building Act. 2010. http://www.notisum.se/rnp/sls/lag/20100900.htm [accessed 1 January 2013].

Runhaar, H., van Laerhoven, F., Driessen, P. and Arts, J. 2013 Environmental assessment in the Netherlands: Effectively governing environmental protection? A discourse analysis, *Environmental Impact Assessment Review*, 39, pp. 13–25.

SFS. 1977:439. *Law on municipal energy planning 1977*. http://www.notisum.se/rnp/sls/lag/19770439.htm [accessed 15 January 2013].

SFS. 1987:10. *Planning and Building Act*. https://www.notisum.se/rnp/sls/lag/19870010.htm [accessed 11 March 2014].

SFS. 1998:808. *Environmental Code*. http://www.notisum.se/rnp/sls/fakta/a9980808.htm [accessed 11 March 2014].

Swedish EPA. 2013. *Environmental Objectives*. http://www.miljomal.nu/sv/Environmental-Objectives-Portal/Undre-meny/About-the-Environmental-Objectives/1-Reduced-Climate-Impact/ [accessed 25 January 2013].

Talen, E. 1997. Success, failure and conformance: An alternative approach to planning evaluation. *Environment and Planning B*, 24, pp. 573–87.

Tett, A. and Wolfe, J.M. 1991. Discourse analysis and city plans. *Journal of Planning Education & Research*, 10(3), pp. 195–200.

Trosa kommun. 2010. Översiktsplan 2010 – Trosa kommun. http://www.trosa.se/Global/Samhällsbyggnadskontoret/Plan/Gammalt%20ÖP%20och%20relaterat/ÖP%202010%20antagen/Översiktsplan%202010%20i%20sin%20helhet.pdf [accessed 15 January 2013].

Uppsala kommun. 2010. Översiktsplan 2010 för Uppsala kommun. http://www.uppsala.se/sv/Boendemiljotrafik/Stadsutveckling--planering/Oversiktsplanering/Oversiktsplan-2010/ [accessed 15 January 2013].

Voogd, H. 1998. 'The communicative ideology and ex ante planning evaluation'. In N. Lichfield et al. 1998. *Evaluation in Planning. Facing the Challenge of Complexity*. Dordrecht: Kluwer, pp. 113–26.

Värmdö kommun. 2012. Översiktsplan 2012–2030 – Värmdö kommun. https://www2.varmdo.se/Resource.phx/plaza/publica/omkommunen/oversiktsplan/index.htx [accessed 15 January 2013].

Örnsköldsviks kommun. 2012. Översiktsplan 2012 – Översiktsplan för Örnsköldsviks kommun. http://www.ornskoldsvik.se/download/18.3ef2cb051386e42ae532ff/1341989689490/ÖP12+kap+1-6+utställning+webb.pdf [accessed 15 January 2013].

Chapter 18

An Evaluation of the Budapest District VIII and IX Renovation Areas Based on Property Prices, Inspection and Narratives

Tom Kauko

Introduction

The development of cities, in general, and urban regeneration, in particular, is an increasingly difficult topic to pin down in Central and Eastern Europe (CEE). More often than not, urban neighbourhoods undergo a complex mixture of decay, inertia and revitalization. To be able to evaluate such multifaceted processes and outcomes we need new tools which go beyond cost-benefit analysis. In more global terms, evaluation tools have developed from urban economic efficiency and entropy to the current idea of sustainable development (see Foxon et al., 2012), and some have expressed the view that further development towards a 'happy healthy cities' paradigm is occurring (see LUDEN, 2012). Within this general movement, the contemporary theoretical discourse suggests a number of paths towards reaching urban sustainability.

When discussing the evaluation of any of the sustainability aspects of urban regeneration/renovation areas, a question arises of how the property development and the other rehabilitation of the blocks is organized at any given time and place. How do actual property development projects manage to connect private-public stakeholders (investor-developers and planner-policymakers) with best practice and in relative depressed circumstances? This chapter presents a case study within the latter context (Budapest, Hungary) to illustrate this. We observe that contemporary urban property developments in Budapest are privately driven – thus the era of public private partnership (PPP) is over, at least for the time being (cf. Keresztély and Scott, 2012). This chapter documents property price effects in two adjacent Budapest neighbourhoods under regeneration where a PPP approach was attempted: central sections of districts VIII (Józsefváros) and IX (Ferencváros). The former case comprises two regeneration areas: Corvin Promenade (mixed use: large-scale block-by-block demolition and new build) and Magdolna (public areas refurbishment). The latter is managed as one coherent whole (exclusively residential, piecemeal).

This contribution conducts an evaluation in a meta-framework of *urban sustainability*, with a particular focus on the *value stability* of the homes sold after and during the planning, building and monitoring stages of the renovation projects studied. The research question addresses the extent to which the different kinds of development processes are considered sustainable in light of the physical and social improvement achieved as a result of the property development projects. It is a multi-method study, where expert interviews, fieldwork and comparable sales analysis (target vs. comparable sales) are combined with a self-organizing map (SOM) analysis of statistical data on house transactions over a 10-year period (2000–2009). The study looks at how market price changes are associated with various urban and neighbourhood improvements in physical, social, cultural and functional terms.

Planning and policies, in addition to purely market-related factors, can affect property prices in a local and spatial sense (see Kauko, 2003, 2009a). This chapter also builds on a prior study by D'Amato and Kauko (2008), which considered the applicability of a 'triangulated price modelling approach' within a planning evaluation context. That study, like the present one, located the price changes in a framework where various effects are isolated in relation to markets, amenities and externalities, and political, policy-based and administrative processes and procedures. What the present study adds is to frame the evaluation within established sustainability criteria and within sub-criteria of general sustainability. The findings suggest a normal market effect, where improved amenities and a well-managed and marketed project –rather than any 'artificial' or 'institutional' effect – keep prices reasonably stable amid falling prices in the wider neighbourhood and city. More broadly, the findings suggest that the evaluation of neighbourhoods under revitalization is about trade-offs between their perceived attractive and unattractive features, which may or may not also be internalized as amenity or nuisance effects of the property prices in the area.

The text is organized as follows. The first section outlines the theoretical context of the chapter, which pertains to two distinct concepts: general urban sustainability and property value stability. The second section follows this conceptualization by outlining the methodology, again focusing on the two different variables under study: the value stability and the general sustainability. The third section then switches the focus to the study area. In the fourth section, the empirical analysis is presented, while the fifth section concludes the chapter.

Sustainability and Value Stability of Urban Regeneration and Property Development

Sustainability

Sustainable development can be defined in many ways, also within an urban housing or real estate context (cf. Støa, 2009; Manzi et al., 2010a, b; Colantonio and Dixon, 2011; Talen, 2011). Real estate is crucial for our understanding of the discourse on urban sustainability: resource conservation is probably the most common element within the sustainability discourse, in general, and the urban real estate discourse, in particular, although more recently the innovativeness and inclusion discourse has added to the complexity of this picture (Manzi et al., 2010a, b). The possibilities offered by real estate due to its centrality to improving the environmental and ecologic state of the built environment are the traditional foci of attention. Furthermore, today the problems of urban decay, the lack of social cohesion and disadvantaged neighbourhoods in many cities are central issues. It is without doubt essential to understand the importance of the interactions between environmental concerns and social and economic domains (see Keivani, 2009).

This discussion has already moved beyond the narrow concept of sustainability that is recognized by NCE and defined simplistically in terms of the balance between costs and benefits. Some authors recognize that certified sustainable buildings are likely to benefit not only the tenant but also the landlord, through reduction in the costs of upkeep, improved public image and security of rent, assuming a market where tenants vote with their feet to find the best balance between costs and benefits, which in turn means less risk for the owner (cf. Lorenz et al., 2008; Eichholz et al., 2009; Fuerst and McAllister, 2011; Warren-Myers, 2011). Kohler and Hassler (2002) promote the role of refurbishment over new build to explain how to maintain the sustainability of the building stock. Dixon (2007) addresses the need to establish buyer-driven markets and niche developments to transition towards a sustainable urban real estate sector (see also Dixon and Earnes, 2013). According to Støa (2009), we should be pragmatic and plan in relation to what we think sustainability is, even if the concept itself is subject to obscurity and debate (see also Pareja Eastaway and Støa, 2004). These are just a few examples that illustrate the richness of the discussion within this realm. By the same token, these examples suggest an apparent need for a further standardization of criteria for urban real estate sustainability.

The company *igloo* in the UK is an example of 'best practice', illustrating sustainability theory from the urban (western/westernized) context – the first self-proclaimed 'sustainable investment and regeneration fund' and a socially responsible investment (SRI) actor (see Colantonio and Dixon, 2011). The success of *igloo* is often attributed to its innovative management processes. It is probably reasonable to assume that, today, economic innovativeness and its subsequent management is a vital precondition for embarking on sustainable development paths. In this context, when times are bad and competitive market processes are accentuated, innovativeness is likely even to become a necessity. Here the discoveries related to the financial crisis and sustainable development must also be regarded as belonging to the same overall paradigm shift in long-term administrative behaviour (via an institutional approach), long-term market behaviour (heterodox economics approach) and human behaviour in general (behavioural approach). We argue that the damage done by the crisis will have lasting consequences for producers, consumers and intermediaries earning a living in the real estate industry. There will ostensibly be fundamental changes in the attitudes, demand and supply structures, markets, regulatory measures and policy initiatives, and most certainly the directions taken by technological progress (e.g. Joss, 2011).

Twelve broad issues related to urban land use and the real estate sustainability agenda can be identified from the literature (the list is not meant to be exhaustive). These are listed here starting from the most localized and continuing to the widest scale:

1. Energy efficiency in buildings (during their lifecycles);
2. Use of renewable energy in buildings (during their lifecycles);
3. Pollution control in buildings (during their lifecycles);
4. Real estate quality;
5. Real estate affordability;
6. Real estate diversity;
7. Optimal density for a block/neighbourhood;
8. Public transport availability (functional issue);
9. Traffic pollution (ecological issue);
10. Social cohesion in the neighbourhood/city;
11. Communicativeness (i.e. collaboration with relevant actors) in local/regional planning (governance transparency);
12. Innovativeness of the region (economic sustainability, including the financial transparency of corporations).

It is important to note that points 11 and 12 are related to process, whereas all other points are related to outcomes. For the sake of manageability, this study deals mostly with points 1 to 10 (points 11–12 are side issues for the evaluation alone). Below, points 4 to 6 are explained separately, as these represent explicit criteria for the evaluation of value stability – a more narrowly defined concept than the general sustainability concept.

Value Stability

An explicit connection between property prices and certain aspects of urban sustainability can be established in a narrow sense in terms of particular criteria. Such criteria include the aforementioned value stability as an analytic economic sustainability concept which assists in the logical identification of situations when prices increase more steeply than quality and/or affordability indicators (see Kauko, 2010a). Obviously, other kinds of economic criteria and social, institutional and physical criteria can also be formulated in relation to the land use preferences of households requiring dwellings, or firms requiring floor space (see Kauko, 2010b). For the evaluation performed in this study, value stability has been selected as an analytical tool complementary to the general sustainability definitions discussed above.

Table 18.1 shows the connections between points 4 to 6 and the dimensions recognized in established conceptualizations of sustainable development. ‘Quality’ and 'affordability' are easily accommodated within mainstream economics. Aspects of quality are often approached using hedonic price models, whereas affordability is often addressed using a welfare economics framework. The innovation here concerns the third class of concept – the 'differentiation' of the product range to allow for innovation in terms of quality or affordability, so as to accommodate an evolutionary perspective.

Table 18.1 Connections between the three sustainability dimensions and three sustainability criteria applied in this study

	Environment	Economy	Social
Quality	X		X
Affordability		X	X
Differentiation	X	X	X

The concept of value stability is defined here as the evaluation of sustainability based on property value. Property value – often approximated as the price of homes sold in an area – is a handy tool, even when used more generally in analyses of the interface between the real estate and planning fields. To illustrate with an example, a hedonic modelling study of Bari, Italy by D'Amato and Kauko (2008) arrived at a proposition as to how to identify rent bubbles from panel house price data series. This offered a relatively novel applicability within a planning evaluation context, the authors suggested. The main points of that study can be summarized as follows:

- Planning evaluation is best performed using the methodology of the ex-post tradition (i.e. evaluating the outcome of an implemented planning policy measure based on data in relation to already realized development). In this case statistical analysis of the housing market outcome data along with a more qualitative approach enables the detection of abnormally high prices or rents (localized 'bubbles'), which subsequently inform the planning of new and existing residential areas.
- A rent bubble was found in Bari using hedonic modelling and five-yearly data cross-sections (1999–2003).
- The implication for planning evaluation is that this approach offers a more rigorous treatment of the interaction between 'market' and 'planning' compared to the ad hoc or ideological practices often applied. In this particular case flexible solutions were developed to diagnose and combat price bubbles in a local rental housing market context.

Overall, D'Amato and Kauko (2008) draw two conclusions: (1) it is feasible to apply an empirical price modelling approach to identify a market/nonmarket related 'bubble' and thereby a real economic and social problem; (2) contextual knowledge and theoretical formalism can be combined when investigating whether the plausible cause of this problem is related to market constraints, amenity increase or speculation (in a policy and political sense). The present project approximates a 'bubble'-free situation relying on the value stability concept.

An Integrative Theoretical Framework

As explained, this study is about the evaluation of urban renovation areas on two distinct criteria:

- Sustainability, in the sense found in the literature;
- Value stability, as a subset of the overall sustainability criterion.

These criteria overlap, as sustainability includes value stability. In other words, value stability is always a part of sustainability (but not the other way around). However, a more formal and standardized reason for this two-criteria approach to sustainability has not been established (at least not in the wider academic literature on urban regeneration). The current research combines various methods to 'triangulate' a result which would be a reasonable approximation of such a judgement:

- Visualized nonlinear analysis of statistical property price data (the SOM);
- Comparability analysis of the same data (targets vs. comparables);
- Field inspection (site visits to the regeneration areas and their comparable counterparts);
- Expert interviews (partly stakeholders).

The cases cover the central sections of districts VIII and IX in Budapest. The former comprises two reasonably homogeneous regeneration areas: Corvin Promenade (at the edge of VIII, adjacent to IX) and Magdolna. The latter is managed as a single coherent project, even though its block composition is quite fragmented.

Methods and Expectations

To determine where higher price areas are situated and to evaluate their sustainability, we required field inspection and in-depth expert interviews, while to evaluate their value stability, we needed more formal and numerical approaches. Unfortunately, the extent to which price data and narratives can be combined remains a controversial issue in the social and economic sciences. This is commonly known as *the positivist vs. non-positivist debate*, with both camps arguing that the two approaches cannot be combined on account of their fundamental differences, and then going on to champion their own approach as the only viable one (the former relying on a model from physics and quantitative data, the latter eschewing statistical deduction and relying on qualitative – often discursive – methodologies). However, from a more pragmatic point of view, this is not the case: triangulation between several methods and datasets will increase the quality of our information for our purposes. This study thus uses the prices and other features of the two urban renovation areas in Budapest. Four specific types of approaches are employed, starting with the more quantitative ones based on how the price variable is related to value stability, and then moving to the qualitative ones, where sustainability is evaluated based on material other than price alone. What is of importance to this study is its pragmatic aim rather than subscribing to a particular theoretical school or interpretative community.

The present study employed a multi-method approach. Data on building location, price and other characteristics were obtained from the Hungarian statistical office (*KSH*) from 2008 and 2009, and on prices and sales volumes aggregated at street-level (from the *KSH* CD-ROM *Ingatlanadattar* for the 10-year period 2000–2009), and combined with interview and site-visit material about districts VIII and IX. (Henceforth: VIII refers to the central section of district VIII and IX refers to the central section of district IX.) This enabled the identification of the extent to which various physical, social, cultural and financial issues affect the micro-level property price development in the selected areas. The many price effects are expected to be more closely related to microeconomic amenity effects or macroeconomic cyclic effects than to planning policy measures, even though this relationship was observed by Kauko (2009a) [see also Kauko (2003); more on these studies below].

Self-Organizing Map

The self-organizing map (SOM) is essentially a measurable similarity-based clustering and classification technique organized as a set of neural network algorithms invented by Teuvo Kohonen (in 1982). Explained briefly, the main principle of the clustering is that the input data is 'won' by predefined output nodes when a response is measured in Euclidean distance in n-dimensional space. In other words, the output nodes receive 'hits' by one or more observations with strong resemblance in terms of input variables. This technique works in three steps. First, the surface is predefined in terms of the number of potential clusters (nodes) and the parameters for adjustment of this map-like surface (feature map). Second, the map is drawn using a dataset of m observations (cases) measured as n variables (map layers). Third, the resulting feature map is examined in terms of the similarities between nodes and the intensities of the relationships between the nodes on a particular map layer. The similarity and intensity of any given node can also be traced across all the map layers if its location is defined and fixed across these layers: this yields a landscape of the dataset for a given cross-section (see Kauko, 2006, 2007 and 2009b, for further description of this methodology).

The SOM can be characterized as a visual approach to quantitative and nonlinear modelling; it enables the detection of clusters of similar cases which are distinct from clusters of different cases, when the similarity is measured by Euclidean distance in n-dimensional space. The visualization of intensities is arguably a more user-friendly alternative to presenting a numerical output; note that the same information, shown here as shades of grey, could also be presented numerically.

The aim is to use the SOM to analyse two datasets: (1) concerning districts VIII and IX in their entireties, using all sales data for 2008 and 2009, including from cases bordering these areas; (2) the *Ingatlanadattar* from the Hungarian statistical office (*KSH*). In the latter case the outcomes for VIII and IX were analysed and compared using a fixed time-window for each cross-section in the period 2000–2009, following Kauko (2009b). The fixed time-windows method is illustrated in Figure 18.1. The key is to identify cases which are inside and outside the urban renovation areas for each district, and to observe from the SOM output surface how the situation changes over the 10-year period.

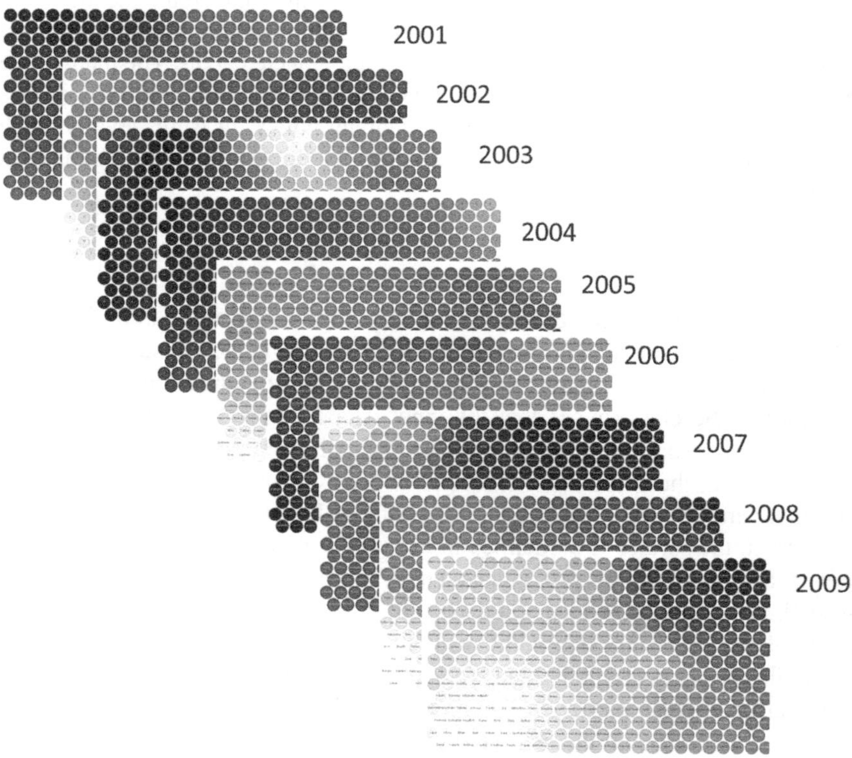

Figure 18.1 The fixed time-windows method

Targets vs. Comparables

Kauko (2003) suggests that conflicts and bottlenecks generate transaction costs, which in turn generate 'artificial' price increasing elements in the property stock of the affected area. On this point, the conceptualization of the process is a counterfactual one: namely, that in the absence of such problematic situations, the partnerships between developers and construction firms proceed smoothly over the long term, which in turn, without such transaction costs, will keep costs reasonable and enable the affordability of high quality for new housing developments, as UK evidence shows (Jones and Evans, 2013, p. 25). Apparently, this is a contextual issue: in some cases artificial elements are found in prices, in others prices are managed to remain on par with general cost and quality developments.

Kauko (2009a) investigated this empirically by comparing prices between supposedly affected and unaffected areas. The findings indicated that, during 1997–2002, such 'artificial' price increases were found in VIII, whereas the price increases in the neighbouring IX were based more on improvement in quality. In any case, price rises faster in VIII than in IX. What will be the corresponding finding in 2008–2009? Is there still an 'artificial' price premium in VIII or did the fact that the development project went smoothly lead to the equalization of prices between VIII and IX? Furthermore, how should the effect of the current financial crisis on property prices be isolated?

If we compare target and comparable sales in four defined areas (using the ideas of Kauko 2009a): within affected area IX (area A), outside affected area IX (B), inside affected area VIII (C) and outside affected area VIII (D), we may be able to again identify a steeper price increase (or a smaller price decrease due to the market downturn) in VIII than in IX. However, if the opposite is the case – the price increase is steeper or the decline smaller – in IX than in VIII, this could indicate economic efficiency, that is to say, an aforementioned 'artificial' element in IX. According to this line of conceptualization, the smoother price trajectory in VIII, if found, will be due to the avoidance of cost-generating conflicts and through having niche market characteristics in terms of specific economic PPP resemblance to IX, where the private investors sold all of their shares to the district when the crisis kicked in.

Obviously, quality changes both increase and decrease prices – which is, in fact, the likeliest explanation for any such price developments. If we compare the two areas, the question is whether environmental or social amenities matter more. Accordingly, price increases can be regarded as a favourable or unfavourable economic development. There could be different possible explanations for the price differences between VIII and XI:

- If prices in VIII > prices in IX, then social (cultural?) factors > environment factors. Or the Corvin Promenade project (VIII) is less affected by the crisis than other projects;

- If prices in IX > prices in VIII, then environmental factors continue to enjoy a scarcity premium. Or, as speculated above, bottlenecks and conflicts have generated 'artificial' price increases in IX, compared to the smoothness of the price development in VIII.

Only cases situated within the boundaries of the central sections of districts VIII and IX are selected as 'target cases' (see Kauko 2009a,b). Nearby situated cases in both districts are 'comparable cases'. For the target vs. comparative sales analysis, only clear cases are selected, not, for example, homes situated along boundary roads.

Interviews

Expert interviews of the following were performed during 2011 and 2012:

- *District IX project*: The general manager of SEM IX;
- *Corvin Promenade project* in district VIII: A senior research fellow, also a former member of the local government of district VIII; the chief Architect of Rév8; a manager from the investor Futureal; one of the two shareholders of the Grund cultural centre;
- *The Magdolna project* in district VIII: the senior research fellow above; an employee of the VÁTI Hungarian Public Non-profit Company for Regional Development and Town Planning.

Inspection

The task was to compare the improvements achieved in various neighbourhood amenities. For district IX, the anticipated benefits were environmental and economic, not social, because of the displacement of social housing tenants which occurred during the project's early stages. For district VIII the benefits were expected to be at least social and cultural.

Interim Conclusions: Towards an Integrative Methodological Framework

The concepts introduced in the introductory section were followed at an operational level by the selection of methods above. The first two subsections in this section introduced the methods based on a technical outcome relating to value stability and Table 18.1. After that, the next two subsections introduced methods that draw on the holistic and multidimensional view of sustainability in a more general sense. Sustainability is considered as a broad concept and value stability as a narrow one, with the former definition here including the latter one. More specifically, sustainability in this study is split into 12 issues on various spatial scales (listed above), whereas value stability only pertains to a subgroup of sustainability comprising three site-specific issues: quality, affordability and diversity.

Background of the Study Area: Urban Regeneration in CEE and Budapest

General Tendencies

When examining historical developments, the post-socialist context offers a potentially fertile ground to analyse the sustainability of urban real estate developments and markets. Scott and Kühn (2012) describe general patterns of (1) *urban regeneration* attempts with relation to post-socialist governance and institutional change; and (2) *urban development* challenges amidst sociospatial change in these circumstances. When evaluating the success of planning and policies, these authors suggest that typical failures are due to obstacles being created by conflicts of interest. However, in the specific case of Budapest, a few qualifications could be added to their analysis in relation to the actual paths and consequences mentioned by Scott and Kühn (cf. Kauko, 2012):

- Apart from a few exceptions, such as Corvin Promenade and the central District IX, PPP was never standard practice in Hungary, even if it may have been so in other CEE countries (p. 1099);
- The statement that the 'lack of national policy frameworks specifically target urban development problems' certainly applies to the previous neo-liberalist regime in Hungary (p. 1103). However, the situation has arguably improved recently. That 'many of the institutional and political conditions necessary for implementation of an "urban social dimension" are only now emerging' is thanks to the political regimes elected into power in both Hungary and in Budapest in 2010 (p. 1104);
- The main reason for the degradation of many inner-city neighbourhoods is not really the transformation but rather the total neglect of these areas during the socialist period (p. 1104).

In the same vein, Keresztély and Scott (2012) list a number of harsh realities in relation to the limits and obstacles of urban regeneration/renovation of inner Budapest. Their evidence points to a number of problems which are characteristic for regeneration strategies in the post-socialist context. Participation processes, for instance, are considered only in their weak form at the most. To implement a comprehensive planning system capable of acting as a counterweight to rapid private investments will not be possible as long as the city government has considerably less say over matters than the district government. Overall, old elites and structures still matter in development in a way which continues to reproduce a strongly socialist-shaped reality (cf. Kulcsar and Domokos, 2005). Nevertheless, hope still remains due to the opportunities, inclusion and shifts in discourse: as Keresztély and Scott note, the massive changes in the urban fabric also include some success stories.

In the context of the interview material collected for the current study, a few issues noted by Keresztély and Scott (2012) deserve further critical scrutiny. To begin with, Keresztély and Scott name political agendas as the key detriment, but the most serious problem here is simply the lack of funds. At a district-specific level it should also be noted that district VIII did not benefit from the Rehabilitation Fund due to the mayor at the time not releasing the promised funding. Lastly, when Keresztély and Scott harshly criticize district VIII renovation, they arguably fail to see the main positive point: that, unlike its district IX counterpart, the housing situation of all displaced households improved considerably. This critique of district VIII renovation could therefore be considered illegitimate. This is, however, not to claim that renovation processes in this district have been complete success stories; in fact, far from it, as will now be discussed.

Urban Renovation Areas of Budapest Districts IX and VIII

Two major renovation projects were conducted in Budapest after the transition from socialism: district IX's (residential project) began in the mid 1980s, while adjacent district VIII's (mixed project) only began in earnest after the turn of the millennium. These two neighbourhoods have undergone rapid social and structural changes during the last decade and can be characterized as particularly dynamic areas in terms of the physical and market processes under study. District VIII is stigmatized in all discourse, although it comprises a great variety of micro-locations and housing stock; district IX in turn, is perceived as a partly gentrified area and undoubtedly the most dynamic neighbourhood in the city.

In the 1980s, the centrally located Ferencváros was designated as an urban renovation area, as the area had become one of the slum neighbourhoods of the city. Three policies were implemented:

1. To keep existing blocks when the buildings were of artistic value, but refurbish the apartments and the facades, and landscape the area within the block.
2. To demolish and build in accordance with decorative 'post-modern' design principles (or leave open, as was the case with a block which was converted into a park).
3. To build completely new buildings and fill in the blocks.

When the first dwellings were completed in 1992, these provided by far the best quality housing in the whole of Pest, and a unique niche market with small-scale, neo-traditional housing blocks with closed gardens for the upper urban middle-class population, found nowhere else in Budapest.

The morphology of district VIII differs from that of district IX in many ways. First, it has a significant proportion of valuable public buildings. Furthermore, unlike the relatively homogeneous and small area of central Ferencváros, the central section of district VIII is composed of several distinguishable areas with different ethnic compositions and styles of building. Moreover, district VIII was originally a village, which was enveloped by the city, unlike district IX. Finally, this district is among the poorest in the city, and its district government struggles to change its unfavourable characteristics.

The central Józsefváros programme is considered a less promising venture than that of the adjacent Ferencváros. Indeed, it represents a great challenge, as the substandard buildings are still extremely dilapidated, and the number of vacant and single-storey houses is high. Curiously, the urban renovation of Budapest actually started in the central section of district VIII in the late 1970s. However, unlike district IX, both economic and aesthetic problems emerged: too many buildings needed to be demolished in relation to the ones retained, and new buildings could not be fitted alongside the old ones. The project did not therefore progress beyond the planning stage. Since its recommencement in 1995, the project has changed management twice: in 1998 and 2002, and has become more complex. Local government has a much wider role in district VIII than in district IX.

On the other hand, the preconditions for a comprehensive demolition-oriented business strategy were good only in the Corvin-Szigony part of district VIII: due to its central location, the prospects of selling the end products were good. This would also decrease the number of poor people in the area, as these would be relocated to other parts of district VIII.

Value Stability in Hungary

The value stability of Budapest real estate was evaluated more widely in Kauko (2012). On the basis of around 20 expert interviews (from the private and public sectors, NGOs and academia) it was found that there are four kinds of value stability related problems in the context of Hungarian property value creation and price setting (see Kauko, 2012):

1. Concerning the building surroundings in the residential sector and elsewhere, there is too little price discounted for the lack of sustainable characteristics (i.e. energy efficiency, renewable energy, and availability of community and infrastructure, as listed above). This is due to scepticism or ignorance in the market actors, which is presumably a consequence of the socialist past.
2. Price-setting by the sellers is unrealistic in the consumer market (falling demand, oversupply). New developments are almost completely seller-driven, which generates tendencies towards value instability in the price-setting. The prevailing 'anything goes' mentality contradicts the establishment of buyer-driven frameworks that both value stability and sustainability would require.
3. Changes in land ownership and land use involve political and lobbying practices – corruption too – which are extremely destructive in the long run. The general strategy (which is myopically a win/win one for both parties

to the transactions) is as follows: a private individual buys agricultural land which a politician, without any physical changes having been made, then redefines as land suitable for development – making it far more valuable. It is then sold for many times its original price with the seller returning a part of the profit to the politician. The result of this sequence of events does not in any way promote value stability (and obviously even less so sustainability).

4. A problem arises from the mismatch between the prices paid for land during the market peak and the prices expected from the sale or lease of the completed developments. In the spring of 2011, 90 per cent of office developments were on hold due to the crisis and many of them will not be continued until 2015–2016. If interest costs need to be paid on the financing of a site over that period, reselling it without considerable economic losses becomes difficult.

The problem types listed above can in principle be seen as variations on the same theme, of price increases having both favourable and harmful consequences on value stability. The first problem, the lack of price discount for a projects unsustainable attributes, could be solved by using a validated pricing model (e.g. an estimated hedonic price equation) which penalizes such features. The second problem, the seller-driven agenda and overpricing of new developments regardless of demand considerations, is in turn more difficult to solve, as this would require the comprehensive upgrading of the existing building stock for urban and suburban areas alike towards the quality levels of new buildings. As for the third problem, detrimental politics and lobbying practices, this would require anticorruption legislation. There is probably no solution to the fourth problem. As the banks are hardly likely to relent, developers who own land (and indirectly also those who already have acquired ownership of the new flats or offices) have the option of immediate panic sale or to wait for the market situation to improve – with both cases resulting in economic losses for years to come. The implication of points 2 and 4 is that, in order to cut costs, the projects might have to limit the luxury attributes they had planned to include, which in itself is not a bad thing from a sustainability point of view. The isolation of the wealthy widens their social distance from poorer groups, which, in the recent literature on social sustainability is characterized unequivocally as a negative development in that it counteracts 'sense of community' agendas (see Manzi et al., 2010a, 2010b; Jones and Evans, 2013).

From this discussion we observe that the first two problems above concern quality in relation to value stability, whereas the latter two problems pertain to the affordability side of value stability.

Findings

The SOM

The district IX project was considered relatively successful in Kauko (2009b), as the quality of the building stock and the neighbourhood in general was improved, whereas its district VIII counterpart was in dead-lock. When the targeted areas are analysed with the SOM for each cross-section during an observed six-year period from 1997 to 2002, two types of dynamics can be identified: the housing market dynamics in the selected 'slice' of the urban area has differentiated micro-locations in terms of actual changes in the housing quality and urban structure in district IX, and in terms of expected house price escalations unrelated to the fundamentals in district VIII. Accordingly, the identified positive change in price levels was of two different natures, depending on whether the change was tangible (district IX) or intangible (district VIII). However, around 10 years have passed since the time of writing, during which the physical changes have been breath-taking.

The new SOM results are shown in figures 18.2 and 18.3. Higher prices are shown as a lighter shaded area on the SOM output in these figures. The analysis using the SOM was performed using two datasets. First, the analysis of the entirety of districts VIII and IX is based on individual data and 10 variables for the years 2008–2009. In year 2008 the most expensive cases were located within the target area of district VIII (C), whereas the following year they fell within district IX, both inside (A) and outside (B; RX indicates the River Danube waterfront) the target area. The interesting findings for 2008 can be listed as follows (top of Figure 18.2):

- The most expensive categories are inside the VIII renovation area;
- Most of these cases are small, old and of low quality and comfort level;
- No relationship between quality and other variables can be ascertained.

As for the corresponding results for 2009 (lower part of Figure 18.2), the interesting findings are that:

- The most expensive categories are now in district IX;
- Most of these cases are old and of low quality and comfort level, situated on a low floor;
- No relationship between quality and other variables can be ascertained.

A second analysis was conducted for each cross-section in 2000–2009, using data aggregated at street and district level (in total for the 10-year period: N = 1658), and six variables in a similar manner to, but not exactly replicating, the analysis carried out in Kauko (2009b). The images in Figure 18.3 show the total

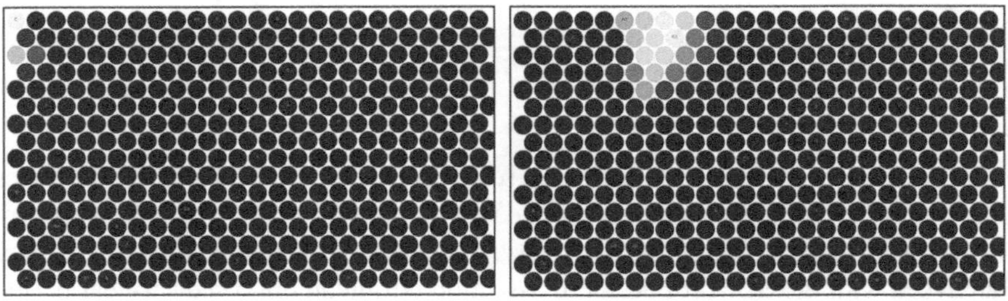

Figure 18.2 Ten variables, individual data (2008, N = 1449; 2009, N = 1088)

price layer for three of the 10 runs. Again, the highest priced areas vary from year to year. For the three years selected, these cases are situated within the district VIII renovation area (see images in Figure 18.3). We can tell this from the identification labels, which are street names. For the three years the situation was as follows:

- In 2002 the highest prices were outside and inside the VIII urban renovation area and inside the IX urban renovation area;
- In 2005 the highest prices were outside and inside the IX urban renovation area and inside the VIII urban renovation area;
- In 2009 the highest prices were inside both the IX and VIII urban renovation areas (in fact, Nagytemplom Street is adjacent to or part of Corvin Promenade).

The interim conclusions based on the SOM runs results with the two datasets (noted above) are as follows:

- Both district VIII and IX renovation areas have expensive flats in old inner-city blocks; this is principally due to their central location rather than their quality (i.e. maintenance and/or comfort and utility);
- A minority of the blocks in the district VIII renovation area are housing estates: these flats are cheap, small and of various qualities;
- For each year in 2000–2009, the most expensive cases can be found in all four 'zones' – even within the district VIII renovation area:
  - those outside are mostly new (the Hungarian type of gated community, 'residential parks');
  - those inside are mostly refurbished – good quality too.

We cannot thus maintain the previous study's conclusion that house prices in the district IX renovation area were categorically found to be higher but more stable (in relation to quality) than in its district VIII counterpart.

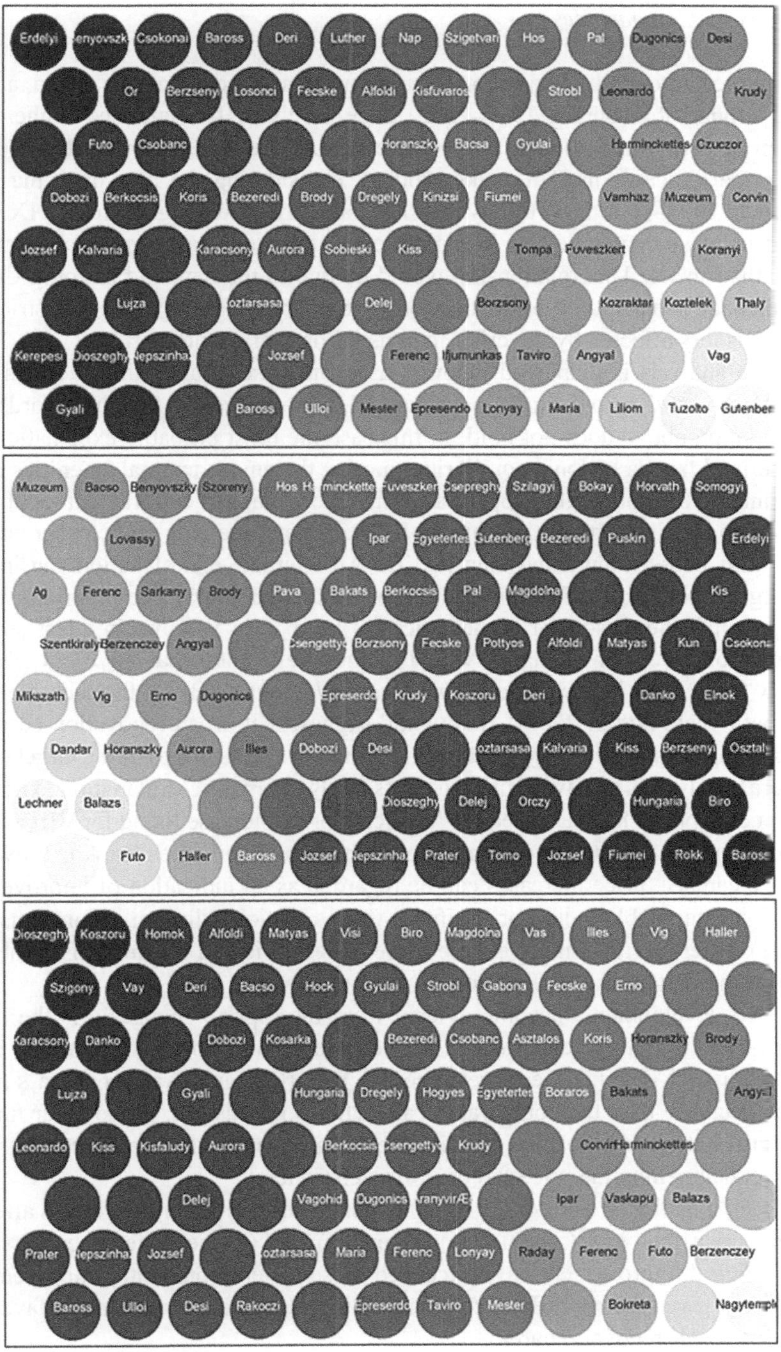

Figure 18.3 Six variables, aggregate data (2002, 2005 and 2009)

Targets vs. Comparables

Using the same method based on a target group and a comparable group, a prior study found that during 1997–2002, price levels in IX remained higher than the price levels in VIII, but at the same time, price increases in VIII exceeded the price increases in IX (Kauko, 2009a). Considering the new data, the price levels in 2008–2009 in VIII may then have already grown higher than in IX. That price increase may have also been steeper (or price decreases more moderate) in VIII than in IX. This relationship would be due to the new and ongoing Corvin Promenade project (as it forms the border to district VIII, adjacent to district IX) which is less affected by the crisis than its IX project counterpart.

The analysis is displayed below. Based on these figures, the prices in VIII (C, D) are still below those in IX (A, B) for both years. In other words, the speculation above does not hold. In this case the most probable explanation can be derived from a hedonic/equilibrium model: the environmental amenities of IX continue to be more highly appreciated than the social and cultural factors of VIII. It is particularly worth noting that, in IX the streetscape is in fact cleaner and the buildings better maintained than in VIII. An alternative explanation can in turn be argued based on a disequilibrium model: an 'artificial' price premium would suggest that the reason for higher prices in IX than in VIII lies in a less smooth renovation process in IX than in VIII. The figures also show that *the prices of target cases* (A, C) *are below those of the comparable cases* (B, D) for each area and for both years (2008, 2009). This would suggest that urban renovation based on improving quality has not yet substantially increased the prices of the target cases. Alternatively, it is also worth noting a less likely alternative explanation exists: the target case prices have avoided the 'artificial' price increase found for VIII and the predominantly private developments elsewhere in the city (see Kauko, 2009a, b). As we can see, price increases can be regarded as an indication of improvement (equilibrium: stable value) or 'artificial value creation' (disequilibrium: unstable value), where improvement can be more or less oriented towards sustainability, depending on the nature of the improvement.

Two developments become evident from these price changes: (1) the target cases of IX (A) have experienced a steep fall relative to the comparable cases in the district (B); (2) the target cases of VIII (C) have experienced a smaller price decrease than the comparable cases (D) in the district. Is the latter finding a manifestation of a positive Corvin effect? In other words, is a successful large project dampening the price fall occurring elsewhere?

If so, we offer two possible explanations: (1) the Corvin project is less affected by the crisis than its IX counterpart (plausible); and (2) the preferences for social and cultural improvement have exceeded those for environmental improvements (unlikely, given the big difference in the environmental improvements in favour of the district IX renovation area).

Finally, we should note here that of the four areas, a price increase was only observed in XI's comparable area (B) due to its luxury and composition (i.e.

residential parks), in line with average new housing developments in the modern market. On the other hand, this price increase could also be due to a knock-on/spill-over effect from the target area or the nearby Corvin Promenade project. These are the reasons areas B and C are highlighted in Table 18.2: the former includes luxury cases, some of which are situated on the River Danube waterfront and the latter includes the Corvin project, with a buffer effect against market downturn. Given the data at our disposal, a method to separate these two effects remains elusive – hedonic regression-based or a more qualitative one.

Table 18.2 Prices and price changes in targeted and comparable cases

2008 means			**N**	**Price changes 2008–2009 as a percentage compared to 2008**	
	A	15,053,999	165	A	-9.55
	B	15,097,130	157	B	18.15
	C	9,749,967	479	C	-8.48
	D	12,539,513	71	D	-22.48
2009 means			**N**		
	A	13,616,851	46		
	B	17,837,281	48		
	C	8,922,699	73		
	D	9,720,588	17		

Interviews

The main findings are summarised in Table 18.3. The rest are described in the neighbourhood change narratives below.

Mid District IX renovation

SEM IX is a public company owned by the municipality and formed in 1991. At that time, it was jointly owned by a French developer, *OTB* Bank and a developer, *Quadrat*. In 2005 the ratio between public and private shares in the project was 1:5, in 2007 it was 1:8, while in 2010 it was 1:10 (thus for the so-far completed project: HUF 8 billion public investment vs. HUF 80 billion private investments over the 20 years). However, due to the crisis, in 2010 the private investors began pulling out and by now *SEM IX* has acquired complete ownership. As a consequence, some of the projects are on hold. The most visible examples of this include the dormitory in Lenhossak utca, planned for 900 students from the nearby University of Medicine, and the ruined office building on Tüzoltó utca owned by a chemical company.

Table 18.3 Comparison of the urban regeneration areas

The projects	Who are the investors?	What sustainability dimensions were considered?	How can these be achieved?	Why were they incorporated into the business strategy?	Was there a land/property value increase or is a land/property value increase anticipated?	What was the fit with the existing land use regulations?	What was the fit with the rest of the neighbourhood?
District IX urban regeneration project	SEM IX; formerly also including private investors.	Environmental and economic. The social sustainability is less clear.	An expectation of interaction between the neighbourhood and its inhabitants. The environment is bound to become cleaner with fewer beggars and homeless people. Green areas will be developed.	They were a financially sound investment as the plots adjacent to green areas could be sold for much higher prices. This is about economic sustainability.	Yes, in 20 years the value has increased in relation to permitted residential area (this is the standard measure here: land price correlates with how many floors can be built on a plot).	Due to the local government-initiated nature of the project, the fit was good. Changes only had to be made to higher density and residential space.	The idea was to fit new buildings and other landscape elements with the existing ones. New buildings are integrated well with the rest of the neighbourhood.
District VIII Corvin Promenade project (Corvin and Szigony neighbourhoods)	District VIII PPP represented by Rév8 (project management organization funded by District VIII) first sold the land to a consortium of several developers; later Futureal bought all of it.	Primarily social, but with economic elements too.	Large-scale redevelopment of homogeneous blocks with social focus. Adaptability of offices and smaller apartments is possible.	The only viable strategy to stop the ghettoization, get rid of the part of the portfolio that was difficult to manage due to old pipes etc. substandard technical quality of the stock, improve the reputation and increase the house prices to a level where selling would be worth considering.	Yes, the Corvin-Szigony area has increased nearby land prices with 20 per cent. The prices set for new flats = market prices of old flats + 30 per cent.	Due to the local government initiated nature of the project the fit with the existing land use regulations was good.	Mixed results.

District VIII Magdolna project	Rév8, with EU funds (unrelated to the Corvin PPPs of the late 1990s and early 2000s, other than that the profits from that project were used).	Social and environmental.	Renovation so that residents can remain in their flats. Safety and educational aspects were targeted: landscaping, technical improvement, a community centre, variation in the courtyards and buildings.	Social upgrading of the neighbourhood instead of direct 'business' strategy.	Yes, rent and value increases. However, a considerable share (40–50 per cent) of the dwellings comprises social housing.	Due to the local government-initiated nature of the project, the fit was good.	The environment of the Magdolna Quarters did/does not change much.

A total of 6,860 new condominiums were built by 54 developers and 936 condominiums were refurbished (by summer 2011; on top of this, 1,107 condominiums had already been refurbished before the project began). The new or refurbished flats were bought by young people with middle incomes, either small families or the parents of university students. The flats are small and priced at around HUF 300,000–350,000 per square metre, which is average for Budapest. For example, a 45 m^2 flat can be bought for HUF 15 million.[1] The reality is that there is not much turnover in the market for old flats. A large part of the housing stock is also owned by the municipality. The most derelict stock is retained as social housing (with rents of only HUF 2000–4000 per month on Gát utca). Some of the stock has, however, been improved by the municipality (5000 partly refurbished flats) and subsequently offered as upgraded social housing for the same pool of tenants at a higher rent (HUF 20,000 per month). The rest of the refurbished flats were put up for sale.

The first areas to be completed had 24-hour surveillance, while today the situation varies in this respect: one park is closed while another is open, for instance. A unique characteristic of this neighbourhood is that it does not have residential parks or luxury residences, and that the municipality wants to maintain all pedestrian access.

A brochure about the renovation project emphasizes the increase in green space, starting from almost none, and progressing to a point where there are trees, bushes and lawns everywhere. A park has even been developed out of a brownfield site.[2] This is anticipated to increase the land and house prices. As can be seen in Table 18.4, the price of land (approximated through the floor-space efficiency) increased almost fivefold in seven years.

Table 18.4 The development of floor-space efficiency in the district IX renovation area

Year	Residential floor-space efficiency per m^2
1993	6,767
2001	16,796
2010	33,000
2011	30,000

1 At the time of the interview in 2011; in 2012 these prices declined somewhat (from HUF 15 to 12 million).

2 This is known as 'the Leipzig model' after the 'perforated' urban fabric in many cities in the former East German republic where the population declined and the reuse of sites at lower efficiency becomes an acute issue.

We notice here a small decrease between 2010 and 2011: this is due to the financial crisis. Until then, however, the trend was strongly upwards. This is partly due to a general increase in Budapest and the micro-level improvement in green spaces along with the construction of better quality buildings. The green or sustainability features for the location are expected to generate a price premium when the residential market, along with that for dormitories and hotels, picks up again.

An additional issue here concerns the decreasing population. Over the last 20 years the number of registered inhabitants has fallen. However, the density in the area has increased (see tables 18.3 and 18.4), which suggests that many inhabitants are registered elsewhere. Apart from the fact that the number of flats has increased, many new buildings have four or five stories, whereas the majority of the old ones have only one or two. The reason for more efficient building is that the average flat size has increased.

Problems:

- The Roma ghetto on Gát utca constitutes an area of unrefurbished municipal stock which remains social housing. Here, however, EU funding was received for two blocks.
- The main problems concern social sustainability: the supply of flats is small in terms of area and developers cannot be forced to build bigger flats. Residential turnover in the area is therefore rapid which is socially unsustainable, given the criteria for retaining a sense of community. This is often mentioned in this context.
- Another problem concerns economic sustainability: the municipality forces developers to build parking space underground, but in the current market this is impossible to sell. The empty underground parking space is expected to be sold by 2021 at the earliest.

According to the original plans, the project was supposed to end in 2018 once the renovation frontier reached the outer limits of the area:[3] depending on the finances the frontier is expected to move somewhat more slowly, so that the new projected completion date is 2020.

Central district VIII renovation, Corvin Promenade

In 1997 *Rév8* was founded as a development agency, with a more public than private character. This development agency had less power than its Western counterparts, as the ultimate decision-making power remains with the district municipality. *Rév8* consists of an interdisciplinary team (architects, sociologists,

3 Namely, Haller utca – except for the Mihalkovics housing estate, which is private and omitted from the project.

social workers, geographers, etc.). It mediates between local government and investors, and to a lesser extent with the municipality of Budapest.

In 2000–2001 an open tender was announced. It was won, however, by a person who was unable to commence the project. In 2003 four construction companies bought the land, but none of these had experience in managing a development process. In 2004 *Futureal* invested in the project, buying the land (350,000 m^2 above ground and a garage of 500,000 m^2). The project curiously also incorporated the protection of the corner site which included a cultural centre (*Grund*). In 2005 the investor rethought the project's functions: the aim was to provide a mixed area, not just residential. It was a courageous step for *Futureal* – as the largest residential developer in Hungary – to get involved, despite the huge political risk involved. The aim was to steadily improve the image of the area. The local government then contacted *Rév8* to prepare a plan. In 2002 a regulatory plan was drawn up for the area. Up to 10 per cent of the plan was agreed at workshops involving senior local (i.e. district government), Budapest and national architects.

Rév8 then advertised and bought the flats in question 'at going market rates', which quickly raised the market prices of the flats in the area. A consequence of this was a growing demand for empty building sites in other (non-Corvin) areas of Józsefváros.

One third of the total costs were incurred by the investor, the rest by local and state government. A total of 1,100 apartments were demolished in this neighbourhood, 800 of which would have had to be demolished anyway due to their unacceptably poor state. Financing was obtained as indicated in Table 18.5 (2002 figures).

Table 18.5 Financing the district VIII project

Total cost of project: HUF 12 billion
1/3 from private developers: new buildings (doubling of permitted floor space), new promenade, widened streets.
1/3 from local municipality budget.
1/3 as follows: HUF 1.3 billion from the state and HUF 2.7 billion from the Budapest municipality (over many years) to cover sewage and other infrastructure. However, the municipality paid only 15 per cent of the promised HUF 2.7 billion. Eventually, these losses had to be covered by increasing the density.

The problem, however, was that the Budapest city government did not (for some undocumented reason) provide the promised funding. As a consequence, *Futureal* had to take over some of the infrastructure provision which was originally supposed to be provided by the public sector. *Futureal* thus took responsibility for developing the public area. In return, *Futureal* was permitted to build an extra level to the block.

This is in fact the only on-going large development in Budapest at the moment. The size of the site is 22 ha, 8 ha of which is residential land. From a starting point of around 90 per cent social housing in this area, the new flats command a price range of 420,000–500,000 HUF/m² for normal condominiums (arguably, a kind of residential park but lacking security systems). The cheapest flats (30 m²) sell for HUF 14 million. Rental apartments are included, starting from around HUF 60,000 per month plus HUF 20,000 per month service charge.

When compared to the central district IX renovation area, where development was realized piecemeal and on small plots, a big advantage here was that the development was achieved comprehensively in a single block. Furthermore, the project creates *sign value* – an intangible element in value – and a stronger sense of place (see Kauko, 2004).

Problems:

- The increase in land prices has led to land sales for political purposes. Subsequently, a legal challenge was brought against the local government sales of plots at the original prices to the original consortium of developers, who later sold them on at market prices despite price control agreements (i.e. the typical problem about corruption and insider dealing already discussed).
- To compensate for the financial losses the area had to be built more densely.[4] While the architecture can be criticized, there was really no other option from an economic perspective: district VIII was a loss-making project under the urban renovation strategy due to bad relations with the then mayor of Budapest.
- The closed courtyards are considered socially dubious as they isolate the residents from their surroundings (contrast the positive experiences of open courtyards in the district IX renovation area).

As for the sustainability aspects, they were environmental: some of the new buildings have geothermal heating systems. Moreover, the future market could have an element of healthy innovativeness associated with a price premium The Corvin Promenade brochure is quite optimistic, mentioning geothermal water and heating, as well as reinvestments in parks, schools, cultural buildings, heritage building renovation and even social/affordable housing. This is, however, the self-proclaimed view of the developer-investor. We must speculate as to the reality – or at least a more balanced view. How much Corvin really enhances sustainability is unclear. According to an interviewed expert, geothermal heating is insignificant since such heating has not been a success in inner-city Budapest. It is solely a

4 Here it should however be noted that, according to one expert involved in the planning stages, this project would never have been profitable on strictly business grounds.

marketing gimmick. However, the provision of social housing and the cultural aspect and community agenda is a plus, according to this interviewee.

A total of 1,100 families were displaced (a third of these Roma). They were free to remain in an apartment in the district or move elsewhere. Each family was given several options, one of which was always a home in district VIII. Families with their own financial resources could also get loans based on the market value of their old apartments. The important thing here is that everybody who wanted to could move into another apartment within the district. This was mostly not new housing, but bought from existing stock. The last 200 apartments were the exception: these were new-build (see inspection below).

The poorest of the inhabitants got new apartments, whereas money rather than replacement housing was offered only to those who could obtain normal bank loans. This is in sharp contrast to the corresponding strategy in the district IX renovation, where money was offered to poor people, which according to the interviewee was wrong from a social perspective. Furthermore, increasing the average flat size from 27 m^2 to 38 m^2 and improving comfort levels (notably, giving all flats their own toilets) also served social goals.

Central District VIII Renovation, the Magdolna Project

Once the Corvin project had generated profits, the aim was to socially rehabilitate the Magdolna and Orczy neighbourhoods. The ghettoization had generally slowed in central district VIII – largely because the substantial proportion of lower middle-class inhabitants who remained there.

This project was therefore started as an offshoot of the Corvin project in the sense that the small-scale renovation was financed from the funds generated by land sales revenue from the district. This is known as 'complex integrated renovation' without demolition. In some cases demolition would have been a better strategy though.

EU money cannot be used for the renovation of flats: therefore very few flats were renovated. This is a problem with EU regulations: there is no funding available for the systematic refurbishment of flats. In cases where residents have to leave temporarily, 90 per cent move back.

Problems:

- This plan was a pilot. There are no similar cases in Budapest in other rundown areas.
- While this can be considered as locally sustainable, there is a risk that the area will deteriorate again.
- The social improvement was a fact, but the project had already ended in 2011. Since then, those involved have remained hopeful, however.

Field Inspection

In November 2013 the district IX renovation area looks almost complete. Modest new buildings and renovated old ones can be observed, with pedestrian passages, trees and bushes, cafes and so forth. By and large, this creates a pleasant atmosphere (although drivers in search of parking spaces might disagree). Simultaneously, the district VIII renovation area remains under construction. Here, the focus appears to be more on parking and car accessibility than in its IX counterpart. Moreover, of the two areas, the environmental amenities and abatement of negative externalities is far more extensive in the IX area, with the unfortunate exception of the previously mentioned Gát utca, with its dirty streets and pavements, rundown building stock, social housing and high concentration of socioeconomically deprived Roma.

A year earlier, site visits were conducted at several specific locations in the context of a wider study of Budapest (see Kauko, 2013). Some of the locations visited were also situated within the present study area and exemplify some of the sustainability arguments brought up in this study. An example of two different locations within B (comparable cases from district IX, distance around 200 m) provided given below.

Lechner Ödön fasor This is a special case because of its explicit function as a new high-rise type of gated community overlooking the river. It can be characterized as 'bling' mixed development blocks, or a global standard development for the upper-tier market – the kind with a concierge (a kind of 'snobbier' janitor service). They offer luxury without real innovation: glass palaces called 'gardens' offering various services (including wine-cellar and 'boy' services) for the residents, and others for the public. While some parts are still under construction, the first flats were completed around 2005, and sold to wealthy urbanites as well as investors, many of whom are foreign (Irish, for example). Many of the flats remain empty, which suggests that this is not a popular location for the buyer segment it is intended for. Despite the river view this does not appear to be a pleasant location to live in, especially as the lower level flats are exposed to noise from either the suburban railway or a motorway immediately outside the block. Ostensibly, there are no sustainability elements here, even the nearby Cultural Palace was debated and it is part of the same waterfront complex. This development was also known for suspicious land dealings.

Drégely utca This is an altogether anomalous case of relatively high sales prices. It is a ghetto for poor people, mainly Roma. The buildings are mainly old and derelict, with some decent ones, including a large block with long balconies. There is at least one large empty site here. The only conceivable explanation for the higher-than-average prices and turnover here is therefore that in 2007, a speculator acting on behalf of a developer or investor bought all the flats in the block, which always increases the prices, as this kind of manoeuvre is bound to generate an artificial price premium (see Kauko, 2009a). This chain of events is unsustainable by definition, as already discussed.

District VIII – A rare instance of social housing construction in Budapest Near the end of the Corvin area relocation process discussed above, the prices of the existing flats rose so much that it became cheaper to build five new buildings. Building these new houses was also important for local revitalization. They were built at different locations in the district. Field inspection found that these five sites lie within 1.5 km of the Corvin site. All the buildings have four to five stories, with individual designs, and were simple and functional. They do not substantially differ from most moderate new private developments nearby, although they do of course contrast markedly with the rundown neighbourhood overall. The assumption is that well-designed social housing can indeed contribute to a socially and economically sustainable environment.

Summary of the Evaluation

This section empirically followed the methodology outlined. The evaluation of the urban renovation areas in districts IX and VIII in relation to criteria 1–10 is presented in Table 18.6. The criteria are provided in the first column and the sustainability of each renovation area is evaluated with respect to each criterion in the other three columns. A ticked box means an item is sustainable (or in the case of the real estate quality, affordability and diversity criteria, only that value stability is found), ticked with parentheses means it is possibly sustainable, not ticked means unsustainable – in all cases with respect to the specific criterion in question.

Table 18.6 Evaluation of the sustainability of the three renovation areas

Criteria	IX	VIII/Corvin	VIII/Magdolna
Energy efficiency		(X)	
Renewable energy use		(X)	
Pollution control		(X)	
Real estate quality	X	X	
Real estate affordability	X	X	
Real estate diversity	X		
Optimal density for a block/neighbourhood	X		
Public transportation availability	X	X	X
Traffic pollution			
Social cohesion in the neighbourhood		X	X

Conclusions

The evidence collected in the previous section enables us to draw conclusions about the different findings: first, with price in relation to value stability; and with other aspects related to sustainability. The analyses based on the price variable (the SOM and target vs. comparable cases) showed that both the district IX area and the Corvin part of district VIII show signs of value stability with regard to quality and affordability. However, only the district IX case is also found to exhibit reasonable diversity in the housing stock. The analyses based on more qualitative material (expert interviews and site visits), in turn, indicated that only the buildings in the Corvin area possess possible ecological sustainability characteristics, that the density is optimal only in the district IX area, and that social cohesion is achieved only in the VIII renovation cases. While these findings indicate some sort of segmentation between the three projects, unfortunately more detailed conclusions proved impossible with the current research design.

Kauko (2009a, b) serves as a point of reference for the new empirical material documented in this chapter. In the prior study it was found that prices in district VIII were below those of district IX, it was also found that the rate of increase was faster in district VIII than in district IX. From a value stability perspective, the renovation in district IX was deemed more favourable than its counterpart in district VIII. We must observe, however, that the social element was not present in the prior evaluation. The findings from the present study however suggest that the displacement of some of the original residents in district IX is a rather unsustainable outcome from a social point of view.

When considering the present study, the SOM results comprise a 'mixed bag' insofar as now the urban renovation of district VIII also contains expensive cases, and the location of the expensive cases changes from one year to another. Accordingly, putting the two trends together, it is plausible that the increased prices in VIII are due to refurbishment or redevelopment. The more qualitative part of the study subsequently does find evidence of a more normal market effect (and not an effect based on value creation), as improved amenities and a well-managed and marketed project are keeping prices reasonably stable, amid falling prices elsewhere. Corvin Promenade in Budapest district VIII has provided a buffer against the market downturn, insofar as that prices in the central district VIII urban regeneration area have not fallen as much as would be expected compared to the trend in corresponding areas elsewhere – especially in the adjacent central IX renovation area. Overall, based on this evidence, it seems as though the district VIII case does include sustainable elements after all, or at least has experienced a stable value development during its redevelopment and rehabilitation. The clearest conclusion of this research is that Corvin Promenade is a rare case in Hungary: it has a niche market character despite its huge size. It can, after all, be considered a reasonably successful redevelopment idea from both a business and a social perspective.

This study's findings thus suggest a more mixed and equal evaluation than the previous study, where the renovation of district IX was found to be more sustainable overall, as well as to have a more stable value development than its district VIII counterpart. Overall, based on our evidence, it appears as if the district VIII case does includes elements that contribute to sustainability after all, and that it is quickly catching up with district IX. The point here is that the two areas are different, but they are nonetheless not evaluated that differently. Whereas the previous studies suggested, rather unequivocally, that the district IX regeneration area is more sustainability oriented – both as a whole and in the narrower sense of value stability – than its district VIII counterpart, the current study is more cautious in its overall evaluation. This is because, since the end of the previous study period (1997–2002), socially unsustainable development has been found in the former area, while the latter area has become more attractive in social and economic terms.

References

Colantonio, A. and Dixon, T. 2011. *Urban Regeneration and Social Sustainability: Best Practice from European Cities*. RICS Research, Wiley-Blackwell, chs 1–4 and 11.

D'Amato, M. and Kauko, T. 2008. 'Hedonic indexes on residential rents in the real estate market of Bari: A contribution to urban planning'. In Miller, D., Khakee, A., Hull, A. and Woltjer, J. (eds) 2008. *New Principles in Planning Evaluation.* Aldershot: Ashgate, pp. 143–65.

Dixon, T. 2007. The Property Development Industry and Sustainable Urban Brownfield Regeneration in England: An Analysis of Case Studies in Thames Gateway and Greater Manchester. *Urban Studies*, 44(12), pp. 2379–400.

Dixon, T. and Earnes, M. 2013. Scaling up: The challenges of urban retrofit. *Building, Research & Information*, 41(5), pp. 499–503.

Eichholz, P., Kok N. and Quigley J. 2009. *Why Do Companies Rent Green? Real Property and Corporate Social Responsibility*. RICS Research, London.

Foxon, T.J., Köhler, J., Michie, J. and Oughton, Ch. 2012. Towards a new complexity economics for sustainability. *Cambridge Journal of Economics*, advance access, November 23, pp. 1–22.

Fuerst, F. and McAllister, P. 2011. Green Noise or Green Value? Measuring the Effects of Environmental Certification on Office Values. *Real Estate Economics*, 39(1), pp. 45–69.

Jones, Ph. and Evans, J. 2013. *Urban Regeneration in the UK*. London: SAGE.

Joss, S. 2011. Eco-cities: The mainstreaming of urban sustainability – key characteristics and driving factors. *International Journal of Sustainable Development Planning*, 6(3), p. 268.

Kauko, T. 2003. Planning processes, development potential and house prices: Contesting positive and normative argumentation. *Housing, Theory and Society*, 20(3), pp. 113–26.

Kauko, T. 2004. Sign value, topophilia and house prices, *Environment and Planning A*, 36(5), pp. 859–78.

Kauko, T. 2006. Urban housing patterns in a tide of change: Spatial structure and residential property values in Budapest in a comparative perspective. *DUP Science Publication, Delft*.

Kauko, T. 2007. A heterodox economic analysis of the housing market structure in Budapest using neural network classification. *Journal of Real Estate Literature*, 15(1), pp. 85–124.

Kauko, T. 2009a. Policy impact and house price development at the neighbourhood-level – a comparison of four urban regeneration areas using the concept of 'artificial' value creation. *European Planning Studies*, 17(1), pp. 85–107.

Kauko, T. 2009b. The housing market dynamics of two Budapest neighbourhoods. *Housing Studies*, 24(5), pp. 587–610.

Kauko, T. 2010a. Value stability in local real estate markets. *International Journal of Strategic Property Management*, 14, pp. 191–9.

Kauko, T. 2010b. Sustainable urban property development and neighbourhood dynamics, 'mini-review'. *The Open Urban Studies Journal*, 3, pp. 103–111 [online]. http://www.benthamscience.com/open/tousj/openaccess2.htm.

Kauko, T. 2012. End in sight? On the (un)sustainability of property development in the Budapest region. *International Journal of Strategic Property Management*, 16(1), pp. 37–55.

Kauko, T. 2013. 'A demonstration of sustainability arguments using house price data', ERES Conference, 3–6 July 2013, Vienna, Austria [online]. http://library.eres.org/eres2013/paperupload/146.pdf [accessed 18 November 2013].

Keresztély, K. and Scott, J.W. 2012. Urban Regeneration in the Post-Socialist Context: Budapest and the Search for a Social Dimension. *European Planning Studies*, 20(7), pp. 1111–34.

Keivani, R. 2009. A review of the main challenges to urban sustainability. *International Journal of Urban Sustainable Development*, 1(1–2), pp. 5–16.

Kohler, N. and Hassler, U. 2002. The building stock as a research object. *Building Research & Information*, 30(4), pp. 226–36.

Kulcsar, L.J. and Domokos, T. 2005. The Post-Socialist Growth Machine: The Case of Hungary. *International Journal of Urban and Regional Research*, 29(3), pp. 550–63.

Lorenz, D., D'Amato, M., Des Rosiers, F., et al. 2008. Sustainable Property Investment & Management – Key Issues & Major Challenges, *RICS* [online]. http://www.rics.org [accessed 22 February 2011].

LUDEN. 2012. Towards a new paradigm for Local Urban Development [online]. http://qeceran.cluster003.ovh.net/documents/Towards-a-new-paradigm-for-Local-Urban-Development.pdf [accessed 24 November 2012].

Manzi, T., Lucas, K., Lloyd-Jones, T. and Allen, J. (eds) 2010a. *Social Sustainability in Urban Areas. Communities, Connectivity and the Urban Fabric*. London and Washington, DC: Earthscan.

Manzi, T., Lucas, K., Lloyd-Jones, T. and Allen, J. 2010b. 'Understanding Social Sustainability: Key Concepts and Developments in Theory and Practice'. In Manzi, T., Lucas, K., Lloyd-Jones, T. and Allen, J. (eds) 2010. *Social Sustainability in Urban Areas*. London and Washington, DC: Earthscan, pp. 1–28.

Pareja Eastaway, M. and Støa, E. 2004. Dimensions of housing and urban sustainability. *Journal of Housing and the Built Environment*, 19, pp. 1–5.

Scott, J.W. and Kühn, M. 2012. Urban Change and Urban Development Strategies in Central East Europe: A Selective Assessment of Events since 1989. *European Planning Studies*, 20(7), pp. 1093–109.

Støa, E. 2009. 'Housing in the sustainable city – issues for an integrated approach'. In Holt-Jensen, A. and Pollock, E. (eds) 2009. *Urban Sustainability and Governance*. Hauppauge, USA: Nova Publishers, pp. 31–48.

Talen, E. 2011. Sprawl retrofit: Sustainable urban form in unsustainable places. *Environment and Planning B*, 38, pp. 952–78.

Warren-Myers, G. 2011. Valuing Sustainability: A new challenge for the profession. *Australia and New Zealand Property Journal*, September, 3(3), pp. 157–64.

Chapter 19

Evaluating the Occurrence of Natural Extreme Events on Infrastructure in Italy

Caterina De Lucia, Atif Kubursi and Dino Borri

Introduction

Over the last decade there has been increasing interest in the international community in studying and assessing catastrophic events in the light of recent developments in climate change policies at the global scale. According to the International Disaster Database, our planet has experienced an increase in the frequency of extreme natural disasters and in the economic losses and damages that these events produce, giving us pause to consider the magnitude of the climate change – that is, the climatic response to increasing greenhouse gas emissions and environmental degradation – and its future effects in this century.[1]

A closer look at these statistics shows, however, that most of the damage costs resulting from natural disaster events are related to a territory's intrinsic characteristics. In other words, disasters would usually be defined according to the characteristics of a physical agent or by the physical effects of this agent in terms of damage to property and persons, adopting an interpretative approach and technical and physical-engineering intervention strategies. The problem of analysis of the pre-impact conditions and the reasons for vulnerability of a system to natural disasters is not captured fully. Furthermore, the increasing interest in managing the risks from catastrophic events, in particular through insurance policies, has now shifted from agriculture to other economic sectors such as infrastructure. It would appear that the evaluation and study of extreme climatic events is crucial not only to catalogue the frequency and magnitude of events but also, and in particular, to understand the exposure and sociophysical vulnerability of those territories characterized by high population and infrastructure density.

In this chapter we aim to describe the effects of natural disasters on infrastructure and illustrate the relevant policy implications. To this end, we will consider a case study in the Italian context of the phenomenon of landslides. In particular, the chapter will: (i) describe some issues and considerations of infrastructural exposure to climate change; (ii) briefly review the concept of vulnerability linked to natural disasters; (iii) illustrate the modelling approach of our Italian case study

1 See: http://directory.eoportal.org/res_p1_Naturalhazardsanddisasters.html.

using a logit model and discuss some relevant policy implications; and (iv) make some concluding remarks.

Infrastructural Exposure to Climate Change: Some Issues and Considerations

The economy, population and climate intersect and interact to define the likely impacts on a community's economic and physical infrastructure of climate change and extreme weather events. The magnitudes of these impacts depend on the size of the population and the ratio of this population likely to be exposed to severe climatic events and their aftermath. Equally important in defining the magnitude of the impact is the state of the economy, the state and magnitudes of the physical infrastructure exposed to climatic conditions, and their resilience. A vibrant and high-growth economy is likely to lose more from an adverse climate than a dormant or contracting economy. Enjoying mature, diversified and flexible infrastructure could mean a community risks losing heavily due to the exposure of its infrastructure to climatic conditions. The lines of causation in Figure 19.1 suggest that there are feedback effects on climate change from the economy, population, and the state of the economic and physical infrastructure. These feedback effects are of second order magnitudes but they demand recognition. More important is the extent of a community's exposure to climatic conditions and its resilience, as exemplified in a diversified economy and adaptable infrastructure.

Climate change impacts are not conceived in a vacuum; rather they interact with different assumptions on the economic and physical infrastructure and population growth as depicted in Figure 19.1.

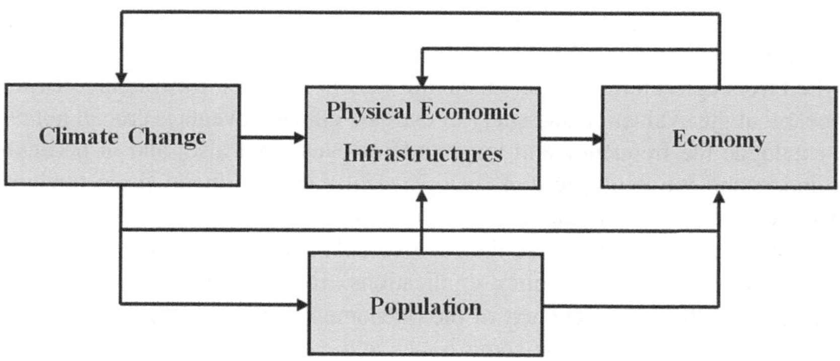

Figure 19.1 Transmission mechanism of climate change impacts on a community's physical and economic infrastructure, population and economy

Source: Economics of Climate Adaptation Working Group (2009).

To help local decision makers address the potential risk posed by climate change, we consider three scenarios for 2035 developed by the Economics of Climate Adaptation Working Group (2009), in which it is possible to model the economic loss expected from different levels of climate change and their impact on the hazards applicable to the cases. Drawing on the available local and global data and numerous interviews with experts and local decision makers, the three (and in some cases more) plausible climate change scenarios for 2035 include:

1. Today's climate: a scenario which assumes no impact of climate change, modelling a continuation of historical weather patterns alongside expected economic and population growth to 2035. We note that other factors unrelated to climate change cannot be kept constant.
2. Moderate change: A scenario built on the average forecasted climate change for the particular hazards in the location under study, drawn from all available studies and expert interviews.
3. High change: A scenario built on the basis of the outer limits of forecasted climate change considered possible in 2035 by existing studies and experts.

In the test cases the 'high change' scenario is developed using the upper end of the conditions described in the majority of IPCC reports and global circulation models published and accepted as likely by the scientific community. Scientific evidence is increasingly being accumulated that suggest that past 'high change' scenarios might be exceeded.

To calculate the economic loss resulting from each scenario, the hazard impact of the scenarios is assessed against the location's total asset values (agricultural production, asset values of residential, commercial and public infrastructure, water quality or quantity) as well as the vulnerability of those assets to the hazard in question (based for example on records of historical weather impacts and associated insurance data).

This process enables the decision maker to arrive at a robust comparison of the probable loss that would occur under each of the three climate change scenarios broken down by asset class or hazard (see Figure 19.2). Different climate scenarios are listed on the X-axis in Figure 19.2. The Y-axis presents a class of assets that are climate sensitive and the likely impact (share of the asset exposed to loss) in the special application depicted in Figure 19.2. All of these figures are adapted from the Economics of Climate Adaptation Working Group (2009) study and are used here for illustration purposes.

Any given location faces three categories of future climate-related loss: (i) those caused to existing assets (infrastructure and residential and non-residential assets) and income levels by a continuation of today's weather patterns; (ii) those brought about by economic growth which exposes greater asset values to climate risk; and finally (iii), any additional loss which may be caused by the exposure of a larger population.

The estimation of expected losses due to different climate scenarios is simple and compelling. The first step in this methodology is to identify the most relevant local climate hazards such as flood, wind, drought or climate zone shift and the sectors and groups most at risk from those hazards in the location under study. The potential loss within these areas is then estimated using three inputs (see Figure 19.3).

1. Hazard: frequency and severity scenarios are developed for most relevant hazards, and a map is generated of the impact of those hazards, for example, on public infrastructure or residential, commercial or agricultural assets. Hazard expresses the proportion of the value assets (see the definition below) expected to be lost on occurrence of the hazardous event. The value assets include a degree of vulnerability (expressed as a percentage of the damaged asset) to the hazard.
2. Value: the value of the assets at risk in the area is quantified in terms of population, private assets, public infrastructure and income value. To arrive at this output the area's population and economic performance are projected to 2035.
3. Vulnerability: the vulnerability of population, assets, infrastructure and incomes to the hazard is determined through the use of 'vulnerability curves' which define, for asset classes such as agriculture, residential and industrial/commercial, the percentage of value damaged by hazards of various severities.

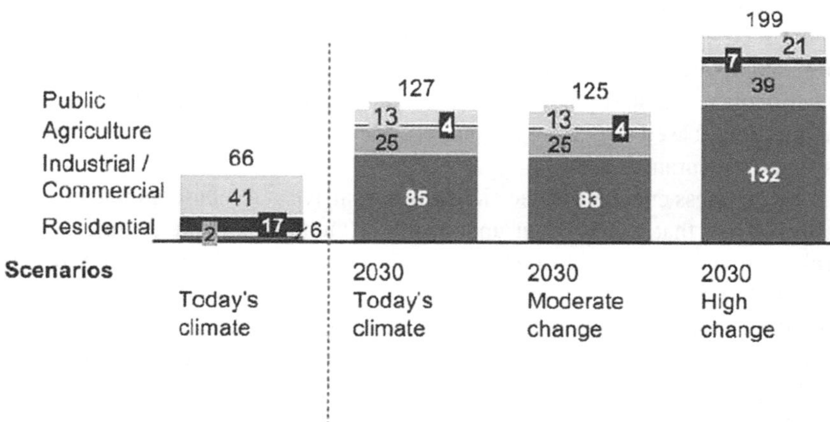

Figure 19.2 Annual expected losses differentiated by major economic sectors or asset types under different climate scenarios (USD m)

Note: This is an example from Georgetown, Guyana.

Source: Economics of Climate Adaptation Working Group (2009), p. 98.

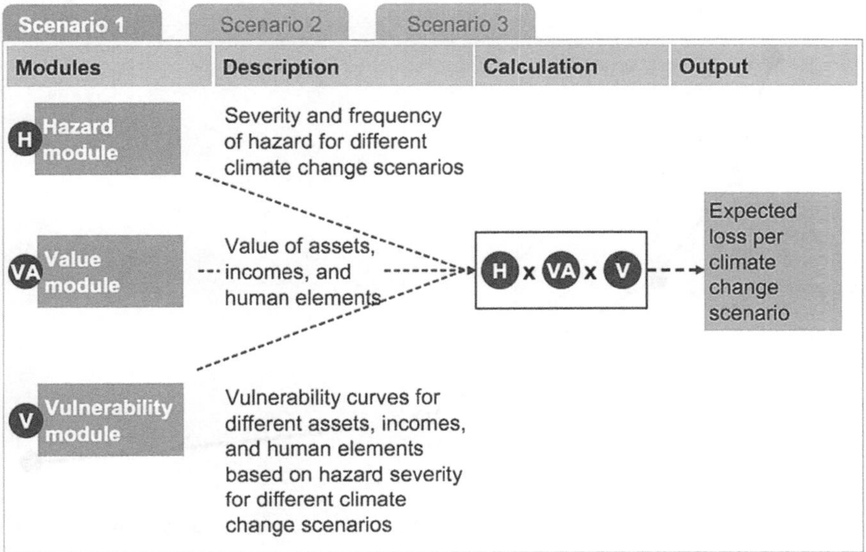

Figure 19.3 The methodological framework for quantifying expected losses for various climate scenarios

Note: Expected loss under each climate change scenario depends on the severity of the hazard, the value of the assets and the vulnerability of assets exposed.

Source: Economics of Climate Adaptation Working Group (2009), p. 128.

The estimation of expected losses based on the probability of occurrence is depicted in Figure 19.4. The area under the curve represents the magnitude of the expected loss at different probability values displayed on the X-axis, whereas the expected values of the losses are read from the Y-axis. Figure 19.4 is used as an illustration of how exceedance curves can be calculated. The magnitudes are not real but illustrative.

Expected loss is a statistic which provides an appreciation of the magnitude of climate hazard impacts. An expected loss figure is a powerful piece of information in that it allows a decision maker to compare accumulated losses across multiple timeframes and to determine how much damage is expected to increase between now and some point in the future, as well as compare the loses from climate hazards to those from other risks (Economics of Climate Adaptation Working Group, 2009).

A simple scheme or approach is presented below to be followed in determining the expected losses from climate change.

First, define the assets at risk by asset type with a special emphasis on public infrastructure. In a majority of cases the major component of the assets at risk is made up of residential, commercial, industrial and public physical assets. Although the economic loss due to damage to physical assets is certainly an important driver

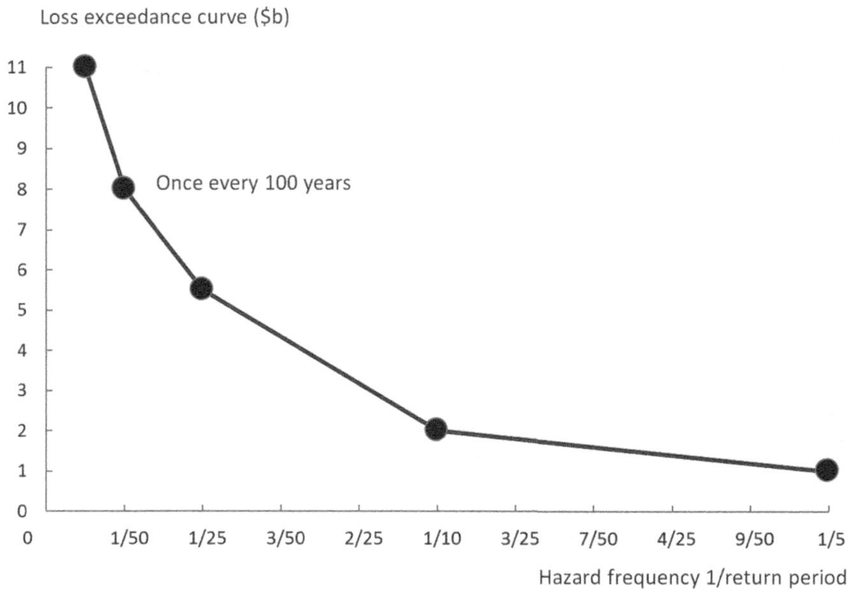

Figure 19.4 Expected losses calculated at different probabilities of occurrence

Source: Economics of Climate Adaptation Working Group (2009), p. 132.

of the total climate hazard, when implementing the methodology at a community scale, a broader range of potential valuable assets should be defined. These could include residential, agricultural, industrial and infrastructural assets.

Second, determine the value and distribution of assets. In many cases data on the value of assets exists and quite often it includes income and revenue streams. The geographic location of assets is a crucial variable: where the assets are located makes a difference but so does their composition and concentration in different areas and their nature, resilience and capacities. This is why the use a Geographic Information Systems (GIS) is of great help for research outputs. Where data is limited, asset values can be estimated from other sources, for example records of recent public investments, historical crop yields and disease mortality rates, etc. The spatial distribution of these assets can then be estimated, for example, using population density as a proxy. In all cases, a variety of sources – and strong local engagement for verification – is required to estimate the value and distribution of assets and infrastructure. A possible feedback effect is embedded in these calculations. A flexible, adaptable and resilient infrastructure (for example, sewers of a particular capacity can minimize the flood damage and other consequences of extreme weather events). When projecting the likely value of assets into the future for the selected test case communities, it makes sense to derive the exogenous GDP growth based on the best available estimates in each location. These top-down estimates do not inherently include climate risks in their models. This is where climate scenarios are useful in the estimation of losses between different scenarios. These losses can be associated with physical damage to infrastructure or assets or other associated social costs.

Third, create vulnerability curves for the value of the assets and infrastructure at risk to events of different severities. The third step marries hazard risk with asset values. The result of this step is a set of vulnerability curves which plot the percentage of value damaged by hazards of different severities (Figure 19.4). A vulnerability curve shows the correlation between event severity and asset loss, where asset loss is presented as a proportion of total asset value (as in Figure 19.4, it could also be presented as a percentage). Event severities are specified by different metrics (for example, height for flood and wind speed for storms) and assessed in earlier steps in the methodology.

Vulnerability in the Context of Natural Disaster: A Brief Review

The vulnerability of territories is one of the issues most debated within the international scientific community. In this section we will briefly review the definition of vulnerability in the context of natural disasters.

Birkmann (2006) recognizes vulnerability's reduction as being the first and perhaps the most important step towards a 'disaster resilient society'. According to this author, vulnerability underlines different aspects which are, on the one hand, linked to events with various risk degrees, and on the other related to the spatial dimension and the nature of territories, including their morphological, geographical and geological compositions. The term vulnerability reveals that its use has been adopted in different contexts and, as a consequence, described according to how it is perceived. We distinguish (i) a particular condition or state of a system before an event triggers a disaster, described in terms of criteria such as susceptibility, limitations, incapacities or deficiencies, that is the inability to resist the impact of the event (resistance) and the inability to cope with an event (coping capacities); (ii) a direct consequence of the exposure to a given hazard; and (iii) the probability or possibility of an outcome in the system when exposed to an external event associated with a hazard, expressed in terms of potential losses such as fatalities or economic losses (Alwang et al., 2001; Brooks, 2003).

In the context of natural disasters, the term vulnerability first appeared in 1972 in a report prepared by the Office of Emergency Preparedness of the Executive Office of the President. The report recognizes vulnerability as the predisposition of an economy (communities and economic sectors) to be affected by a natural disaster (Office of Emergency Preparedness, Executive Office of the President, 1972). While the document recognizes hazards and system vulnerabilities as factors affecting disasters, it does not provide an explicit definition of vulnerability. It contextualizes vulnerability to the geopolitical territory of the United States only.

Chambers (1989) introduces a more systematic definition of vulnerability targeting communities and their livelihoods, identifying two aspect of the definition: (i) external vulnerability, which is the exposure to external shocks and stresses; and (ii) internal vulnerability, which is the incapacity to cope without damaging losses. Cardona (2003) presents a more holistic view and defines vulnerability to natural disasters as the predisposition of an element, a system or a community to be affected by or susceptible to damage. In this definition vulnerability originates as a consequence of three factors: (i) exposure of human settlements to natural or social phenomena due to their particular spatial location in an hazard area; (ii) socioeconomic fragility, meaning the tendency of a particular area to suffer harm due to marginalization, social segregation in human settlements, and poverty; and (iii) lack of resilience, due to the limitations of access and mobility of resources, and incapacity to absorb the effects of a disaster. It can be linked with the economic conditions of underdevelopment and the lack of risk-management strategies.

Watts and Bohle (1993) extend Chamber's definition of vulnerability by considering the term as multidimensional in a social space defined by the political, economic and institutional capabilities of people in specific places and times. The relation between Chamber and Watts and Bohle's models can be interpreted in terms of exposure and coping mechanisms for natural disasters. The exposure side of vulnerability is in turn influenced by (i) population dynamics and capacities to manage the environment; (ii) the inability of people to obtain or manage assets; and (iii) social inequalities. On the other hand, the coping mechanism is affected by: (i) how people interact to each other either by free will or as a result of societal, governmental or economic constraints; (ii) how people mitigate their vulnerability through access to various assets; and (iii) management of resources and assets, the capacity to manage crisis situations and resolve conflicts. As in Chamber's conception, Watts and Bohle recognize the relationship between vulnerability, coping capacities and assets (economic, sociopolitical, infrastructural, ecological and personal). The more assets people control, the less vulnerable they appear because such assets increase their capacities to cope with risks and disasters. Watts and Bohle conclude that the most vulnerable people usually only control social assets. Therefore, the control over assets becomes a relevant issue in case of crisis and conflicts. Alexander (2000) defines vulnerability as 'the potential for casualty, destruction, damage, disruption or other forms of loss with respect to a particular element' Alexander (2000, p. 13) and recognizes the degree of its perception as a factor which increases or reduces the interventions (response) of policymakers.

The *International Strategy for Disaster Reduction* at UN considers the term vulnerability as 'the set of conditions and processes resulting from physical, social, economic, and environmental factors, which increase the susceptibility of a community to the impact of hazards' (United Nations Office for Disaster Risk Reduction, 2004, p. 24). Dilley et al. (2005) 'use the term vulnerability to represent the apparent weakness of the physical and social systems to particular hazards' (Dilley et al., 2005, p. 24). The vulnerability of the physical system refers to the fragility curves for infrastructure and is considered a function of hazard intensity.

On the other hand, the vulnerability of a social system refers to a complex function of social, economic, political and cultural variables, and the estimates obtained generally express economic losses. Finally, due to the increasing frequency of extreme climatic events in recent years, the international scientific community has established an Expert Working Group with the aim of ensuring continuous dialog across communities and territories in vulnerability research.

Case Study: Landslides and the Italian Context

The Italian morphological, lithological, geological and structural compositions make the whole territory vulnerable to landslide phenomena such that the risks from an isolated or coupled event are particularly high. The latest Italian report on landslides (Institute for Environmental Protection and Research, 2008) argues that the occurrence of such events is widespread throughout the territory, causing a large number of deaths across communities. The significant increase in anthropogenic activities, along with the expansion of urban and transport infrastructure, has put pressure on the already unstable areas of the territory. In this framework landslides represent a diffuse phenomenon of social, economic and environmental relevance. It is true, nonetheless, that a landslide, like any other process affected by mountainside erosion, can be regarded as a natural process responsible for producing the sediments necessary for the formation of floodplains. These, in turn, are vital constituents of various terrestrial ecosystems, including those supporting many human activities.

The morphology of the Italian territory is characterized by mountains (80 per cent), including uplands up to an altitude of 400–800 m. One of the studies of landslides conducted by the National Council of Geologists reported over 3,000 instances of slope instabilities involving over 200,000 ha of land area since 1970 (Mazziotti, 1991). Over the centuries, the Italian territory has undergone a transformation process which has witnessed landslides changing from being a 'natural process' to a 'natural disaster' as a result of the diffusion of human and urban settlements and the development of infrastructural networks in extremely vulnerable areas, inefficient land use for inadequate agricultural practices, and excessive deforestation. Furthermore, these side effects also represent causes which, along with the geological, geomorphological and climatic features of the Italian territory, have exacerbated other hydrogeological instability processes.

The extreme vulnerability to landslides of the Italian territory also impacts on the national urban heritage. Generally, the risk of landslide events is high in the Apennines. This area is in fact characterized by an irregular morphology which leaves urban settlements and developments vulnerable to such events. The history of landslide events in Italy (ISPRA, 2008) also teaches us that, given the wide range of geological formations (for example, structural, geomorphological and geotechnical), landslide occurrences in urban settlements are of two types (i) *rapid falls* caused by anthropic interventions (i.e. excavations, overloading, groundwater seepage, intensive urbanization processes, etc.) and natural phenomena from small entities of strong intensity (i.e. earthquakes, heavy rain etc.). These types of landslides are relevant in terms of the loss of life and damage to infrastructure; and (ii) *slow falls* with prolonged impacts. These are mostly localized in areas on clay soils and their effects in terms of mortality appear less significant.

The main literature (D'Elia, 1980; Cotecchia, Del Prete, 1984; Lanzo et al., 2004) considers the nature of the landslides and the degree of vulnerability of the urban systems as the main factors affecting the risk of landslides in the Italian territory. In other words, landslide vulnerability in sites characterized by slow falls, with densely populated urban centres that are rich in monuments and cultural heritage, would be different to the vulnerability of sites with low population density and rapid falls. Therefore, the landslide vulnerability of urban settlements is affected by the quality and degree of the slope instability and also by a series of other factors more or less correlated to each other, such as: (i) the number of buildings; (ii) the lack of adequate planning policies for buildings and infrastructure; (iii) the lack or inefficient realization of civil protection mechanisms; (iv) the lack of adequate water support, drainage and disposal devices; (v) tectonic or volcanic uplifts, deposition loading slopes or its crests, and fluvial and wave erosion modifying the hydrographical network; and (vi) lack of adequate materials for building construction.

The Dataset

Our sample includes 10,769 Italian sites affected by landslide events during the period 1918–2001. Data are from the AVI Project (Italian Vulnerability Areas) database[2] and the information pertains to 18,500 sites with a number of relevant variables, such as: (i) number of landslides; (ii) type of affected site; (ii) geological and geotechnical structure; (iii) data sources; (iv) land use; (v) agricultural damage; (vi) topography; (vii) event type; (viii) movement type; (ix) potential causes; (x) damage to infrastructure and buildings; (xi) number of lives lost; (xii) zoo-technical damages; (xiii) indirect and direct effects; (xiv) soil type; and (xv) monetary estimate of damage (for some sites).

Variables description For this study we have elaborated the information contained in the AVI Project and considered the following variables suitable for our analysis, as shown in Table 19.1. The dependent variables are the damage to infrastructure such as civil, industrial and network infrastructures. The independent variables are a list of direct causes of landslides such as water infiltration, rain, earthquakes, erosion, slope oscillation, excavation, lack of site protection, underground extraction activities, surface extraction activities, frost weathering, deposition,

2 For details see http://avi.gndci.cnr.it/welcome_en.htm.

melting snow, agricultural practices, indirect effects variables such as floods and water stagnation, and land use variables such as human settlements and forests.

Table 19.1 List of variables

Damages to infrastructures from landslides	Public buildings	Dependent variable
	Industrial buildings	Dependent variable
	Network infrastructures	Dependent variable
Triggering causes of landslides	Water infiltration	Independent variable
	Rain	Independent variable
	Earthquakes	Independent variable
	Erosion	Independent variable
	Slope oscillation	Independent variable
	Excavation	Independent variable
	Frost weathering	Independent variable
	Deposition	Independent variable
	Snow melting	Independent variable
	Agricultural practices	Independent variable
Indirect effects of landslides	Floods	Independent variable
	Water stagnation	Independent variable
Land use	Human settlements	Independent variable
	Forests	Independent variable

Models results We estimated three logit regression models to infer the probability that the damage from landslides occurred in public buildings, industrial buildings and network infrastructures depend on a series of independent variables as listed in Table 19.1. For each model we ran a stepwise logit regression which drops variables as not statistically significant at 90 per cent confidence interval. Estimated results for damage to public buildings[3] are shown in Table 19.2.

A change (in this case from 0 to 1) in the occurrence of water infiltration, for example, causes the odds ratio in favour of damage to buildings being caused to increase by: 236% – 100% = 136%. To best interpret this result, we need to compute the predictive margins for each variable, which is the effect on the conditional mean of the damage caused to public buildings by landslides of a change from 0 to 1 in the categorical variable of interest, keeping all other variables at their mean. Table 19.3 shows the predictive margin at means values.

3 Estimated results for the industrial building and network infrastructure models are available on request from the authors.

Table 19.2 Logit model results for damage from landslides to public buildings

	Odds ratio	SE	P>¦z¦	[95% CI]
Constant	0.13	0.00	0.00	0.12 0.14
Forests	1.90	0.45	0.00	1.20 3.02
Floods	2.52	0.50	0.00	1.71 3.72
water_stag	2.98	0.39	0.00	2.30 3.86
water_filtration	2.36	0.13	0.00	2.12 2.63
Earthqk	.744	0.13	0.00	0.53 1.04
Erosion	1.27	0.06	0.08	1.15 1.40
slope_oscil	2.18	0.30	0.00	1.66 2.86
Excavations	0.74	0.10	0.00	0.57 0.97
lack_protection	1.93	0.29	0.03	1.44 2.60
extractions_ug	1.61	0.43	0.00	0.95 2.72
extractions_su	0.30	0.10	0.07	0.16 0.57
Depositions	0.07	0.02	0.00	0.04 0.14
snow_melting	2.18	0.44	0.00	1.47 3.23
N. of Obs	10769			
Log likelihood	-5288.80			
LR chi2(13)	740.04			
Prob>chi2	0.00			

Note: Independent variables are categorical.

In this case the marginal effect at mean (MEM) for water infiltration of 0.13, for example, tells us that the predicted probability that a landslide will occur is 0.13 greater than when water infiltration is absent. To confirm this, we can easily compute the predicted probabilities for the presence or absence of water infiltration (shown in Table 19.4) where the difference is: 0.26–0.13 = 0.13. From Table 19.4 we can also argue that water stagnation along with floods, snow melting, lack of site protection and forests would favour damage to public buildings due to landslides events. It is interesting to note that the presence of forests in terms of a territory's land use would cause landslides to occur with a probability of 32 per cent. This finding seems to contrast with current literature results.

However, we should consider the geological formation of the Italian territory. From the Alps to Sicily it is mostly calcareous, which lets water infiltrate deep underground, that diminishes the potential barrier effect of forests to landslides. Table 19.5 shows predicted probabilities for landslides damaging industrial buildings and network infrastructure conditional on the occurrence of the variables of interest (those which are statistically significant at 95 per cent confidence interval).

Table 19.3 Predictive margins at means, Delta method

	dy/dx (margins)	SE	P>¦z¦	[95% CI]
Forests	0.12	0.05	0.02	0.02 0.22
Floods	0.18	0.05	0.00	0.09 0.28
water_stag	0.22	0.03	0.00	0.16 0.29
water_filtration	0.13	0.01	0.00	0.11 0.14
Earthqk	-0.04	0.02	0.06	-0.09 0.00
Erosion	0.04	0.01	0.00	0.02 0.05
Slope_oscil	0.15	0.03	0.00	0.09 0.21
Excavations	-0.04	0.02	0.02	-0.08 -0.01
lack_protection	0.12	0.03	0.00	0.06 0.19
extractions_ug	0.09	0.05	0.11	-0.02 0.19
extractions_su	-0.13	0.02	0.00	-0.17 -0.09
Depositions	-0.19	0.01	0.00	-0.21 -0.18
snow_melting	0.15	0.04	0.00	0.06 0.25

Table 19.4 Predicted probabilities of landslides damaging public buildings

	Pr(Y=1¦X, x=1)	Pr(Y=1¦X, x=0)
Forests	**0.32**	0.20
Floods	**0.38**	0.20
water_stag	**0.42**	0.19
water_filtration	0.26	0.13
Earthqk	0.16	0.20
Erosion	0.22	0.18
Slope_oscil	**0.34**	0.20
Excavations	0.16	0.20
lack_protection	**0.32**	0.20
extractions_ug	0.28	0.20
extractions_su	0.07	0.20
Depositions	0.02	0.21
snow_melting	**0.35**	0.20

Note: Values are statistically significant at 0.05 per cent.

Table 19.5 Predicted probabilities for landslides damaging industrial buildings and network infrastructures

	Predicted probabilities of landslides damaging industrial buildings conditional on X=1 and X=0		Predicted probabilities of landslides damaging network infrastructure conditional on X=1 and X=0	
	Pr(Y=1¦X, x=1)	Pr(Y=1¦X, x=0)	Pr(Y=1¦X, x=1)	Pr(Y=1¦X, x=0)
forests	0.04	0.01	0.06	0.01
water_stag	0.03	0.01	0.03	0.01
Rain	0.01	0.01		
erosion	0.02	0.01		
excavations	0.00	0.01		
lack_protection	0.06	0.01	0.10	0.01
extractions_ug	0.11	0.01		
extractions_su	0.12	0.01		
snow_melting	0.05	0.01	0.06	0.01
agr_practices	0.04	0.01		

Note: Blank cells correspond to variables not statistically significant at 95 per cent. These have been dropped from the logit models.

Policy implications of the Italian case study According to the national conference on hydrogeological risk held in Rome in 2013, vulnerability to landslides is one of the issues most debated across Italian municipalities. The debate is focused on the persistent lack of adequate prevention policies at all governance levels. Over the last years, the frequency of the events and their occurrence even in non-urban areas (i.e. those characterized by low numbers of human and industrial settlements and infrastructures), has led policymakers to consider efficient planning and prevention strategies at the national level and large scale information for the majority of the population. Eighty-two per cent of Italian municipalities are at hydrogeological risk, and the magnitudes of the risks are mostly endogenous and structural in nature. There is a need for a national regulation which defines the principles of adequate territorial governance in terms of prevention of landslides and flood risks. An accurate land use policy will also valorise and preserve hydrographical basins, as highlighted by the governance strategies proposed by the Water Framework Directive (2000/60/EC). Finally, the potential benefits deriving from establishing a national regulation for hydrogeological risk would not only cause an increase in safety measures to protect the local population and communities but also favour job creation (due to the demand for highly skilled labour) in vulnerable territories.

Concluding Remarks

The present chapter has developed and discussed vulnerability to extreme climatic events due to climate change. We considered the case study for Italy which centred on the effects of landslide events on infrastructure.

In particular, we assessed the effects of landslide events on infrastructure by means of a logit model. We estimated three logit models for public buildings, industrial buildings and network infrastructure. In the first model, we noted how water stagnation and floods, melting snow, lack of site protection and forests would exacerbate damage to public buildings. We also pointed out how the presence of forests in terms of a territory's land use would cause landslides at a probability of 32 per cent. We found this figure in contrast with current literature results. However, given that the geological formation of the Italian territory is mostly calcareous, this allows water to infiltrate deep underground, which diminishes the potential barrier effect of forests to landslides. We also found that the other two models are particularly relevant to our analysis. In terms of policy implications, we argued that there is a need for policymakers to consider the implementation of efficient planning and prevention strategies aimed at the preservation of land use, the conservation of vulnerable territories and population safety at the national as well as local level. These policies will serve to valorize hydrographical basins as elements of territorial governance practices and as economic drivers to enhance employment creation in vulnerable territories.

References

Alexander, D. 2000. *Confronting Catastrophe: New Perspectives on Natural Disasters*. Edinburgh: Dunedin Academic Press Ltd.

Alwang, J., Siegel, P.B. and Jorgensen, S.L. 2001. *Vulnerability: A View from Different Disciplines*, Social Protection Discussion Paper No. 0115. Washington, DC: The World Bank.

Andrey, J. and Mills, B. 2003. 'Climate Change and the Canadian Transportation System: Vulnerabilities and adaptations'. In J. Andrey and K. Christopher (eds) *Weather and Road Transportation*. Department of Geography Publication Series, Monograph 55. University of Waterloo, Waterloo, Canada.

Birkmann, J. 2006. 'Measuring vulnerability to promote disaster-resilient societies. Conceptual frameworks and definitions'. In J Birkmann (ed.) 2006. *Measuring Vulnerability to Natural Hazards*. United Nations University.

Brooks, N. 2003. *Vulnerability, Risk and Adaptation: A Conceptual Framework*. Norwich: Tyndall Centre Working Paper No. 38.

Cardona, O.D. 2003. 'The need for rethinking the concepts of vulnerability and risk from a holistic perspective: A necessary review and criticism for effective risk management'. In G. Bankoff, G. Frerks and D. Hilhorst (eds)

2003. *Mapping Vulnerability: Disasters, Development and People*. London: Earthscan Publishers.

Chambers, R. 1989. *Vulnerability, Coping and Policy*. IDS Bulletin 20, Institute of Development Studies, Sussex.

Cotecchia, V. and Del Prete, M. 1984. 'The reactivation of large flows in the parts of southern Italy affected by the earthquake of November 1980, with reference to the evolutive mechanism'. In *Proceedings, International Symposium on Landslides*, Toronto.

D'Elia, B. 1980. 'Processi di Ammorbidimento di Un'argilla Marnosa con Struttura a Scaglie'. In *Proceedings, 14th Convegno Nazionale di Geotecnica*, Firenze.

Intergovernmental Panel of Climate Change IPCC. 2007. *Fourth Assessment Report* [online]. http://www.ipcc.ch [accessed 20 January 2014].

Economics of Climate Adaptation Working Group. 2009. *Shaping Climate Resilient Development. A Framework for Decision Making* [online]. http://www.rockefellerfoundation.org/uploads/files/d453d8f9-7aff-473d-b414-5a8add2a5d0b.pdf [accessed 14 January 2014].

Institute for Environmental Protection and Research. 2008. *Landslides in Italy*. Special Report, 83, Rome, Italy.

Lanzo, G., Olivares, L., Silvestri, F. and Tommasi, P. 2004. 'Seismic Response Analyses of Historical Towns Rising on Rock Slabs Overlying a Clayey Substratum'. In *Proceedings 5th International Conference on Case Histories in Geotechnical Engineering*, Paper 3.25, New York.

Mazziotti, L. 1991. *Landslides Instabilities in Italy* (*I dissesti franosi in Italia*). ANTINCENDIO.

Office of Emergency Preparedness, Executive Office of the President of the United States (OEP-EOP). 1972. *Disaster Preparedness*. Report to the Congress, US Government Printing Office.

Peng, C.-Y.J., Manz, B.D. and Keck, J. 2001. Modeling categorical variables by logistic regression. *American Journal of Health Behavior*, 25(3), pp. 278–84.

Shaw, J., Taylor, R.B., Solomon, S., et al. 1998. Potential impacts of sea level rise of Canadian coasts. *The Canadian Geographer*, 42(4), pp. 365–79.

United Nations Office for Disaster Risk Reduction. 2014. *Living with Risk: A Global Review of Disaster Reduction Initiatives* [online]. http://www.unisdr.org/we/inform/publications/657 [accessed 13 January 2014].

Watts, M.J. and Bohle, H.G. 1993. The Space of Vulnerability: The Causal Structure of Hunger and Famine. *Progress in Human Geography*, 17(1), pp. 43–67.

Index

For Product Safety Concerns and Information please contact our EU representative GPSR@taylorandfrancis.com Taylor & Francis Verlag GmbH, Kaufingerstraße 24, 80331 München, Germany